Springer Series in

ADVANCED MICROELECTRONICS 8

Springer
Berlin
Heidelberg
New York
Barcelona
Hong Kong
London
Milan
Paris
Tokyo

Springer Series in

ADVANCED MICROELECTRONICS

The Springer Series in Advanced Microelectronics provides systematic information on all the topics relevant for the design, processing, and manufacturing of microelectronic devices. The books, each prepared by leading researchers or engineers in their fields, cover the basic and advanced aspects of topics such as wafer processing, materials, device design, device technologies, circuit design, VLSI implementation, and subsystem technology. The series forms a bridge between physics and engineering and the volumes will appeal to practicing engineers as well as research scientists.

1 **Cellular Neural Networks**
Chaos, Complexity and VLSI Processing
By G. Manganaro, P. Arena, and L. Fortuna

2 **Technology of Integrated Circuits**
By D. Widmann, H. Mader, and H. Friedrich

3 **Ferroelectric Memories**
By J. F. Scott

4 **Microwave Resonators and Filters for Wireless Communication**
Theory, Design and Application
By M. Makimoto and S. Yamashita

5 **VLSI Memory Chip Design**
By K. Itoh

6 **High-Frequency Bipolar Transistors**
Physics, Modelling, Applications
By M. Reisch

7 **Noise in Semiconductor Devices**
Modeling and Simulation
By F. Bonani and G. Ghione

8 **Logic Synthesis for Asynchronous Controllers and Interfaces**
By J. Cortadella, M. Kishinevsky, A. Kondratyev, L. Lavagno, and A. Yakovlev

Series homepage – http://www.springer.de/phys/books/ssam/

J. Cortadella M. Kishinevsky
A. Kondratyev L. Lavagno A. Yakovlev

Logic Synthesis for Asynchronous Controllers and Interfaces

With 146 Figures

Springer

Prof. J. Cortadella
Department of Software
Universitat Politecnica de Catalunya
Campus Nord. Modul C6
08034 Barcelona, Spain

Dr. M. Kishinevsky
Intel Corporation
JF4-211, 2111 N.E. 25th Ave.
Hillsboro, OR 97124-5961, USA

Dr. A. Kondratyev
Cadence Design Systems
2001 Addison Avenue
Berkeley, CA 94704, USA

Prof. L. Lavagno
Dipartimento di Elettronica
Politecnico di Torino
Corso Duca degli Abruzzi 24
10129 Torino, Italy

Prof. A. Yakovlev
Department of Computing Science
University of Newcastle upon Tyne Claremont
Tower, Claremont Road
Newcastle upon Tyne, NE1 7RU, U.K.

Series Editors:

Dr. Kiyoo Itoh
Hitachi Ltd., Central Research Laboratory, 1-280 Higashi-Koigakubo
Kokubunji-shi, Tokyo 185-8601, Japan

Professor Thomas Lee
Stanford University, Department of Electrical Engineering, 420 Via Palou Mall, CIS-205
Stanford, CA 94305-4070, USA

Professor Takayasu Sakurai
Center for Collaborative Research, University of Tokyo, 7-22-1 Roppongi
Minato-ku, Tokyo 106-8558, Japan

Professor Doris Schmitt-Landsiedel
Technische Universität München, Lehrstuhl für Technische Elektronik
Theresienstrasse 90, Gebäude N3, 80290 München, Germany

ISSN 1437-0387
ISBN 3-540-43152-7 Springer-Verlag Berlin Heidelberg New York

Library of Congress Cataloging-in-Publication Data applied for.
Die Deutsche Bibliothek – CIP-Einheitsaufnahme
Logic synthesis for asynchronous controllers and interfaces / J. Cortadella .... – Berlin ; Heidelberg ; New York ; Barcelona ; Hong Kong ; London ; Milan ; Paris ; Tokyo : Springer, 2002 (Springer series in advanced microelectronics ; 8) (Physics and astronomy online library) ISBN 3-540-43152-7

Springer-Verlag Berlin Heidelberg New York
a member of BertelsmannSpringer Science+Business Media GmbH

http://www.springer.de

Printed in Germany

Typesetting by the authors using a Springer TEX makro package.
Final processing by Steingraeber Satztechnik GmbH Heidelberg.
Cover concept by eStudio Calmar Steinen using a background picture from Photo Studio "SONO". Courtesy of Mr. Yukio Sono, 3-18-4 Uchi-Kanda, Chiyoda-ku, Tokyo
Cover design: *design & production* GmbH, Heidelberg

Printed on acid-free paper SPIN: 10832637 57/3141/mf - 5 4 3 2 1 0

To Cristina, Viviana, Kamal,
Lena, Katie, Gene, Masha, Anna,
Mitya, Paola, Maria, and Gregory,

who sometimes waited patiently at home,
sometimes joined in the fun.

Preface

This book is the result of a long friendship, of a broad international cooperation, and of a bold dream. It is the summary of work carried out by the authors, and several other wonderful people, during more than 15 years, across 3 continents, in the course of countless meetings, workshops and discussions. It shows that neither language nor distance can be an obstacle to close scientific cooperation, when there is unity of goals and true collaboration.

When we started, we had very different approaches to handling the mysterious, almost magical world of asynchronous circuits. Some were more theoretical, some were closer to physical reality, some were driven mostly by design needs. In the end, we all shared the same belief that true Electronic Design Automation research must be solidly grounded in formal models, practically minded to avoid excessive complexity, and tested "in the field" in the form of experimental tools. The results are this book, and the CAD tool petrify. The latter can be downloaded and tried by anybody bold (or desperate) enough to tread into the clockless (but not lawless) domain of small-scale asynchronicity. The URL is `http://www.lsi.upc.es/~jordic/petrify`.

We believe that asynchronous circuits are a wonderful object, that abandons some of the almost militaristic law and order that governs synchronous circuits, to improve in terms of simplicity, energy efficiency and performance. To use a social and economic parable, strict control and tight limits are simple ways to manage large numbers of people, but direct loose interaction is the best known means to quickly and effectively achieve complex goals. In both cases, social sciences and logic circuits, speed and efficiency can be conjugated by *keeping the interaction local*. Modularity is the key to solving complex problems by preserving natural boundaries.

Of course, a good deal of *automation* is required to keep any complex design problem involving millions of entities under control. This book contains the theory and the algorithms that can be, and have been, used to create a complete synthesis environment for the synthesis of asynchronous circuits from high-level specifications to a logic circuit. The specification can be derived with relative ease from a Register Transfer Level-like model of the required functionality. The circuit is guaranteed to work regardless of the delays of the gates and of some wires.

Unfortunately asynchronous circuits must pay something for the added freedom they provide: algorithmic complexity. Logic synthesis can no longer use Boolean algebra, but must resort to sequential Finite State Machine-like models, which are much more expensive to manipulate. Fortunately the intrinsic modularity of the asynchronous paradigm comes to rescue again, since relatively small blocks can be implemented independent of each other, without affecting correctness.

It is not clear today whether asynchronous circuits will ever see widespread adoption in everyday industrial design practice. One serious problem is the requirement to "think different" when designing them. One must consider *causality relations* rather than fixed steps. However, this book shows that once this conceptual barrier is overcome, due e.g. to the rise of unavoidable problems with the synchronous design style, it is indeed possible to define and implement a synthesis-based design flow that ensures enough designer productivity.

We all hope, thus, that our work will soon be used in practical designs. So far it has been a very challenging, fruitful and exciting cooperation.

The list of people and organizations that deserve our gratitude is long indeed. First of all, some colleagues helped with the formulation and the implementation of several ideas described in this book. Alexander Taubin (Theseus Research) contributed heavily to the topic of synthesis from STGs. Shai Rotem initiated and Ken Stevens and Steven M. Burns (all from Intel Strategic CAD Lab) contributed to the research on relative timing synthesis. Enric Pastor (Universitat Politecnica de Catalunya) cooperated on logic decomposition and technology mapping. Then, we would like to acknowledge support from Intel Corporation and various funding agencies such as: ESPRIT ACiD-WG Nr. 21949, EPSRC grant GR/M94366, MURST research project "VLSI architectures", CICYT grants TIC98-0410 and TIC98-0949. Finally, all members of the asynchronous community, both users and opponents of the STG-based approach, and in particular the users of petrify, provided us with excellent feedback.

Barcelona, Spain,	*Jordi Cortadella*
Hillsboro, Oregon,	*Michael Kishinevsky*
San Jose, California,	*Alex Kondratyev*
Torino, Italy,	*Luciano Lavagno*
Newcastle upon Tyne, United Kingdom,	*Alexandre Yakovlev*
October, 2001	

Contents

1. Introduction

This book is devoted to an in-depth study of logic synthesis techniques for asynchronous control circuits. These are logic circuits that do not rely on global synchronization signals, the clocks, to dictate the interval of time at which other signals are sampled. The difficulty with their design is well known since the late 1950's. Asynchronous circuits cannot distinguish between combinational behavior, governed by Boolean algebra, and sequential behavior, governed by Finite State Machine (FSM) algebra. This separation, together with static timing analysis to compute minimum clock cycles, is essential to modern synchronous logic design. Asynchronous circuits can still, by means of appropriate delay models, abstract away most complex electric and timing properties of transistors and wires. However, both synthesis and analysis of asynchronous circuits must consider everything as *sequential*, and hence are subject to the *state explosion* problem which plagues the sequential world.

Our goal is to show that, despite the increase in complexity due to the need to consider sequencing at every step, *asynchronous control circuits can be synthesized automatically*. The very property that makes them desirable, i.e. *modularity*, makes their analysis and synthesis also *feasible* and *efficient*. We thus claim that, while it is unlikely that asynchronous circuits will become part of any mainstream ASIC (Application-Specific Integrated Circuit) design flow in the near future, they do have a range of (possibly niche) applications, and (locally) very efficient synthesis algorithms. Although testing and datapath design are not dealt with explicitly in this book, we provide references to the main publications in these areas.

1.1 A Little History

Asynchronous circuits have been around, in one form or another, for quite some time. In fact, since they are a proper superset of synchronous circuits, one may even claim that they never disappeared. Note that we do not consider aspects of individual flip-flops such as asynchronous reset as true manifestations of asynchronicity, since they are handled very differently from truly asynchronous logic. They are just a nuisance to synchronous CAD tools, and can be handled with simple tricks.

It is interesting, however, to analyze the history of asynchronous circuits, and that of their synchronous counterparts, in order to identify the origin of problems and misconceptions.

The clock, which is a means to dictate timing globally for a digital circuit, was considered an essential device for synchronizing a digital computer by Turing, who claimed that asynchronous circuits are hard to design[1]. On the other side of the Atlantic, Von Neumann approximately at the same time showed that addition using a linear carry chain has on average a logarithmic carry propagation length, with respect to the linear worst-case. This is often claimed to be one of the main reasons for claiming the superiority of asynchronous over synchronous circuits: better performance by exploiting the average case [1, 2]. An asynchronous adder with a circuitry, completion detection, that identifies when the carry propagation is finished has potentially the same average performance as a theoretically much more expensive carry lookahead adder.

We will see later that this is not often exploited in the data path, since only few designers can today afford the heavy area and delay cost of completion detection for arithmetic. The size increase due to completion detection is 2×–4× (linear) and the delay is logarithmic in the number of inputs, but even this is considered excessive. See, however, [3] and [4] for ways of at least partially exploiting the performance advantage of average case versus worst case performance with acceptable overheads.

Early computers were either asynchronous, or had significant asynchronous components. This was due to the fact that computer design was still mostly an art, and that algebraic abstractions for electronics were not as well developed as today. Moreover these abstractions did not match well the physical reality of the underlying devices, and this is becoming very important today again, with Deep Sub-Micron technology. Classical logic design books (e.g. Unger's famous book [5]) devoted a lot of attention to asynchronous design.

The current demise of asynchronous circuits as a mainstream implementation platform for electronic devices is due, in our opinion, to two factors.

1. Static CMOS logic implements an excellent approximation of the ideal properties of Boolean gates,
2. Logic synthesis, based on Boolean algebra, is essential to implement both datapath and control logic [6, 7, 8].

The former factor will probably become less significant, due to unavoidable changes in the technology [9]. The latter, on the other hand, seems very

[1] "We might say that the clock enables us to introduce a discreteness into time, so that time for some purposes can be regarded as a succession of instants instead of a continuous flow. A digital machine must essentially deal with discrete objects, and in the case of ACE this is made possible by the use of a clock. All other computing machines except for human and other brains that I know of do the same. One can think up ways of avoiding it, but they are very awkward."

well entrenched into both engineering practice, e.g. EDA (Electronic Design Automation) tools and flows, and business practice, e.g. design handoff.

This is due to the fact that the combinational Boolean abstraction, enabled by the use of the clock to separate functionality from timing, has become an essential mechanism used by the modern designer. It is thus a deeply rooted assumption in any digital implementation flow. If one looks inside modern commercial logic synthesis tools, *sequential optimizations*, such as state encoding, retiming, etc., are much less exploited, and much more carefully used, than purely combinational ones, despite the potential advantages. The size of modern datapath, control and interfacing circuitry is such that any sequential optimization becomes automatically too complex to be useful. Global optimizations, which are possible only in the combinational domain, and even there often only for algebraic techniques [7, 10], are the only means to satisfy area, testability, delay and power constraints.

1.2 Advantages of Asynchronous Logic

1.2.1 Modularity

It is becoming increasingly difficult to satisfy the assumptions that made synchronous logic synthesis so successful, namely:

- all the logic on a chip works with one basic clock, or a few clocks rationally related to each other,
- clocks can be transmitted to the whole chip with reasonable skews,
- communication wires can be implemented more or less independently from logic.

The latter in particular is currently already violated at the chip level, and has caused a flurry of work in areas related to the integration of placement and synthesis [11], and to driver and wire sizing [12].

Clock skew is certainly a serious problem, one which causes considerable design effort and on-chip energy to be spent. Even though an increasing number of routing layers and improved routing techniques may alleviate the problem, asynchronous circuits are very promising in terms of *modularity*. Modularity, that is tightly connected with the first and last assumption above, is becoming increasingly important, especially with the emergency of System-On-Chip integration. The capability to "absorb clock skew variations inherent in tens-of-million transistor class chips" was cited as the key reason for Sun Microsystems to incorporate an asynchronous FIFO into the UltraSPARC IIIi design [13, 14].

Hence asynchronous logic may see increasing application, as a means to achieve better *plug and play* capabilities for large systems on a chip. This is particularly true when there is no possibility to re-do the clock routing and fabrication, or reduce the clock speed if a subtle timing problem is discovered, especially if it is intermittent. This is because tight clocking disciplines, like IBM's LSSD [15] (Level-Sensitive Scan Design), break down and become too inefficient when applied to several millions of gates. Moreover, it is unlikely that they can be used if an integrated circuit is developed by assembling Intellectual Property from several parties. Both organizations, such as the Virtual Socket International Alliance [16], and private initiatives, such as the Re-Use manual [17], are attempting to solve this problem by providing standards and guidelines. However, the appeal of a self-adapting, self-configuring, self-timing-repairing technique such as asynchronous design is obvious.

As usual, everything can be done synchronously, by adopting appropriate guidelines, policies, and devoting enough time to the design, but asynchronous techniques may make it easier, and faster. In the end, they may well become the only *practical* means to complete a design on time and within specs. The same most likely holds, as we will see, for all the other claimed advantages of asynchronicity. Whether this will give them enough momentum to conquer at least some niche areas, e.g. low Electro-Magnetic Emission, or even make them the tool of choice for mainstream design, is still to be seen. But we need to list a few more bright spots of asynchronous logic. The black ones are known to everybody, and are even part of unsubstantiated folk opinions that this book is aimed at eradicating.

1.2.2 Power Consumption and Electromagnetic Interference

A number of asynchronous circuits have also demonstrated advantages of lower energy per computation and better electromagnetic emission spectrum than synchronous ones [18, 19, 20]. Asynchronous control circuits tend to be *quiet*, because they:

- avoid unneeded transition, in particular those of the omnipresent clock signal, thus requiring less energy to perform a given computation, and
- spread out needed transitions without the strict period and phase relationships that are typical of synchronous circuits.

In synchronous circuits, not only clock signals have equally spaced transitions, but also each data and control signal changes value with a more or less constant phase relationship to the clocks. In fact, the departure time from the flip-flops and the logic propagation times are more or less constant for each gate in a given circuit over successive clock cycles. This causes significant correlation between signal edges, and hence an electro-magnetic emission that shows impressive peaks, due both to supply and signal lines, at several clock frequency harmonics. See, for example, Fig. 1.1 (provided by Theseus Logic, Inc.), where the spikes in the supply current, and hence the emitted harmonics of the clock frequency, for two versions of the same micro-processor are much higher in the synchronous case than in the asynchronous case.

Asynchronous circuits, as discussed in [21], have a much better energy spectrum than their synchronous counterparts, one which is generally more

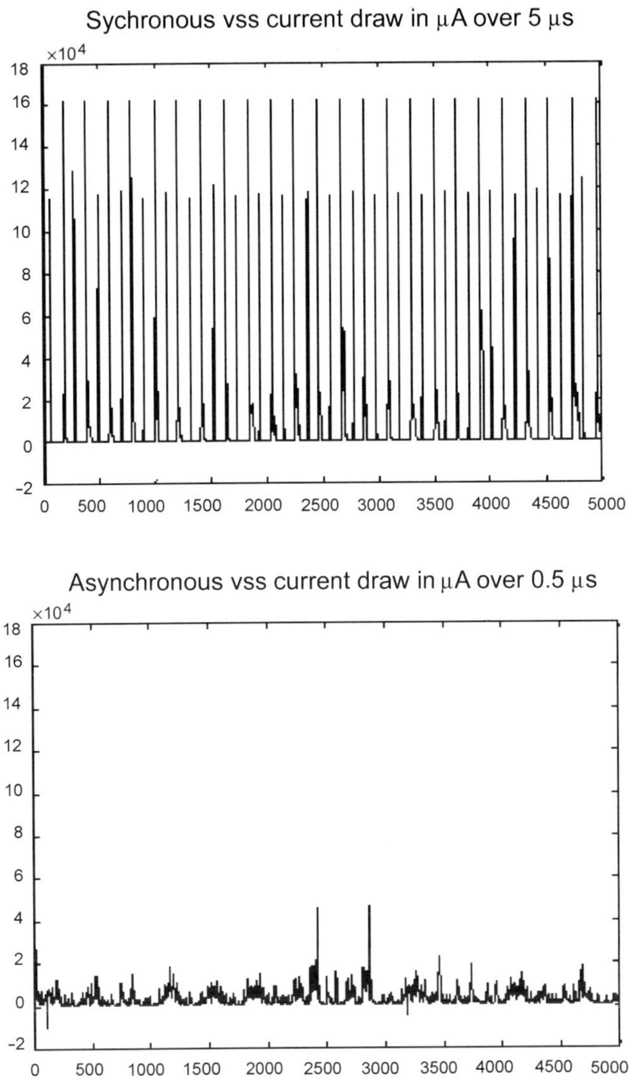

Fig. 1.1. Supply current for synchronous and asynchronous circuit

flat due to the irregular computation and communication patterns. This is important to reduce the needs for expensive Electro-Magnetic Interference shielding, in order to meet regulatory requirements that generally impose a limit on the maximum energy that can be emitted at a given frequency, rather than on the global emitted energy over a large spectrum interval. Reduced electro-magnetic interference was cited as a main motivation for Philips to include asynchronous microcontroller chips into commercial pagers [19].

On the energy saving side, asynchronous control is generally less power-hungry than synchronous control, since there are no long clock signal lines. These lines require a lot of power to be driven with limited skew. Asynchronous control, on the other hand, is much more *local*, and thus requires less energy if the layout is done properly, and handshaking circuits are kept close to each other. Of course, significant power savings in synchronous can be achieved by careful clock gating, however this requires very careful logical and layout design, as any gate on the clock path can lead to timing errors if not handled properly. This requires both:

- A careful manual selection of the gating regions, trading off energy versus performance. The latter can be lost any time logic is inserted in the clock path, because the uncertainty in the clock skew increases.
- Support of complex clocking rules from the synthesis and layout tools.

Asynchronous logic can provide the same or better levels of power savings *inherently* and without special effort [22]. Power saving is cited as one of the main advantages of the Sharp DDMP asynchronous multi-media processor [23, 24].

Of course, these advantages apply only to *dynamic* power consumption. Static power is closely related to area, within a given technology. Hence it is addressed by keeping area overhead low in comparison with synchronous controllers, as we will discuss further in the rest of the book.

1.2.3 Performance

The global performance advantages of asynchronous logics are less obvious than the energy-related ones. Asynchronous modules obviously may exhibit *average case* performance, rather than worst case as synchronous ones [1]. However, it is not clear how often this advantage can be exploited at the system level, due to the fact that the throughput of asynchronous pipelines is limited by the delay of the *slowest stage* [25]. A key performance advantage of asynchronous control circuits is that they can be finely tuned down to the levels of transistor sizing and individual transition delay, which can in turn improve overall system performance, as discussed in [26] and Chap. 7. Moreover, such tunings can be performed independently by teams working on the internals of blocks connected by asynchronous interfaces. Correctness, due to the above mentioned modularity, is not compromised, assuming that the timing optimizations do not affect the interface protocols. Thus asynchronous circuits can achieve impressive *latency and throughput* comparable to or sometimes better than high-speed synchronous control. This is achieved either by aggressive use of local timing optimizations [26, 27] or by the use of low granularity pipelining which hides the reset phase latency [28].

The issue of performance of asynchronous datapath components is a bit more problematic. It is not discussed at all in this book, which is devoted to *asynchronous control* circuits. Truly asynchronous datapath units may exploit the better average case delay of arithmetic operations, rather than the worst-case performance (both among all input patterns and among all manufactured integrated circuits) of many synchronous designs. However, some robust asynchronous design styles require expensive dual-rail or other delay-insensitive encodings and complex completion detection trees, which may result in a globally worse performance and cost than synchronous ones. These problems can be avoided design, either by keeping completion detection out of the critical path [3], or by approximating completion detection [4]. However, many asynchronous designs today use a micro-pipelined architecture with synchronous datapath [29, 30, 31]. The most successful example of an asynchronous datapath is probably Ted Williams' iterative divider that was used in a commercial microprocessor [32].

1.3 Asynchronous Control Circuits

Traditionally, designers distinguish two main parts in the structure of a sequential circuit that are often synthesized by using different methods and tools:

- **data path** that includes wide, e.g. 16-bit or 32-bit, units to perform transformations on the data manipulated by the circuit. Some units perform arithmetic operations, e.g. adders, multipliers, whereas others are used for communication and storage, e.g. multiplexors, registers.
- **control** that includes those signals that determine which data must be used by each unit, e.g. control signals for multiplexors, and which type of operations must be performed, e.g. operation code signals for ALUs.

In synchronous circuits, clock signals determine when the units in the datapath must operate. The frequency of each clock is calculated to accommodate the longest delay of the operations that can be scheduled at any cycle. For this reason, the answer to the question *"when will an operation be completed?"*, typically is *"I do not know, but I know it will be completed by the end of the cycle"*. Thus, there is an implicit synchronization of all units by the end of each cycle. Time borrowing based on the use of transparent latches, skew-tolerant or self-resetting domino brings more elements of asynchrony in the synchronous design [33]. Such asynchrony is however typically localized within a few phases of the clock cycle.

Asynchronous circuits do not use clocks and, therefore, synchronization must be done in a different way. One of the main paradigms in asynchronous circuits is *distributed control*. This means that each unit synchronizes only with those units for which the synchronization is relevant and at the time when the synchronization may take place, regardless of the activities carried out by other units in the same circuit. For this reason, asynchronous functional units incorporate explicit signals for synchronization that execute some *handshake* protocol with their neighbors. If one could translate the dialogue between a couple of functional units into natural language, one would hear sentences like

`U1:` *Here you have your input data ready.*
`U2:` *Fine, I'm ready to start my operation.*
Silence for some delay ...
`U2:` *I'm done, here you have the result.*
`U1:` *Fine, I got it. You can do something else now.*

To operate in this manner, data-path units commonly have two signals used for synchronization:

- **request** is an input signal used by the environment to indicate that input data are ready and that an operation is requested to the unit, and
- **acknowledge** is an output signal used by the unit to indicate that the requested operation has been completed and that the result can be read by the environment.

The purpose of an asynchronous control circuit is that of interfacing a portion of datapath to its sources and sinks of data, as shown in Fig. 1.2. The controller manages all the synchronization requirements, making sure that the data path receives its input data, performs the appropriate computation when they are valid and stable, and that the results are transferred to the receiving blocks whenever they are ready.

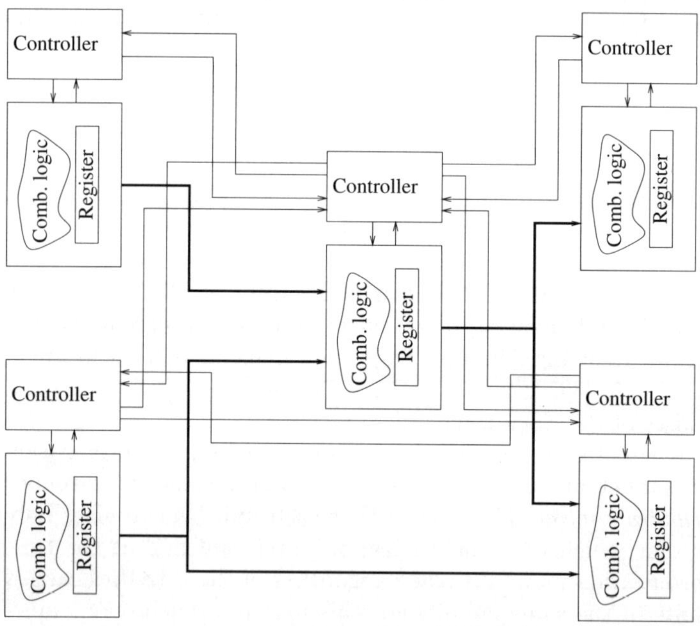

Fig. 1.2. Asynchronous controller and data path interaction

The controller does so by *handshaking* with other controllers, in a totally distributed fashion. Communication with the data path is mediated, as usual, by control and status signals. In addition to the normal interfacing conventions that are common to synchronous datapaths, status signals also carry the acknowledge information. They signal when the datapath is done computing, in reaction to events on command signals from the controller that carry the request information. This information can be derived either by completion detection, or by matched delays.

Completion detection is safer, in that it works independent of the datapath delays, but requires the datapath to be built with expensive delay-insensitive codes, also called self-indicating codes [34, 35]. These codes are redundant, i.e. require more than $\log n$ bits to represent n values. The most common delay-insensitive code, *dual rail*, requires $2 \log n$ bits to represent n values plus an additional *spacer* code.

Matched delays are cheaper, because they rely on a chain of delay elements, carefully sized to exceed slightly the length of the longest delay in the combinational datapath. These delay elements simply delay the request, to generate an acknowledgment event when it is guaranteed that the datapath logic has settled, just like a *local clock*. Of course, they do not result in average case delays, unless one employs the technique of [4] where a set of matched delays, from fast to slow, is attached to the functional unit and an appropriate delay is selected to indicate completion for each operation. In this case, at least average case among input patterns is achieved, while worst-case among manufactured ICs is obviously still required.

Several handshake protocols have been proposed and studied in the past, e.g. [36, 37, 30, 38, 39]. Discussing these protocols is not the purpose of this book.

Design of asynchronous controllers has followed several main paradigms, each based on an underlying *delay and operation model*, on a *specification mechanism*, and on *synthesis and verification algorithms*. Here we will only outline a few of these that are needed to understand the rest of the book, as well as notions on Petri nets, speed-independence and state graphs. We refer the interested reader to, e.g. [40, 41, 42] for further details on other techniques.

Hazards, or the potential for glitches, are the main challenge of designing asynchronous circuits. Informally, a hazard is any deviation from the specified circuit behavior due to the fact that Boolean gates are not instantaneous, and signal communication along wires takes time. In synchronous systems, glitches do not usually affect correct operation, since they are permitted at all times except near the sampling period around a clock edge. In asynchronous systems, in contrast, there is no clock, and clean glitch-free communication is required.

In order to analyze and avoid hazards, a delay model must be used which fairly accurately captures the physical circuit behavior, and the circuit's interaction with its environment must also be accurately modeled.

1.3.1 Delay Models

An asynchronous circuit is generally modeled using the standard *separation between functionality and timing*, at a much finer level of granularity than in the synchronous case. Each Boolean gate is still regarded, following the classical approach of Muller [43, 44], as an atomic evaluator of a Boolean function. However, a delay element is associated with either every gate output, **gate delay model**, or with every wire or gate input, **wire delay model**. The delays may be assumed to be either unbounded, i.e. arbitrary finite delays, or bounded, i.e. lying within given min/max bounds. The circuit must be hazard-free for any delay assignment permitted by the given model.

The gate delay model is optimistic, since it assumes that the *skew* between the delay of wires after a fanout is negligible with respect to that of the receiving gates. The *unbounded* gate delay model is quite robust, and gives rise to the formal theory of speed-independent circuits that is discussed in the rest of this book. It is similar to the quasi-delay-insensitive model [45, 40], that requires the skew to be bounded only for some specific fanouts. However, as discussed in [46, 47], both these assumptions are problematic, especially when the long wires of modern ICs are considered.

Unfortunately, design techniques that use the wire delay model

- either must assume that bounds on the delays of gates and wires are known, and use timing analysis and delay padding techniques to eliminate hazards [48, 49],
- or must use more complex elementary blocks, such as arbiters, sequencers, etc., than the well-known Boolean gates [50]. This is due to the fact that in any delay-insensitive circuit built out of basic gates, every gate must belong to every cycle of the circuit [51].

In this book, except for Chap. 7, we will use the **unbounded gate delay model**. In order to alleviate the problem with the negligible skew assumption, we will rely on a *hierarchical* technique, where the whole system is manually decomposed into a set of modules communicating through *delay insensitive interfaces* [52, 53] or interfaces satisfying timing constraints [54, 26]. This, roughly speaking, means that the wire delays between the modules can be neglected without affecting the functional correctness of the circuit. The estimated size of each module must be such that wire delay skews are negligible within it.

Our synthesis methods then use the unbounded gate delay model within each module. This modularity also helps reduce the execution time of the complex sequential synthesis algorithms that asynchronous circuits require.

Then each module must be placed and routed with good control over wire length, e.g. by using floor and wire planning [55].

This decomposition allows us to preserve modularity, since interfaces are delay insensitive, and efficient implementation, since we can still use the Boolean abstraction and powerful techniques from combinational and sequential logic synthesis to synthesize, optimize and technology map into a *standard library*, e.g. standard cells or FPGA, the specification of each module. This also allows one to play various trade-offs: e.g. larger modules imply better optimization, but also are more expensive in terms of synthesis time and less robust with respect to post-layout timing problems [56].

In Chap. 7 we will use a different delay model, in which gate and wire delay bounds are assumed to be known, or, to be more precise, enforceable during gate sizing and physical design. There we will need to compare the delays of sets of paths in the circuit. Hence we will use the **bounded path delay** model.

1.3.2 Operating Modes

Operating modes govern the interaction between an asynchronous controller and its environment.

In **Fundamental Mode**, based on the work of Huffman [57] and Unger [5], a circuit and its environment, which generally represents other circuits modeled in abstract manner, strictly follow a two-phase protocol that is reminiscent of the way *synchronous* circuits work. First *communication* occurs, when the environment receives outputs from the circuit and sends new inputs to it. Then *computation* occurs, when the circuit reacts to new inputs by providing new outputs. Hence the environment may not send any new input until the circuit has stabilized after a computation. The two phases must be strictly separated and must alternate. Various definitions of when one completes and the next one starts exist.

For each circuit operating in **Huffman mode** [57], one defines two time intervals δ_1 and δ_2, where $\delta_1 < \delta_2$ and both intervals depend on the specific circuit delay characteristics. If input changes are separated by less then δ_1, then they are considered to belong to the same communication phase, i.e. to be simultaneous. If they are separated by more then δ_2, then they are considered to belong to two different communication phases. Otherwise the change is illegal, and the circuit behavior is undefined. There are some special cases of Huffman Mode that deserve special mention and are attributed names of their own:

- **Fundamental Mode**: inputs from the environment are constrained to change only when all the delay elements are stable, i.e. they have the input value equal to the output value. In this case:
  - δ_1 is the minimum propagation delay inside the combinational logic and

  - δ_2 is the maximum settling time for the circuit, i.e. the maximum time that we must wait for all the delay elements to be stable once its inputs have become stable.
- **Normal Fundamental Mode**: only one input is allowed to change in each communication phase.

Burst Mode [49, 58] is often improperly called "Fundamental Mode". In this case, for each state of the circuit there is a set of possible distinct *input bursts* that can occur. Each burst is characterized by some set of changing signals, i.e. no signal is allowed to change twice in a burst, and no burst can be a subset of another burst, because one must be able to tell which one has occurred as soon as it has occurred. In this case, the end of the communication phase is signaled by the occurrence of a burst. The computation phase, on the other hand, lasts at most for δ_2, the maximum settling time for the circuit.

In burst mode, the outputs can be produced while the circuit is still computing the next state signal values, as long as it is known that the environment will be slow enough not to change the inputs while this computation is still in progress. Hence computation and communication are not so strictly separated.

A different definition of Burst Mode, somewhat similar to the Input/Output Mode mentioned below, is also possible. In this case, the end of a computation phase is signaled by the end of an output burst that is defined analogously to an input burst. However, the performance of circuits designed to work in this mode is much lower than both Burst Mode and Input/Output Mode circuits, since it does not permit overlap between output communication, computations done by the environment, and computations done by the circuit.

Input/Output Mode is totally different from the various flavors of Fundamental Mode. In this case computation and communication can overlap arbitrarily, only constrained by the overall protocol between the circuit and its environment.

Generally, circuits operating in Fundamental Mode are specified using some variation of Finite State Machine mechanisms, e.g. Flow Tables [5]. These are quite similar to the corresponding Control Unit models used in the synchronous case, and thus are easier to understand and handle by designers. On the other hand, when the operation of an Input/Output Mode circuit must be specified, the role of fine-grained concurrency is much more significant. Even though there exist state-based modeling formalisms such as State Graphs (Sect. 2.2) that describe the detailed interaction between the circuit and its environment, these models are more suitable for CAD tools than human interaction, due to the huge size of the state space. In this case, truly concurrent specifications such as Signal Transition Graphs (Sect. 2.1) are much more convenient.

2. Design Flow

The main purpose of this book is to present a methodology to design asynchronous control circuits, i.e. those circuits that synchronize the operations performed by the functional units of the data-path through handshake protocols.

This chapter gives a quick overview of the proposed design flow. It covers the main steps by posing the problems that must be solved to provide complete automation. The contents of this chapter is not technically deep. Only some intuitive ideas on how to solve these problems are sketched.

2.1 Specification of Asynchronous Controllers

The model used to specify asynchronous controllers is based on Petri nets [59, 60, 61] and is called *Signal Transition Graph* (STG) [62, 63]. Roughly speaking, an STG is a formal model for *timing diagrams*.

The example in Fig. 2.1 will help to illustrate how the behavior of an asynchronous controller can be specified by an STG. Fig. 2.1a depicts the interface of a circuit that controls data transfers between a VME bus and a device. The main task of the bus controller is to open and close the data

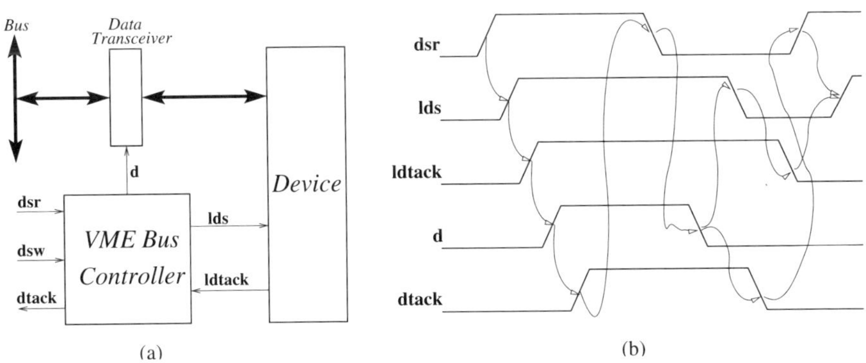

Fig. 2.1. (**a**) Interface for a VME bus controller, (**b**) timing diagram for the read cycle

transceiver through signal `d` according to a given protocol to read/write data from/to the device.

The input and output signals of the bus controller are the following:

- `dsr` and `dsw` are input signals that request to do a read or write operation, respectively.
- `dtack` is an output signal that indicates that the requested operation is ready to be performed.
- `lds` is an output signal to request the device to perform a data transfer.
- `ldtack` is an input signal coming from the device indicating that the device is ready to perform the requested data transfer.
- `d` is an output signal that controls the data transceiver. When high, the data transceiver connects the device with the bus.

Fig. 2.1b shows a timing diagram of the read cycle. In this case, signal `dsw` is always low and not depicted in the diagram. The behavior of the controller is as follows: a request to read from the device is received by signal `dsr`. The controller transfers this request to the device by rising signal `lds`. When the device has the data ready (`ldtack` high), the controller opens the transceiver to transfer data to the bus (`d` high). Once data has been transferred, `dsr` will become low indicating that the transaction must be finished. Immediately after, the controller will lower signal `d` to isolate the device from the bus. After that, the transaction will be completed by a return-to-zero of all interface signals, seeking for a maximum parallelism between the bus and the device operations.

Our controller also supports a write cycle with a slightly different behavior. This cycle is triggered when a request to write is received by signal `dsw`. Details on the write cycle will be discussed in Sect. 2.1.2.

2.1.1 From Timing Diagrams to Signal Transition Graphs

A timing diagram specifies the events (signal transitions) of a behavior and their causality relations. An STG is a formal model for this type of specifications. In its simplest form, an STG can be considered as a causality graph in which each node represents an event and each arc a causality relation. An STG representing the behavior of the read cycle for the VME bus is shown in Fig. 2.2. Rising and falling transitions of a signal are represented by the suffixes `+` and `-`, respectively.

Additionally, an STG can also model all possible dynamic behaviors of the system. This is the rôle of the tokens held by some of the causality arcs. An event is *enabled* when it has at least one token on each input arc. An enabled event can *fire*, which means that the event occurs. When an event fires, a token is removed from each input arc and a token is put on each output arc. Thus, the firing of an event produces the enabling of another event. The tokens in the specification represent the initial state of the system. The exact

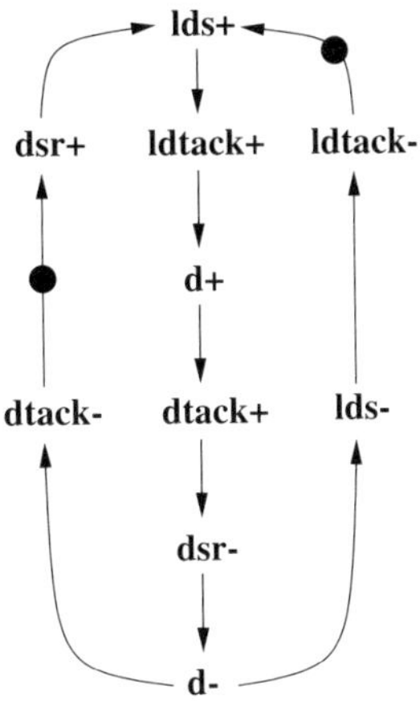

Fig. 2.2. STG for the read cycle of the VME bus

semantics of this token game will be described in Chap. 3, when discussing the Petri net model.

The initial state in the specification of Fig. 2.2 is defined by the tokens on the arcs `dtack-` $\rightarrow$ `dsr+` and `ldtack-` $\rightarrow$ `lds+`. In this state, there is only one event enabled: `dsr+`. It is an event on an input signal that must be produced by the environment. The occurrence of `dsr+` removes a token from its input arc and puts a token on its output arc. In that state, the event `lds+` is enabled. In this case, it is an event on an output signal, that must be produced by the circuit modeled by this specification.

After firing the sequence of events `ldtack+`, `d+`, `dtack+`, `dsr-` and `d-`, two tokens are placed on the arcs `d-` $\rightarrow$ `dtack-` and `d-` $\rightarrow$ `lds-`. In this situation, two events are enabled and can fire in any order independently from each other. This a situation of concurrency, which is naturally modeled by STGs.

2.1.2 Choice in Signal Transition Graphs

In some cases, alternative behaviors can occur depending on how the environment interacts with the system. In our example, the system will react differently depending on whether the environment issues a request to read or a request to write.

Typically, different behaviors are represented by different timing diagrams. For example, Fig. 2.3a and 2.3b depict the STGs corresponding to the read and write cycles, respectively. In these pictures, some arcs have been split and circles inserted in between. These circles represent *places* that can hold tokens. In fact, each arc going from one transition to another has an implicit place that holds the tokens located in that arc.

By looking at the initial markings, we can observe that the transition `dsr+` is enabled in the read cycle, whereas `dsw+` is enabled in the write cycle. The combination of both STGs models the fact that the environment can non-deterministically choose whether to start a read or a write cycle.

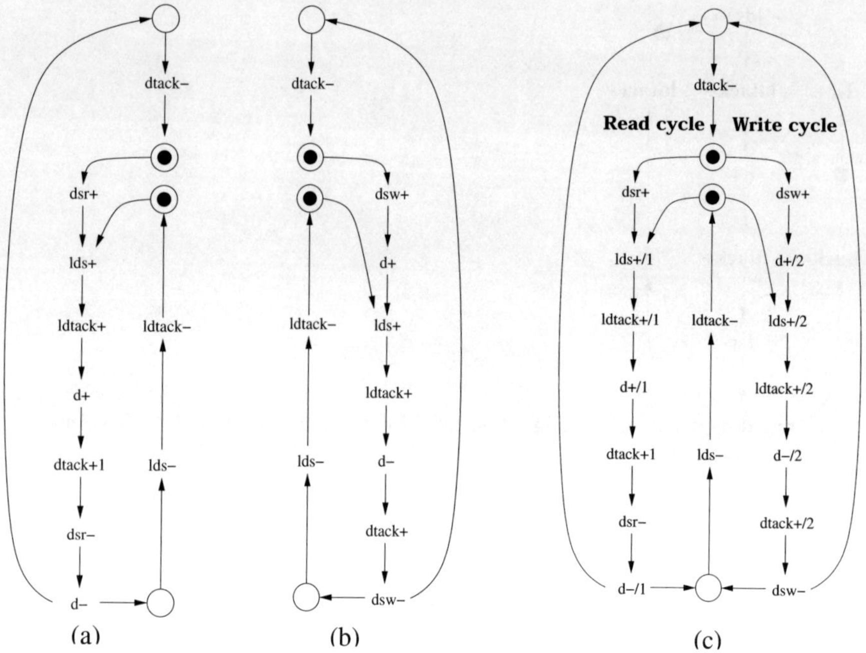

Fig. 2.3. VME bus controller: (**a**) read cycle, (**b**) write cycle, (**c**) read and write cycles

This combination can be expressed by a single STG with a choice place, as shown in Fig. 2.3c. In the initial state, both transitions, `dsr+` and `dsw+`, are enabled. However, when one of them fires, the other is disabled since both transitions are competing for the token in the choice place.

Here is where we can observe an important difference between the expressiveness power of STGs with regard to timing diagrams: the capability of expressing non-deterministic choices. More details on the semantics of STGs in particular, and Petri nets in general, will be given in Chap. 3.

2.2 Transition Systems and State Graphs

2.2.1 State Space

An STG is a succinct representation of the behavior of an asynchronous control circuit that describes the causality relations among the events. However, the state space of the system must be derived by exploring all possible firing orders of the events. Such exploration may result in a state space much larger than the specification.

Fig. 2.4 illustrates the fact that the concurrency manifested by a system may produce extremely large state spaces. The Petri net in Fig. 2.4a describes

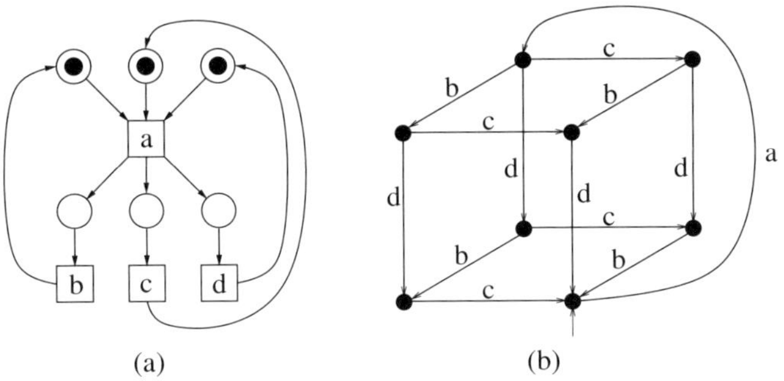

Fig. 2.4. (**a**) Petri net, (**b**) transition system

the behavior of a system with four events. In the initial state, event `a` is enabled. When it fires, three events become enabled, namely `b`, `c` and `d`. They can be fired in any order. After all three have fired, the systems returns to its initial state.

Fig. 2.4b depicts a *transition system* (TS) that models the same behavior. A transition system has nodes (states), arcs (transitions) and labels (events). Each transition is labeled with an event. Each path in the TS represents a possible firing order of the events that label the path. The concurrency of the events `b`, `c` and `d` is represented by a 3-dimensional cube.

A similar system with n concurrent events would be modeled by a TS with an n-dimensional cube, i.e. 2^n states.

Unfortunately, the synthesis of asynchronous circuits from STGs requires an exhaustive exploration of the state space. Finding efficient representations of the state space is a crucial aspect in building synthesis tools. Other techniques based on direct translation of Petri Nets into circuits or on approximations of the state space exist [64, 65], but usually produce circuits with area and performance penalty.

Going back to our example of the VME bus controller, Fig. 2.5 shows the TS corresponding to the behavior of the read cycle. The initial state is depicted in gray.

For simplicity, the write cycle will be ignored in the rest of this chapter. Thus, we will consider the synthesis of a bus controller that only performs read cycles.

2.2.2 Binary Interpretation

The events of an asynchronous circuit are interpreted as rising and falling transitions of digital signals. A rising (falling) transition represents a switch from 0 (1) to 1 (0) of the signal value. Therefore, when considering each signal of the system, a binary value can be assigned to each state for that signal. All those states visited after a rising (falling) transition and before a falling (rising) transition represent situations in which the signal value is 1 (0).

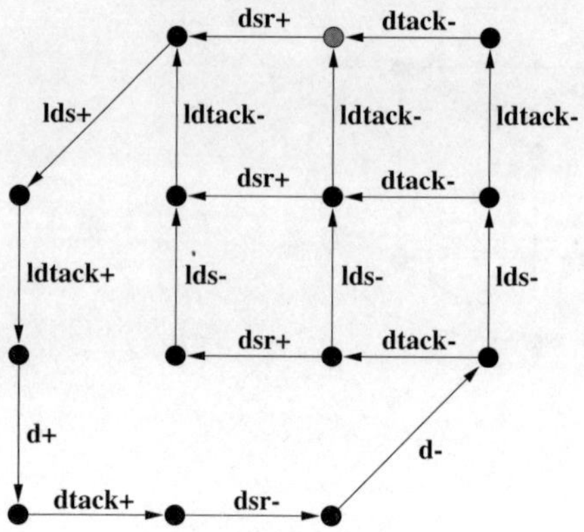

Fig. 2.5. Transition system of the read cycle

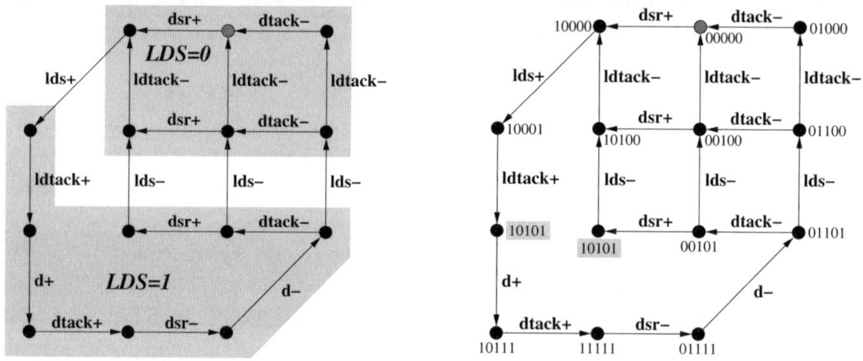

(a) (b)

Fig. 2.6. (**a**) Binary encoding of signal `lds`. (**b**) State graph of the read cycle. States are encoded with the vector (*dsr*, *dtack*, *ldtack*, *d*, *lds*)

In general, the events representing rising and falling transitions of a signal induce a partition of the state space. As an example, let us take signal `lds` of the bus controller. Fig. 2.6a depicts the partition of states. Each transition from `LDS=0` to `LDS=1` is labeled by `lds+` and each transition from `LDS=1` to `LDS=0` is labeled by `lds-`.

It is important to notice that rising and falling transitions of a signal must alternate. The fact that a rising transition of a signal is enabled when the signal is at 1 is considered a specification error. More formally, a specification with such problem is said to have an *inconsistent state coding*.

After deriving the value of each signal, each state can be assigned a binary vector that represents the value of all signals in that state. A transition system with a binary interpretation of its signals is called a *state graph* (SG). The SG of the bus controller read cycle is shown in Fig. 2.6b.

2.3 Deriving Logic Equations

The main purpose of this book is to explain how an asynchronous circuit can be automatically obtained from a behavioral description. This section introduces the bridge required to move from the behavioral domain to the Boolean domain.

We can distinguish two types of signals in a specification: inputs and outputs. Further, some of the outputs may be observable and some internal. Typically, observable outputs correspond to those included in the specification, whereas internal outputs correspond to those inserted during synthesis and not observable by the environment. Synthesizing a circuit means providing an implementation for the output signals of the system.

This section gives an overview of the methods used for the synthesis of asynchronous circuits from an SG.

2.3.1 System Behavior

The specification of a system models a protocol between its inputs and its outputs. At a given state, one or several of these two situations may happen:

- The system is waiting for an input event to occur. For example, in the state `00000` of Fig. 2.6, the system is waiting for the environment to produce a rising transition on signal `dsr`.
- The system is expected to produce a non-input (output or internal) event. For example, the environment is expecting the system to produce a rising transition on signal `lds` in state `10000`.

In concurrent systems, several of these things may occur simultaneously. For example, in state `00101`, the system is expecting the environment to produce `dsr+`, whereas the environment is expecting the system to produce `lds-`. In some other cases, such as in state `01101`, the environment may be expecting the system to produce several events concurrently, e.g. `dtack-` and `lds-`.

The particular order in which concurrent events will occur will depend on the delays of the components of the system. Most of the synthesis methods discussed in this book aim at synthesizing circuits whose correctness does not depend on the actual delays of the components. These circuits are called *speed independent*.

A correct implementation of the output signals must be in such a way that signal transitions on those signals must be generated *if and only if* the environment is expecting them. Unexpected signal transitions, or not generating signal transitions when expected, may produce circuit malfunctions.

2.3.2 Excitation and Quiescent Regions

Let us take one of the output signals of the system, say signal `lds`. According to the specification, the states can be classified into four regions:

- The **positive excitation region** (ER(`lds`+)) includes all those states in which a rising transition of `lds` is enabled.
- The **negative excitation region** (ER(`lds`−)) includes all those states in which a falling transition of `lds` is enabled.
- The **positive quiescent region** (QR(`lds`+)) includes all those states in which signal `lds` is at 1 and `lds`− is not enabled.
- The **negative quiescent region** (QR(`lds`−)) includes all those states in which signal `lds` is at 0 and `lds`+ is not enabled.

Fig. 2.7 depicts these regions for signal `lds`. It can be easily deduced that ER(`lds`+) ∪ QR(`lds`−) and ER(`lds`−) ∪ QR(`lds`+) are the sets of states in which signal `lds` is at 0 and 1, respectively.

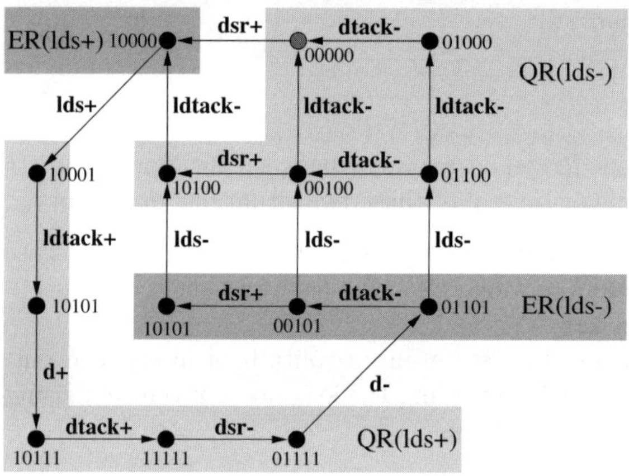

Fig. 2.7. Excitation and quiescent regions for signal `lds`

2.3.3 Next-state Functions

Excitation and quiescent regions represent sets of states that are behaviorally equivalent from the point of view of the signal for which they are defined. The semantics of these regions are the following:

- ER(`lds`+) is the set of states in which `lds` is at 0 and the system must change it to 1.
- ER(`lds`−) is the set of states in which `lds` is at 1 and the system must change it to 0.
- QR(`lds`+) is the set of states in which `lds` is at 1 and the system must not change it.
- QR(`lds`−) is the set of states in which `lds` is at 0 and the system must not change it.

According to this definition, the behavior of each signal can be determined by calculating the *next value* expected at each state of the SG. This behavior can be modeled by Boolean equations that implement the so-called *next-state* functions (see Table 2.1).

Table 2.1. Next-state functions

State region	current value of `lds`	next value of `lds`
ER(`lds`+)	0	1
QR(`lds`+)	1	1
ER(`lds`−)	1	0
QR(`lds`−)	0	0

Let us consider again the bus controller and try to derive a Boolean equation for the output signal `lds`. A 5-variable Karnaugh map for Boolean minimization is depicted in Fig. 2.8. Several things can be observed in that table. There are many cells of the map with a *don't care* (-) value. These cells represent binary encodings not associated to any of the states of the SG. Since the system will never reach a state with those encodings, the next-state value of the signal is irrelevant.

The shadowed cells correspond to states in the excitation regions of the signal. The rest of cells correspond to states in some of the quiescent regions. If we call $f_{\texttt{lds}}$ the next-state function for signal `lds`, here are some examples on the value of $f_{\texttt{lds}}$:

$f_{\texttt{lds}}(10000) = 1$ state in ER(`lds`+)
$f_{\texttt{lds}}(10111) = 1$ state in QR(`lds`+)
$f_{\texttt{lds}}(00101) = 0$ state in ER(`lds`−)
$f_{\texttt{lds}}(01000) = 0$ state in QR(`lds`−)

2.4 State Encoding

At this point, the reader must have noticed a peculiar situation for the value of the next-state function for signal `lds` in two states with the same binary encoding: 10101. This binary encoding is assigned to the shadowed states in Fig. 2.6b.

Unfortunately, the two states belong to two different regions for signal `lds`, namely to ER(`lds`−) and QR(`lds`+). This means that the binary encoding of the SG signals alone cannot determine the future behavior of `lds`. Hence, an ambiguity arises when trying to define the next-state function. This ambiguity is illustrated in the Karnaugh map of Fig. 2.8.

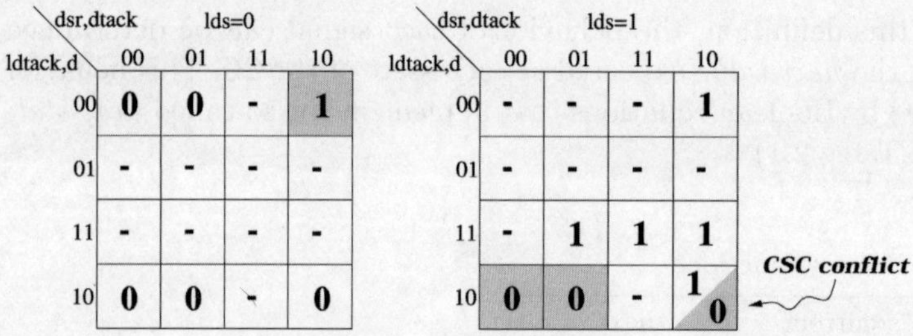

Fig. 2.8. Karnaugh map for the minimization of signal `lds`

Roughly speaking, this phenomenon appears when the system does not have enough memory to "remember" in which state it is. When this occurs, the system is said to violate the *complete state coding* (CSC) property.

Guaranteeing CSC is one of the most difficult problems in the synthesis of asynchronous circuits. Chap. 5 will be completely devoted to methods solving this problem.

Fig. 2.9 presents a possible solution for the SG of the VME bus controller. It consists of inserting a new signal, `csc`, that adds more memory to the system. After the insertion, the two conflicting states are disambiguated by the value of `csc`, which is the last value in the binary vectors of Fig. 2.9.

Now Boolean minimization can be performed and logic equations can be obtained (see Fig. 2.10). In the context of Boolean equations representing gates we shall liberally use the "=" sign to denote "assignment", rather than

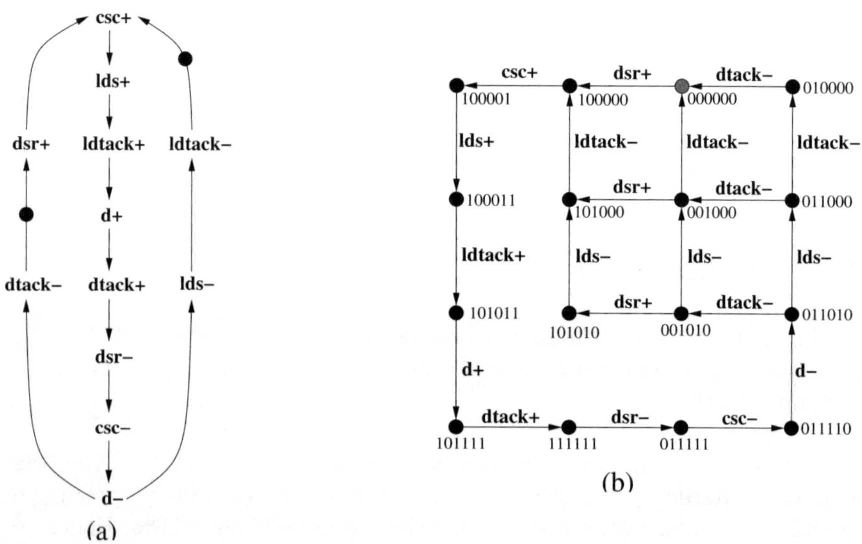

Fig. 2.9. (a) STG and (b) SG with the CSC property

$$
\begin{aligned}
\texttt{lds} &= \texttt{d} + \texttt{csc} \\
\texttt{dtack} &= \texttt{d} \\
\texttt{d} &= \texttt{ldtack} \cdot \texttt{csc} \\
\texttt{csc} &= \texttt{dsr} \cdot (\texttt{csc} + \overline{\texttt{ldtack}})
\end{aligned}
$$

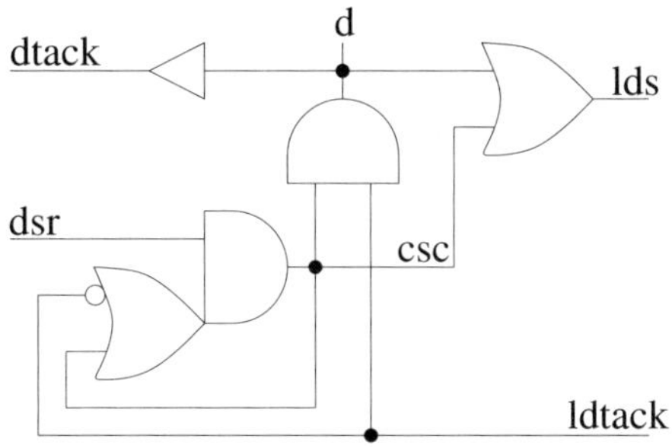

Fig. 2.10. Logic equations and implementation of the VME bus controller

mathematical equality. Hence `csc` on the left-hand side of the last equation stands for the *next* value of signal `csc`, while `csc` on the right-hand side corresponds to its previous value. The resulting circuit is not acyclic. The combinational feedbacks play the rôle of local memory in the system. Chap. 4 will cover other synthesis strategies to derive implementations from STG specifications.

The circuit shown in Fig. 2.10 is said to be speed-independent, i.e. it works correctly regardless of the delays of its components. For this to be true, it is required that each Boolean equation is implemented as one *complex gate.* This roughly means that the internal delays within each gate are negligible and do not produce any externally observable spurious behavior. However, the external delay of the gates can be arbitrarily long.

Note that signal `dtack` is merely implemented as a buffer. One might think that a wire would be enough to preserve that behavior. However, the specification indicates that the transitions on `dtack` must occur always after the transitions on `d`. For this reason, the resulting equation is

$$\texttt{dtack} = \texttt{d}$$

and not vice versa. Thus, the buffer introduces the required delay to enforce the specified causality.

2.5 Logic Decomposition and Technology Mapping

The implementation of a circuit must be usually done under some technological constraints. In semi-custom design, circuits must be built up from cells in a gate library. In custom design, the complexity of the gates is determined by constraints like maximum fanin.

In general, each Boolean equation derived after minimization may not be implementable as one gate. But decomposing such equations into smaller ones that meet the technology constraints may introduce undesired behavior in the system. To illustrate such problem, we will try to decompose the circuit in Fig. 2.10 into another one that only uses 2-input gates. The only gate that does not meet this constraint is the one implementing the signal `csc`.

A possible implementation is presented in Fig. 2.11, where the 3-input gate has been split into two gates. Thus, a new internal signal, `x`, is introduced. Unfortunately, when considering arbitrary delays for the gates, the circuit does not work correctly.

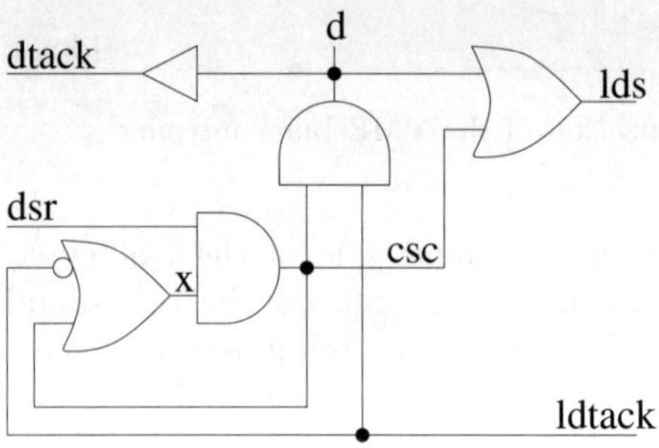

Fig. 2.11. Hazardous implementation of the VME bus controller with 2-input gates

Let us consider the initial state of the specification in which signal `x` is at 1 and the rest of the signals at 0. Let us now consider the following sequence of events:

`dsr+ csc+ lds+ ldtack+ d+ dtack+ dsr- csc-` • `d- dtack- dsr+ csc+ d+`

At the state represented by •, a falling transition of signal `x` is enabled. Let us assume that the gates implementing `x` and `lds` are extremely slow. Under this assumption, and after having fired the events `d-`, `dtack-` and `dsr+`, the internal signal `csc` is ready to rise. This is an event that, although not observable, is not expected in the specification of Fig. 2.9. Unfortunately, this malfunction can also be propagated to an observable output by producing the unexpected event `d+`.

The reader can simply verify that this malfunction cannot occur in the implementation of Fig. 2.10 if it is assumed that the internal delays of the 3-input gate are negligible.

Solving the problem of logic decomposition and technology mapping is another difficult task in the synthesis of asynchronous circuits. Chap. 6 will be devoted to cover this problem.

Meanwhile, a solution for our example is proposed in Fig. 2.12. This solution is obtained by properly re-wiring the new internal signal, so that it cannot manifest any unexpected behavior.

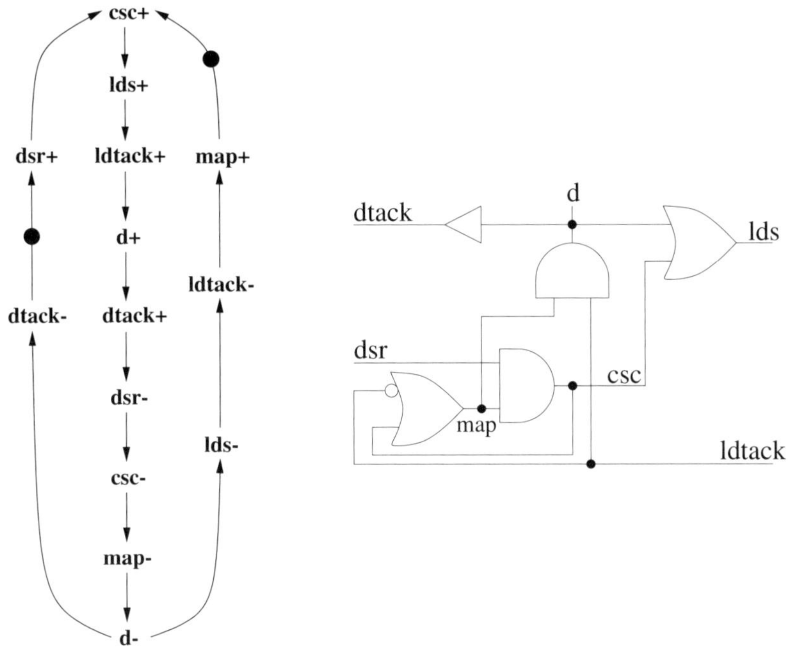

Fig. 2.12. Hazard-free implementation of the VME bus controller with 2-input gates

2.6 Synthesis with Relative Timing

The assumption that the components of a system can have arbitrary delays may be too conservative in practice and lead to inefficient circuits. The knowledge of the timing behavior of some components can be used to simplify the functionality of the system and derive simpler and faster circuits.

Let us illustrate this issue with our example. Let us assume that the response time of the bus is longer than the response time of the device. In particular, this can be translated into the following timing assumption:

> *The falling edge of* `ldtack` *(completion of the read cycle) will always occur before a new request to read is issued* (`dsr+`).

Fig. 2.13a shows the specification of the system with a timing (dashed) arc that represents the previous assumption. Fig. 2.13b depicts the state space after removing those states that become unreachable when considering the timing assumption. In this particular case, the reduction of the state space has a significant impact on the synthesis of the circuit for the following reasons:

- The removal of one of the states, with code 10101, avoids the CSC conflict of the original specification. Therefore, there is no need to insert a new signal to properly encode the SG.

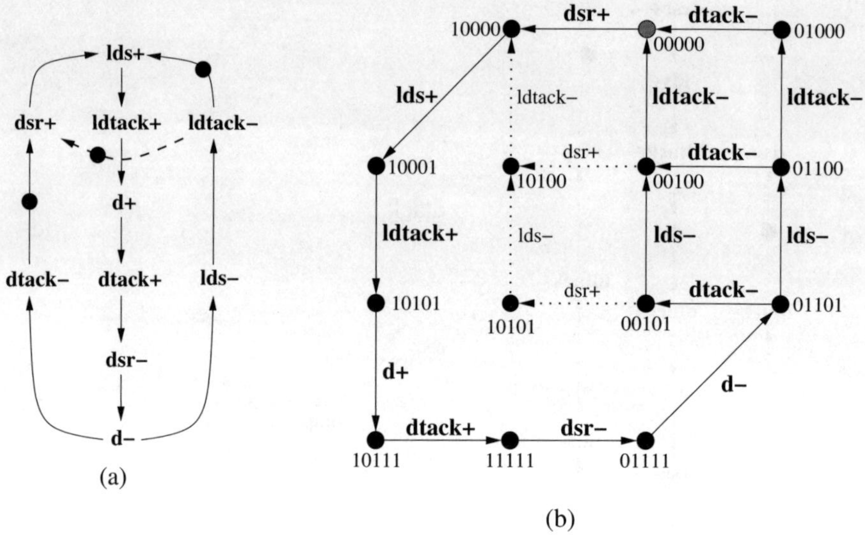

Fig. 2.13a,b. Incorporation of timing assumptions

- The reduction in the number of reachable states enlarges the don't care conditions in the Boolean space. Thus, Boolean minimization is able to derive simpler equations.

The resulting circuit is shown in Fig. 2.14. The correctness of this implementation is however restricted to environments in which the response time of the bus is longer than the one of the device. In other words, the timing assumption used to improve the quality of the circuit now becomes a timing constraint to ensure its correctness.

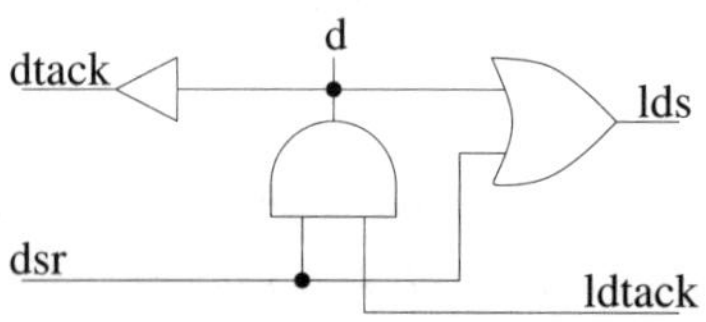

Fig. 2.14. Hazard-free implementation with timing constraints

Chapter 7 will discuss different types of assumptions that can be made to improve the quality of the circuits and strategies to derive those assumptions automatically. It will also describe a method to back-annotate the constraints required to ensure the correctness of the circuit.

2.7 Summary

No clock, distributed control, synchronization through handshakes, ... These are paradigms that make the synthesis of asynchronous controllers significantly different from their synchronous counterparts.

The fact that control is distributed imposes an explicit treatment of the system's concurrency. For this reason, specification models for concurrent systems, such as Petri nets, are frequently used in design flows for asynchronous circuits.

The design flow proposed in this book is depicted in Fig. 2.15. A formal model similar to timing diagrams, STGs, is used for specification.

The synthesis of asynchronous controllers is a difficult and error-prone task. The main purpose of this book is to show that this task can be fully automated. In the forthcoming chapters, methods and algorithms to solve the synthesis problems that appear in the design flow will be proposed and studied.

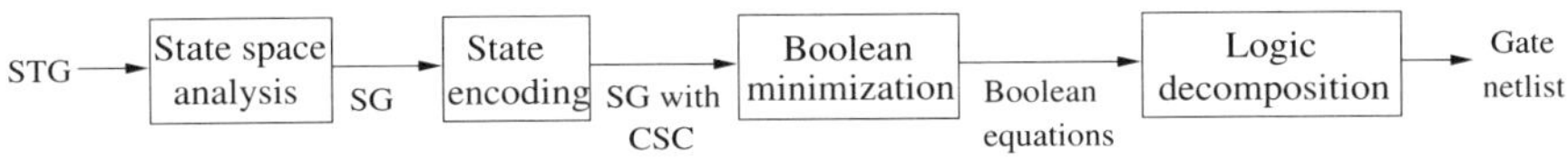

Fig. 2.15. Design flow for asynchronous control circuits

3. Background

In this chapter we will introduce models that are used for the specification, synthesis and verification of asynchronous circuits: Petri Nets and Transition Systems. We will also discuss relationships between the two, and transformations that can be done with each of them. These relationships and transformations are interesting within the context of asynchronous circuit synthesis for several reasons. Firstly, some of the synthesis steps, in particular state encoding as discussed in Chap. 5, can be made more efficient by using concepts, such as regions, that were first define to transform Transition Systems into Petri Nets. Secondly, that same transformation can be used to show in a readable form the result of various synthesis steps to the designer (*back-annotation*). Thirdly, the "new signal insertion" transformation, that is the basis of most synthesis phases, is defined so that it preserves some theoretically important properties of Transition Systems that are discussed in this Chapter.

Petri Nets and Transition Systems have an abstract view of the events and states in the system, without considering their binary encoding. Binary encoded versions of these two models, Signal Transition Graphs and State Graphs, come into play for logic synthesis, and will be discussed in the next chapters.

3.1 Petri Nets

A Petri net [59, 60, 61] is a formal model of concurrent behavior as might arise in asynchronous digital circuits, distributed protocols, operating systems, or multi-processor systems. PNs play the same role for asynchronous digital circuits as automata play for synchronous ones. The major difference between the two is that while automata define system behavior in terms of global states (at any moment a system is in a single state), a PN defines behavior as a composition of local states that might evolve independently influencing each other by synchronization or communication.

Definition 3.1.1 (Nets). *A net is a triple* $N = (P, T, F)$*, where*

- P *is a finite set of places,*
- T *is a finite set of transitions* $(T \cap P = \emptyset)$*,*
- $F : (P \times T) \cup (T \times P) \to \mathbb{N}$ *is the flow function.*

A net does not define the initial state of the system and models behaviors under all possible initial conditions. A Petri Net adds the initial state to the model.

Definition 3.1.2 (Petri Nets). *A Petri net (PN) is a pair* (N, m_0)*, where*

- N *is a net, and*
- $m_0 : P \to \mathbb{N}$ *is the initial marking of the net.*

If the flow of a net is a relation on $(P \cup T)$, i.e. a mapping $F : (P \times T) \cup (T \times P) \to \{0, 1\}$, then the net (and the corresponding PN) is called an *ordinary* net. In this book we will mostly consider ordinary nets and will often talk about the flow relation F instead of the flow function.

It is often necessary to label transitions of a net or a PN by symbols from some alphabet (e.g. by the names of signal transitions in a circuit netlist, as we will often do in the following chapters). Different transitions can be labeled with the same symbol. The labeling function need not be injective, i.e. labeling can be partial (not all of transitions are labeled). Formally,

Definition 3.1.3 (Labeled Nets and Petri Nets). *A Labeled (Petri) net is a triple* $LN = (N, A, \lambda)$*, where*

- N *is a (Petri) net,*
- A *is a finite alphabet,*
- $\lambda : T \to A$ *labels transitions of the (Petri) net with symbols (also called labels) from* A*.*

If the labeling is 1-to-1, i.e. each transition is uniquely labeled by a symbol from the alphabet, then we do not distinguish between the net and the labeled net, since they are isomorphic.

Informally, transitions in a PN represent events in the system (e.g. request to access a memory bank in a multi-processor system, production of a car in a plant, change of a signal value in a digital system, etc.). Places represent placeholders for the conditions for the events to occur or for the resources that are needed. The initial distribution of resources is given by the initial marking m_0. For any place $p \in P$, the initial marking $m_0(p)$ gives the number of *tokens* initially placed into p. Tokens in a place can model the number of data samples required to perform a multiplication, the number of printers in a computer system, or some other resource or condition. Each place in a PN can be viewed as a local state. All marked places taken together form the global state of the system.

The behavior of a PN is defined as a token game - changing markings according to the enabling and firing rules for the transitions.

Definition 3.1.4 (Pre-set and post-set). *The* pre-set of transition t, *denoted by* $^\bullet t$, *is the set of places* p *such that* $F(p,t) > 0$. *The* post-set of transition t, *denoted by* $t^\bullet$, *is the set of places* p *such that* $F(t,p) > 0$. *Symmetrically, the* pre-set of place p, *denoted by* $^\bullet p$, *is the set of transitions such that* $F(t,p) > 0$ *and the* post-set of place p, *denoted by* $p^\bullet$, *is the set of transitions such that* $F(p,t) > 0$. *Informally, the pre-set of a transition (place) gives all its input places (transitions), while its post-set corresponds to its output places (transitions).*

Definition 3.1.5 (Enabling and firing). *A transition* $t \in T$ *is* enabled *at marking* m_1 *if for any place* $p \in^\bullet t : m_1(p) \geq F(p,t)$. *An enabled transition* t *may* fire *producing a new marking* m_2 *such that*

$$\forall p \in P : m_2(p) = m_1(p) - F(p,t) + F(t,p),$$

where $+$ *and* $-$ *are defined component-wise. This is denoted by* $m_1 \stackrel{t}{\rightarrow} m_2$ *or* $m_1 \rightarrow m_2$. *In other words, firing of* t *subtracts* $F(p,t)$ *tokens from each pre-set place* p *of* t *and adds* $F(t,p)$ *tokens to each post-set place* p.

Definition 3.1.6 (Reachability). *The new marking* m_2 *can again make some transitions enabled. We can therefore talk about sequences of transitions that fire in the markings* reachable *from the initial marking* m_0. *Such sequences of transitions will be called* feasible traces *or simply* traces. *The set of all markings reachable from marking* m *is denoted* $[m\rangle$. *The set* $[m_0\rangle$ *of markings reachable from the initial marking of a* PN *induces a graph, called its* **Reachability Graph** *(RG), with nodes corresponding to the reachable markings and edges corresponding to the transitions firing between pairs of markings. Two nodes in the graph are connected with an arc if they correspond to markings* m_i *and* m_j *and* $m_i \rightarrow m_j$ *holds.*

We will further illustrate these notions by example.

3.1.1 The Dining Philosophers

This is a version of the well-known problem [66] of sharing resources (forks) between concurrent processes (philosophers). The system contains m forks for eating spaghetti and n philosophers sitting around the table. To eat spaghetti, a philosopher needs two forks[1]: one from the left and one from the right.

Modeling Concurrency. Fig. 3.1 models the behavior of a single philosopher with two forks, to illustrate how PNs represent **concurrency**. The PN in Fig. 3.1a has four places named fl, fr, hl, hr ("fork left", "fork right", "has left fork", "has right fork") and three transitions named tl, tr, and e ("take left", "take right", and "eat"). Initially, both forks are on the table,

[1] Let us ignore, for the sake of simplicity, whether Italian table manners require this eating protocol.

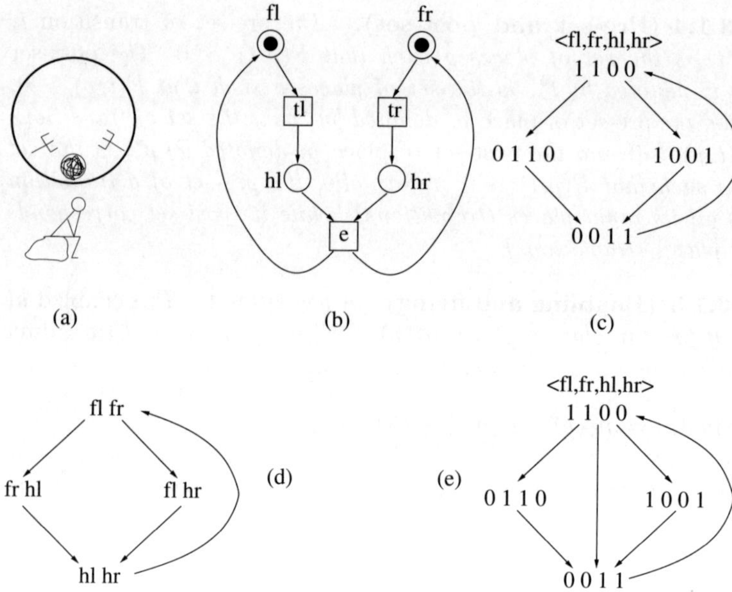

Fig. 3.1. (**a**) Lonely philosopher, (**b**) Petri net, (**c**) reachability graph with vector encoding of markings, (**d**) RG with set encoding of markings, (**e**) RG with true concurrency

and hence places fl and fr are marked with a single token each. Transitions tl and tr are **concurrent** in the initial marking $(1, 1, 0, 0)$, i.e. they are enabled in this marking and can be fired in any order (hands of the philosopher are asynchronous and work independently of each other). Firing one of them does not disable the other, which will also eventually fire. After both forks are taken eating becomes possible (transition e becomes enabled). When e fires the PN returns to the initial marking. Fig. 3.1c shows the reachability graph of this PN. It has four markings and five arcs between the markings, each of which corresponds to the firing of a single transition, namely, $(1,1,0,0) \xrightarrow{tl} (0,1,1,0), (1,1,0,0) \xrightarrow{tr} (1,0,0,1), (0,1,1,0) \xrightarrow{tr} (0,0,1,1), (1,0,0,1) \xrightarrow{tl} (0,0,1,1)$, and $(0,0,1,1) \xrightarrow{e} (1,1,0,0)$.

As shown by this example, markings are denoted by vectors of natural numbers (commas between vector components can be omitted when it does not create confusion). In case of a single philosopher, no place can carry more than one token hence all markings are in fact Boolean vectors with 0/1 components. It is sometimes more convenient to denote markings as multisets (sets in case of Boolean vectors) of marked places as shown in Fig. 3.1d. Set $\{fl, fr\}$ represents the same marking as vector $(1, 1, 0, 0)$. Marking $(2, 1, 0, 0)$ (if reachable) would be denoted as a multiset $\{fl^2, fr\}$. We will use both the vector and the (multi)set notations interchangeably.

A PN is called:

- **k-bounded**, if for every reachable marking the number of tokens in any place is not greater than k (a place is called k-bounded if for every reachable marking the number of tokens in it is not greater than k),
- **bounded**, if there is a finite k for which it is k-bounded,
- *safe*, if it is 1-bounded (a 1-bounded place is called a safe place)

All PNs discussed so far are safe. Fig. 3.2 shows a model for the writer-reader system. It is assumed here that the writer continues publishing and writing books paying no attention to how fast the reader is reading his books. As a result lots of books can be stocked in place p_4. In fact p_4 is unbounded. Fig. 3.3 shows a PN in which a feedback is introduced from the reader to the writer. The writer does not write a book until the previous one has been read. This PN is 2-bounded, since place p_4 (and p_6) can keep up to two tokens. Boundedness of PNs is decidable. Although the complexity of this property is high (in general it requires exponential space [67]), in many applications it can be solved efficiently by using appropriate algorithms to traverse the reachability space, as discussed in Sect 3.3.

Since transitions tl and tr are concurrent, one would expect that they can potentially fire together in the very same instant, as shown by the arc $(1100) \stackrel{tl,tr}{\rightarrow} (0011)$ in Fig. 3.1e. The semantics of PNs that takes into account possible step-transitions when more than one transition fires in the

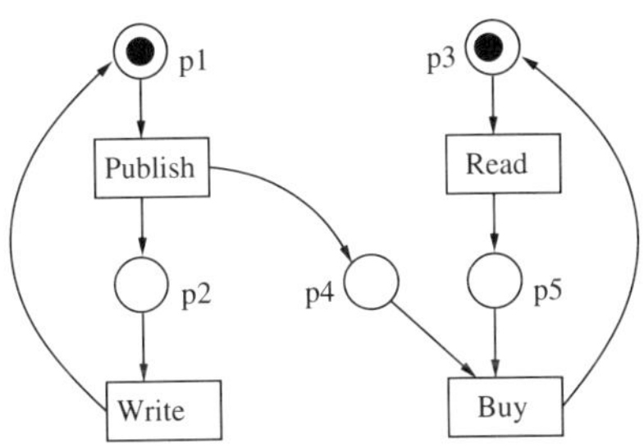

Fig. 3.2. Careless writer - reader: unbounded PN

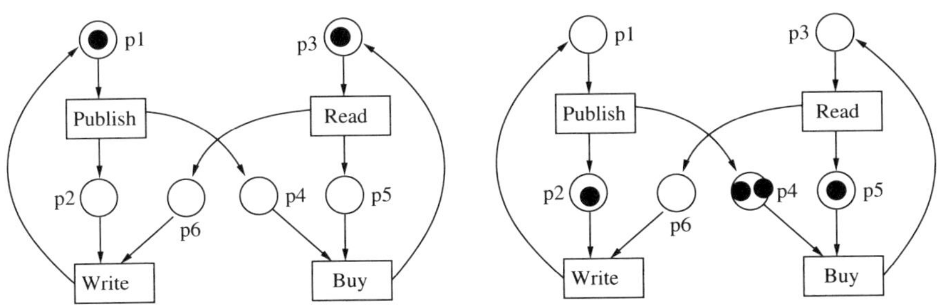

Fig. 3.3. Writer - reader: 2-bounded PN

same instant is called a *true concurrency* semantics. The semantics that abstracts away step-transitions (as shown in Fig. 3.1c) and only allows for single transitions to fire at a time is called an *interleaving semantics*. The true concurrency semantics is more general and allows for handling some subtle properties of concurrent systems that are hard or impossible to see using interleaving semantics [68, 69]. However, it is also more cumbersome as all sets of concurrent transitions should be considered. For the purpose of this book the interleaving semantics suffices, since asynchronous circuit specifications and implementations typically enjoy the same properties for both versions of the semantics. Similarly, the interpretation of the behavior of a single philosopher stays the same under the interleaving and the true concurrency semantics, except that concurrency of transitions tr and tl in the RG is represented by the diamond of four markings and arcs instead of a single step-transition.

Modeling Choice. Let us now consider two philosophers with three forks. Philosophers are in **conflict** over the middle fork, as shown in Fig. 3.4, because place f_2 contains only one token. This place is the only pre-set place for transitions tr_1 and tl_2, both of which are enabled. However, as soon as one of these transitions fires, the other becomes disabled. We say that place f_2 is a **choice** or **conflict** place. The corresponding reachability graph is shown in Fig. 3.5.

Fig. 3.6 adds one more philosopher. Now all three forks are shared. When philosopher i is eating, his left and right neighbor are missing a fork, and they should wait until he finishes and frees the forks. Taking left and right forks for a philosopher are independent actions, e.g. he can take one fork and then wait until the other is free. However, after eating, a philosopher returns both forks simultaneously. Contrary to the two previous examples, the system can

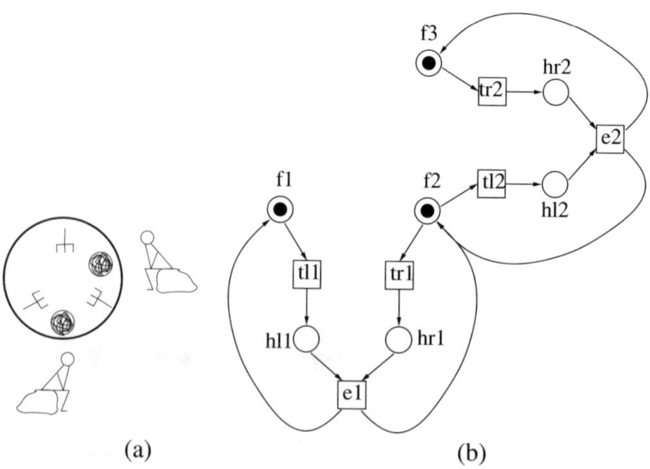

Fig. 3.4. (**a**) Two philosophers with three forks (one shared), (**b**) Petri net

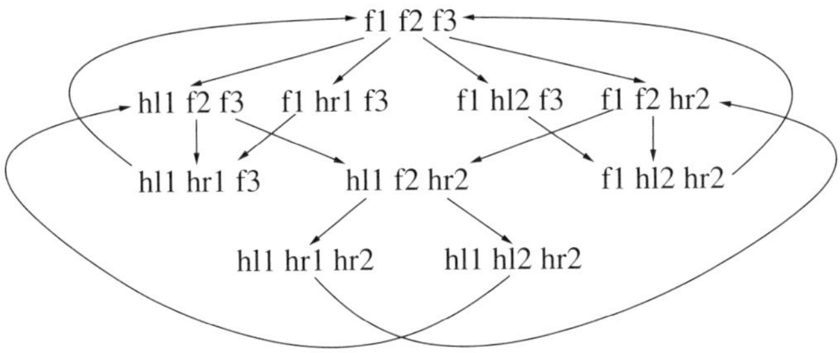

Fig. 3.5. RG for two philosophers with two forks

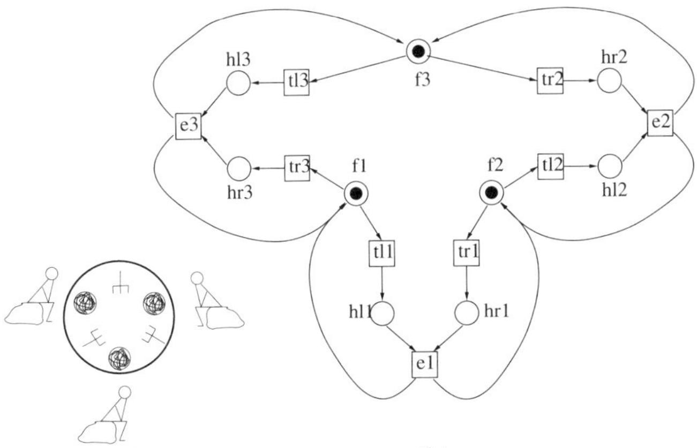

(a) (b)

Fig. 3.6. (**a**) Three philosophers with three forks (all shared), (**b**) Petri net

deadlock. For example, if all three philosophers take their left forks (the PN reaches marking hl_1, hl_2, hl_3) then no transition can ever be enabled. We say that the marking hl_1, hl_2, hl_3 is a *deadlock marking*.

Let us summarize the notions that have been introduced in the above examples.

- **Choice place.** A place is called a choice (or conflict) place if it has more than one output transition.
- **Marked graph and State machine.** A PN is called a *marked graph* (MG) if each place has exactly one input and one output transition. Dually, a PN is called a *state machine* (SM) if each transition has exactly one input and one output place. MGs have no choice. Safe SMs have no concurrency.
- **Free-choice.** A choice place is called *free-choice* if every output transition has only one input place. A PN is *free-choice* if all its choice places are free-choice.
- **Persistency.** A transition t_i is called *non-persistent* if t_i is enabled in a reachable marking m together with another transition t_j, and t_i becomes

disabled after firing t_j. Non-persistency of t_i with respect to t_j is also called *a direct conflict* between t_i and t_j. A PN is *persistent* if it does not contain any non-persistent transition.

- **Pure nets.** A net is called *pure* if for each transition the following condition is satisfied:

$$^\bullet t \cap t^\bullet = \emptyset$$

Interference between Concurrency and Choice. The PN in Fig. 3.1 is a marked graph and is therefore persistent. The PNs in Fig. 3.4 and 3.6 are free-choice. These two PNs are non-persistent (e.g. transitions tr_1 and tl_2 are non-persistent in both nets, since firing one of them disables the other). In free-choice PNs choice and concurrency are separated: choice decisions are made independently from the firing of any other concurrent transition. Fig. 3.7 shows how concurrency and choice can interact. Here each philosopher is not hungry for some time after eating. Only when hunger is approaching and place hi gets a token, the philosopher is allowed to take a shared fork. The other fork can be taken at any time. In the PN on the left, the first philosopher is hungry and can take the shared fork f_2 by firing tr_1. There is no conflict under this marking since transition tl_2 is not enabled. However as a result of firing transition t_2 the second philosopher becomes hungry and transitions tr_1 and tl_2 are in direct conflict, as shown in the figure on the right. Hence firing of t_2, that is concurrent to transition tr_1, leads to potential disabling of tr_1 if the newly enabled transition tl_2 fires faster than tr_1. In the theory of PNs this is called *confusion*, and it cannot occur in free-choice nets. PNs with confusion are in general a more difficult target for the analysis than free-choice nets. Most problems (e.g. finiteness of reachability graph, absence of deadlocks) have a polynomial time solution in the case of confusion-free nets [70].

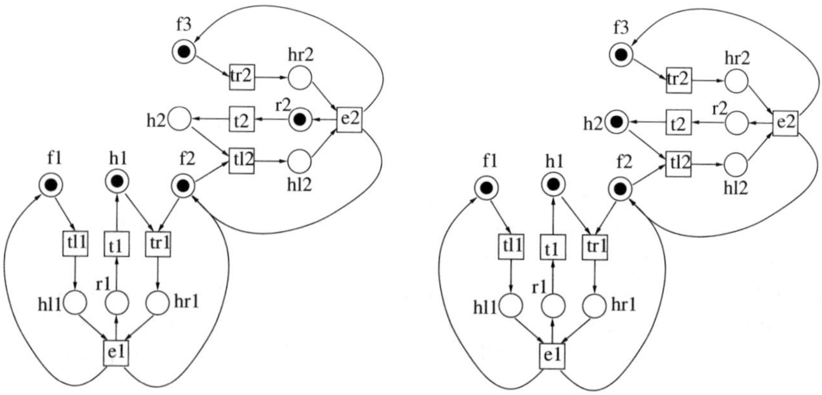

Fig. 3.7. Hungry philosophers illustrating confusion (concurrency-choice interaction) in PNs

3.2 Structural Theory of Petri Nets

Many properties of PNs can be checked by only looking at the underlying structure of the net. This section presents some results of the structural theory of PNs that are relevant for the analysis and synthesis of asynchronous specifications. We refer the reader to more specialized texts for detailed explanations of the structural theory [71, 70, 72, 60].

3.2.1 Incidence Matrix and State Equation

A PN has a bipartite graph as underlying structure and, therefore, it can be represented by a matrix. A PN with p places, $p_1 \ldots p_p$, and t transitions, $t_1 \ldots t_t$, can be represented by a $p \times t$ matrix C, called *incidence matrix* and defined as:

$$C = C^+ - C^-$$

where $C^+_{ij} = F(p_i, t_j)$ and $C^-_{ij} = F(t_j, p_i)$. It can be easily observed that C^+_{ij} and C^-_{ij} represent the number of tokens added to and removed from place p_i when transition t_j fires.

The following incidence matrix represents the net in Fig. 3.8a.

$$C = \begin{bmatrix} -2 & 1 & 1 \\ 1 & -1 & 0 \\ 1 & 0 & -1 \\ 0 & -2 & 2 \end{bmatrix}$$

A marking can also be represented by a p-vector m. Transition t_j is enabled in m if

$$m(i) \geq C^-_{ij}, \quad i = 1, \ldots, p$$

After firing t_j in m, a new marking m' is obtained that can be calculated as follows:

$$m' = m + Cu_j$$

where u_j is a vector with t elements, all of them with value 0 except for the entry corresponding to transition t_j, which has value 1. The previous equation is known as the *state equation*. In general, a firing sequence can be represented by a firing count vector u in which each entry denotes the number of times that transition fires in the sequence. Thus, the following equation holds when firing the sequence from m:

$$m' = m + Cu$$

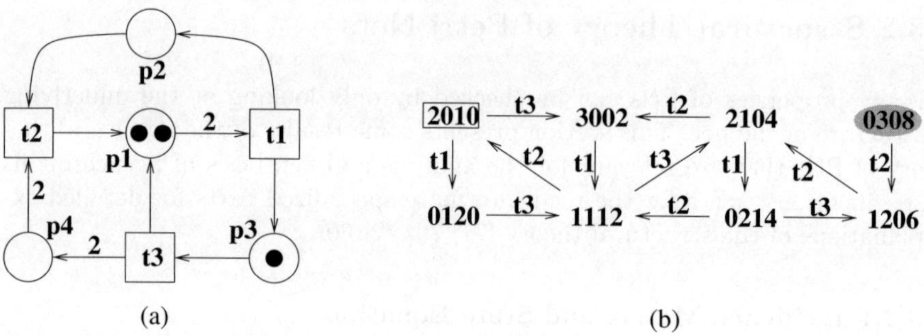

Fig. 3.8. (**a**) Petri net (from [60]) and (**b**) potentially reachable state space. The 4-element vectors represent the number of tokens in $p_1 \cdots p_4$

Consider the example of Fig. 3.8a. After firing the sequence $t_3 t_1 t_3$ from the initial marking $m_0 = (2, 0, 1, 0)$, the following marking is obtained:

$$\begin{bmatrix} 2 \\ 1 \\ 0 \\ 4 \end{bmatrix} = \begin{bmatrix} 2 \\ 0 \\ 1 \\ 0 \end{bmatrix} + \begin{bmatrix} -2 & 1 & 1 \\ 1 & -1 & 0 \\ 1 & 0 & -1 \\ 0 & -2 & 2 \end{bmatrix} \begin{bmatrix} 1 \\ 0 \\ 2 \end{bmatrix}$$

The state equation of a PN gives a necessary condition for a marking m to be reachable from the initial marking m_0. If we call $\Delta m = m - m_0$, a necessary condition for m to be reachable from m_0 is that there exist a non-negative integer vector x such that

$$Cx = \Delta m$$

The markings that fulfill the previous condition define the *potentially reachable set* of markings of the PN. Fig. 3.8b represents the set of potentially reachable markings of the PN in Fig. 3.8a. Note that one of the markings, $(0, 3, 0, 8)$, is not reachable from $m_0 = (2, 0, 1, 0)$. However, the necessary reachability condition holds for $x = (3, 0, 4)$. Therefore, the set of potentially reachable markings can be characterized by algebraic techniques and is an overestimation of the set of reachable markings.

3.2.2 Transition and Place Invariants

An integer solution of the equation $\Delta m = 0$, i.e.

$$Cx = 0$$

is called a transition invariant (T-invariant). Informally, a T-invariant characterizes a set of transitions whose firing leads to the same marking. T-invariants represent repetitive behaviors in PNs. In the previous example, the vector $x = (1,1,1)$ is a T-invariant, since the firing of t_1, t_2 and t_3 from any marking leads to the same marking.

An integer solution of the equation

$$yC = 0$$

where y is a p-vector, is called a place invariant (P-invariant)[2] . A P-invariant denotes a set of places whose weighted sum of tokens remains constant in all reachable markings.

The set of invariants of a PN can be represented by a basis, from which all the other invariants can be obtained by linear combination. Here, we will concentrate on P-invariants, which provide an efficient way of estimating the reachability set of a PN. Calculating a basis of P-invariants or T-invariants can be performed in polynomial time in the size of the net [60].

In the previous example, a basis of all P-invariants is given by the matrix

$$\begin{bmatrix} 2 & 4 & 0 & -1 \\ 1 & 1 & 1 & 0 \end{bmatrix}$$

By calculating the weighted sum of tokens for m_0, the following equations hold for any reachable marking from m_0:

$$\begin{aligned} 2m(p_1) + 4m(p_2) - m(p_4) &= 4 \\ m(p_1) + m(p_2) + m(p_3) &= 3 \end{aligned}$$

These equations exactly characterize the potential set of reachable markings previously defined by the necessary condition of the state equation.

A subnet of a PN that is an SM is said to be a state machine component (SMC) of the PN. Each SMC has a corresponding P-invariant in the PN [70].

In the examples of Fig. 3.2 and 3.3, the sets of places $\{p_1, p_2\}$ and $\{p_3, p_5\}$ define two SMCs. The corresponding invariants are represented by the following equations:

$$\begin{aligned} m(p_1) + m(p_2) &= 1 \\ m(p_3) + m(p_5) &= 1 \end{aligned}$$

As shown in the next section, P-invariants and SMCs are useful objects to efficiently encode the reachability graph. Efficient encoding schemes are crucial for symbolic analysis.

3.3 Calculating the Reachability Graph of a Petri Net

This section briefly explains how sets of markings of a PN can be represented by means of Boolean functions and efficiently manipulated by using Binary

[2] Place invariants are also sometimes called S-invariants.

Decision Diagrams (BDDs) [73, 74]. As an example, it will be shown how traversal of the reachability set of markings can be done symbolically using BDDs.

Given the set P of places of a PN, a set M of *safe*[3] markings over P can be represented by its *characteristic function*, denoted χ_M, that is a Boolean function that evaluates to 1 for each marking in M. In particular, the set of reachable markings of a PN in which a place p_i is marked is denoted by χ_i.

A BDD is a directed acyclic graph with one root and two leaf nodes (0 and 1). Each non-leaf node is labeled with a Boolean variable and has two outgoing arcs with labels 0 and 1. A BDD represents a Boolean function as follows: each variable assignment has a corresponding path that goes from the root node to one of the leaf nodes. The label of the leaf node is the value of the function for that assignment. As an example, the BDD depicted in Fig. 3.9b represents the function $f(v_2, v_3, v_4) = \overline{v_2} + \overline{v_3}\,\overline{v_4}$ corresponding to the reachability set of the PN in Fig. 3.9a. The value of the function for the assignment $v_2 = v_3 = 1$ and $v_4 = 0$ is 0. This assignment corresponds to the path $v_2 \xrightarrow{1} v_3 \xrightarrow{1} 0$. Note that the function has 20 minterms and that each of them corresponds to a reachable marking of the function. Here are some examples of how the markings (represented as sets of marked places) are encoded:

$$(0,1,0,1,1) \rightarrow \{p_1, p_5, p_8\}$$
$$(1,1,1,0,0) \rightarrow \{p_3, p_6\}$$
$$(0,0,1,0,1) \rightarrow \text{unreachable marking}$$

We refer the reader to [74] for further details on how to manipulate Boolean functions efficiently by means of BDDs. With such a representation, the basic operations on sets of states (union, intersection, complement) can be mimicked as Boolean operations on Boolean functions (or, and, not). Moreover, such functions can have a compact representation. In [75] some examples are shown in which graphs with 10^{18} states can be represented with BDDs having 10^3 nodes by using very naive encodings (one Boolean variable per place).

Starting from a simple example of a PN specification shown in Fig. 3.9a, we will explain the following steps.

- Selection of encoding variables for representing markings of individual places. Instead of the simplest naive encoding, one variable per place, we use a more dense encoding based on a state machine decomposition of the original PN.
- Traverse the reachability space of the net starting from the initial marking until the (unique) fixed point is reached. The traversal is done symbolically,

[3] The technique can also be extended to k-bounded PNs by appropriate encoding, but this aspect will not be discussed in the following.

using the BDD representation of Boolean characteristic functions for sets of markings and the transition relation of the net.

- As a result of the computation, we will get a BDD representation of the characteristic function of the reachability set. This function has the encoding variables as its arguments.
- Given this function, we show how different operations with sets of markings can be performed.

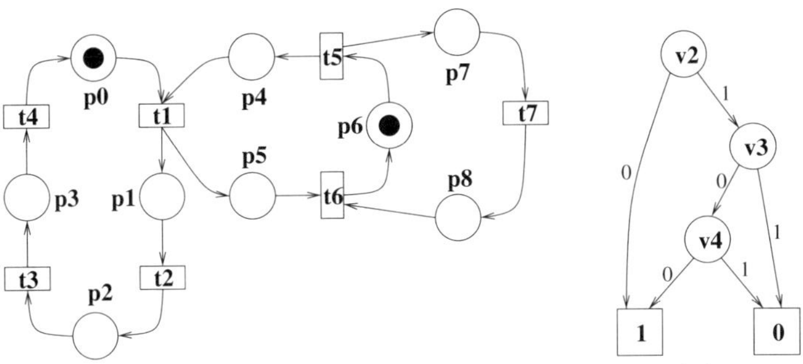

(a) (b)

place	v_0	v_1	v_2	v_3	v_4
p_0	0	0	-	-	-
p_1	0	1	-	-	-
p_2	1	0	-	-	-
p_3	1	1	-	-	-
p_4	-	-	0	0	-
p_5	-	-	0	1	-
p_6	-	-	1	0	-
p_6, p_7	-	-	-	-	0
p_8	-	-	-	-	1

(c)

Fig. 3.9. (**a**) Petri net, (**b**) BDD representing the set of reachable markings and (**c**) place encoding

3.3.1 Encoding

The natural and naive encoding that uses one variable for each place of the PN is not usable in practice, since most traversal algorithms have a complexity that heavily depends on the number of variables. More efficient solutions are possible, as discussed next.

The dense encoding used for the markings of the PN of Fig. 3.9 is based on the observation that the sets of places $P_1 = \{p_0, p_1, p_2, p_3\}$, $P_2 = \{p_4, p_5, p_6\}$ and $P_3 = \{p_6, p_7, p_8\}$ define three state machines components, S_1, S_2, and S_3 with the following sets of transitions $T_1 = \{t_1, t_2, t_3, t_4\}$, $T_2 = \{t_1, t_5, t_6\}$, and $T_3 = \{t_5, t_6, t_7\}$, respectively. This information can be structurally obtained by using algebraic methods, as shown in the previous section. Given the initial marking of the net, at most one of the places of each SMC is marked at each marking. Thus, the following encoding can be proposed. Two Boolean variables (v_0 and v_1) are used to encode the token in S_1 and two Boolean variables (v_2 and v_3) for S_2. Only one Boolean variable (v_4) is sufficient for S_3, since S_2 already uniquely encodes p_6 ($v_2 = 1 \Leftrightarrow p_6$ has a token) and only places p_7 and p_8 must be distinguished. The table in Fig. 3.9c proposes an encoding for the places that leads to the following characteristics functions for places:

place, p	χ_p	place, p	χ_p
p_0	$\overline{v_0}\,\overline{v_1}$	p_1	$\overline{v_0}v_1$
p_2	$v_0\overline{v_1}$	p_3	v_0v_1
p_4	$\overline{v_2}\,\overline{v_3}$	p_5	$\overline{v_2}v_3$
p_6	$v_2\overline{v_3}$	p_7	$\overline{v_4}(\overline{v_2} + v_3)$
p_8	v_4	–	–

Note that since variable v_4 has value 0 both for places p_6 and p_7 and its characteristic function χ_{p_6} depends only on variables corresponding to the second state machine, S_2, the characteristic function for p_7 is $\chi_{p_7} = \overline{v_4}\,\overline{\chi_{p_6}} = \overline{v_4}(\overline{v_2} + v_3)$.

More sophisticated encoding strategies based on the calculation of S-invariants are presented in [76].

3.3.2 Transition Function and Reachable Markings

The methods used for deriving the transition function and calculating the reachable markings of a PN are similar to those used for reachability analysis and equivalence checking of finite state machines [77].

For calculating the transition function let us first introduce the characteristic functions for two important sets related to a transition $t \in T$:

$$\begin{aligned} \mathrm{E}(t) &= \bigwedge_{p_i \in {}^\bullet t} \chi_{p_i}(V) \ (t \text{ enabled}), \\ \mathrm{ASM}(t) &= \bigwedge_{p_i \in t^\bullet} \chi_{p_i}(V) \ (\text{all successors marked}). \end{aligned}$$

Function $E(t)$ ($ASM(t)$) states that all input (output) places of transition t contain a token. For example, for transition t_1 in Fig. 3.9a

$$E(t_1) = \chi_{p_0}(V)\chi_{p_4}(V) = \overline{v_0}\,\overline{v_1}\,\overline{v_2}\,\overline{v_3}$$

and

$$ASM(t_1) = \chi_{p_1}(V)\chi_{p_5}(V) = \overline{v_0}v_1\,\overline{v_2}v_3$$

Let M_N be the set of all possible markings of a PN N. The *transition function* of a Petri net is a function $\delta_N : 2^{M_N} \times T \longrightarrow 2^{M_N}$ that transforms, for each transition t, a set of markings M into a new set of markings M' one-step reachable from M by firing transition t: $M' = \delta_N(M,t) = \{m_2 \in M_N : \exists m_1 \in M, m_1 \xrightarrow{t} m_2\}$.

We illustrate the calculation of the transition function with an example (for a more detailed explanation of the algorithms see [75]). Assume that in the example of Fig. 3.9a we calculate $M' = \delta_N(M,t_1)$ given the set of markings: $M = \{\{p_0,p_4,p_7\},\{p_0,p_4,p_8\},\{p_3,p_6\}\}$ represented by the characteristic Boolean function:

$$M = \chi_{p_0}\chi_{p_4}\chi_{p_7} + \chi_{p_0}\chi_{p_4}\chi_{p_8} + \chi_{p_3}\chi_{p_5}\chi_{p_7} = \overline{v_0}\,\overline{v_1}\,\overline{v_2}\,\overline{v_3} + v_0v_1\overline{v_2}v_3\overline{v_4}.$$

First, by calculating $M \cdot E(t_1) = \overline{v_0}\,\overline{v_1}\,\overline{v_2}\,\overline{v_3}$ one selects those markings from M in which t_1 is enabled. After that we determine all literals that are logically implied by the characteristic functions of input places of transition t_1, i.e. places p_0 and p_4. The following implications hold: $\chi_{p_0} \Rightarrow \overline{v_0}$, $\chi_{p_0} \Rightarrow \overline{v_1}$, $\chi_{p_4} \Rightarrow \overline{v_2}$, and $\chi_{p_4} \Rightarrow \overline{v_3}$. All implied literals should be cofactored from function $M \cdot E(t)$ [4]. The result is $(M \cdot E(t))_{\overline{v_0}\,\overline{v_1}\,\overline{v_2}\,\overline{v_3}} \equiv 1$. Informally this corresponds to removing predecessor places of t_1 from the characteristic function. The final step is adding successor places of t_1 into the characteristic function, which is done by calculating the conjunction of the previous result with $ASM(t_1)$:

$$(M \cdot E(t))_{\overline{v_0}\,\overline{v_1}\,\overline{v_2}\,\overline{v_3}} \cdot ASM(t_1) = \overline{v_0}v_1\,\overline{v_2}v_3.$$

This result is the characteristic function of M', which can be considered as a Boolean version of the marking equation of the PN. The existential quantification of t from the above formula gives the set of markings $\delta_N(M)$ reachable by firing any *one* enabled transition from a marking in M [5].

In such a way, starting from the initial marking, by iterative application of the transition function we calculate the characteristic function of the

[4] The cofactor of $f(v_1, \dots, v_i, \dots, v_n)$ with respect to literal v_i, denoted by f_{v_i}, is $f(v_1, \dots, 1, \dots, v_n)$ and with respect to literal $\overline{v_i}$,$f_{\overline{v_i}}$, is $f(v_1, \dots, 0, \dots, v_n)$. The notion of cofactor can be generalized to a product of literals, e.g. $f_{v_i,v_2} = (f_{v_1})_{v_2}$ [74].

[5] The existential abstraction of $f(v_1, \dots, v_i, \dots, v_n)$ with respect to v_i is $\exists_{v_i}(f) = f_{v_i} + f_{\overline{v_i}}$ [74].

reachability set until the fixed point in the calculation is reached. The resulting function for the reachability set for the example considered above is $f(v_2, v_3, v_4) = \overline{v_2} + \overline{v_3}\ \overline{v_4}$. All calculations can be done using a BDD representation of the corresponding characteristic Boolean functions.

3.4 Transition Systems

The RG of a labeled PN is a particular form of a more abstract state-based model that is called Transition System.

Definition 3.4.1 (Transition system). *A* Transition System *(TS) is a quadruple [78]* $TS = (S, E, T, s_0)$, *where*

- S *is a* non-empty *set of* states,
- E *is a set of* events *(E and S must be disjoint sets),*
- $T \subseteq S \times E \times S$ *is a* transition *relation, and*
- s_0 *is an* initial state.
  The elements of T *are called the* transitions *of TS and will be often denoted by* $s \stackrel{e}{\to} s'$ *instead of* (s, e, s').

A transition system is *finite* if S and E are finite. In the sequel, only finite transition systems will be considered.

Definition 3.4.2 (Determinism). *A TS is called deterministic if for each state* s *and each label* a *there is at most one state* s' *such that* $s \stackrel{a}{\to} s'$.

Otherwise, it is called *non-deterministic*.

The transitive closure of the transition relation T is called the *reachability relation* between states and is denoted by T^*. In other words, state s' is reachable from state s if there is a (possibly empty) sequence of transitions from T: $\sigma = (s, e_1, s_1), ..., (s_k, e_k, s')$. This is denoted by $s \stackrel{\sigma}{\to} s'$ or simply by $s \stackrel{*}{\to} s'$, if the sequence is not important. We also write $s \stackrel{e}{\to}$, $s \stackrel{\sigma}{\to}$, $\stackrel{e}{\to} s$, and $\stackrel{\sigma}{\to} s$ if there is a state $s' \in S$ such that $s \stackrel{e}{\to} s'$, $s \stackrel{\sigma}{\to} s'$, $s' \stackrel{e}{\to} s$, or $s' \stackrel{\sigma}{\to} s$, correspondingly. Each state is reachable from itself, since we allow empty sequences in the definition.

Two states, s and s', are *confluent* if there is a state s'' which is reachable both from s and s' [79]. Note that according to the definition of reachability s'' can coincide with s or s'.

Every transition system $TS = (S, E, T, s_0)$ is assumed to satisfy the following axioms:

(A1) No self-loops: $\forall (s, e, s') \in T : s \neq s'$;
(A2) Every event has an occurrence: $\forall e \in E : \exists (s, e, s') \in T$;
(A3) Every state is reachable from the initial state: $\forall s \in S : s_0 \stackrel{*}{\to} s$.

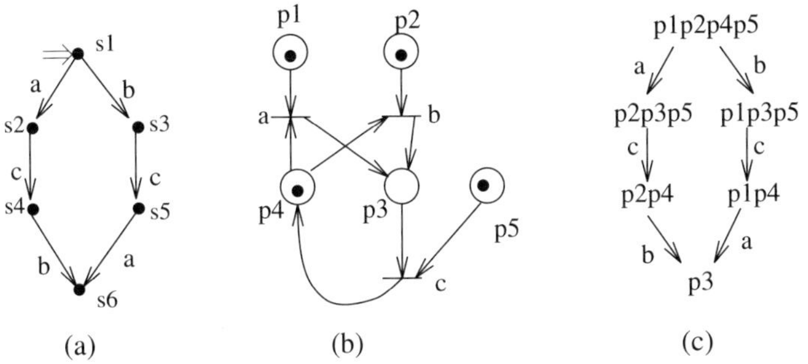

Fig. 3.10. An example of Transition System (**a**), a corresponding PN (**b**) and its RG (**c**)

A TS can be represented by an arc-labeled directed graph. A simple example of a TS without cycles is shown in Fig. 3.10a. A PN expressing the same behavior as the TS in Fig. 3.10a is shown in Fig. 3.10b.

To link the notions of TS and RG, we can say that the *Reachability Graph* (RG) of a PN N, denoted by $RG(N)$, is a transition system in which the set of states is the Reachability Set, the events are the transitions of the net and a transition (m_1, t, m_2) exists if and only if $m_1 \stackrel{t}{\rightarrow} m_2$. One can easily check that the RG in Fig. 3.10c, derived from the PN in Fig. 3.10b, is isomorphic to the TS shown in Fig. 3.10a.

3.5 Deriving Petri Nets from Transition Systems

Section 3.1 showed that for every bounded PN one can build a TS, its reachability graph, that describes the same behavior and hence is equivalent. In this section we address the inverse problem: given a TS derive an equivalent PN. In the context of asynchronous design, this is useful, for example, to generate a PN describing compactly the behavior of an asynchronous circuit ("back-annotation"). The technique for constructing a PN from a TS is based on the theory of state regions, that we sketch below. We will not formally discuss the equivalence relations between PNs and TSs. One can assume that two models are equivalent if they generate the same set of feasible traces, i.e. they generate the same language. In fact, a stronger bisimulation equivalence holds as shown in [80]. The reader interested in the formal theory of PN synthesis is referred to [78, 81, 80].

3.5.1 Regions

Let S' be a subset of the states of a TS, $S' \subseteq S$. If $s \notin S'$ and $s' \in S'$, then we say that transition $s \stackrel{a}{\rightarrow} s'$ *enters* S'. If $s \in S'$ and $s' \notin S'$, then transition $s \stackrel{a}{\rightarrow} s'$ *exits* S'. Otherwise, transition $s \stackrel{a}{\rightarrow} s'$ is *unrelated* with S' (i.e. *does not cross* S'). In particular, if $s \in S'$ and $s' \in S'$, then the transition is said to be *internal* to S', and if $s \notin S'$ and $s' \notin S'$, then the transition is *external* to S'.

Definition 3.5.1. *Let* $TS = (S, E, T, s_0)$ *be a* TS. *Let* $S' \subseteq S$ *be a subset of states and* $e \in E$ *be an event. The following conditions (in the form of predicates) are defined for* S' *and* e*:*

$$\begin{aligned} \texttt{in}(e, S') &\equiv \exists (s, e, s') \in T : s, s' \in S' \\ \texttt{out}(e, S') &\equiv \exists (s, e, s') \in T : s, s' \notin S' \\ \texttt{enter}(e, S') &\equiv \exists (s, e, s') \in T : s \notin S' \wedge s' \in S' \\ \texttt{exit}(e, S') &\equiv \exists (s, e, s') \in T : s \in S' \wedge s' \notin S' \end{aligned}$$

The notion of a *region* is central for the synthesis of PNs. Intuitively, each region corresponds to a place in the synthesized PN, so that there is a 1-1 correspondence between states of the region and markings of the PN in which this place has a token.

Definition 3.5.2 (region). *A set of states* $r \subseteq S$ *in* $TS = (S, E, T, s_0)$ *is called a* region *if the following two conditions are satisfied for each event* $e \in E$*:*

$$\begin{aligned} &\textit{(I)}\ \texttt{enter}(e, r) \Rightarrow \neg\texttt{in}(e, r) \wedge \neg\texttt{out}(e, r) \wedge \neg\texttt{exit}(e, r) \\ &\textit{(II)}\ \texttt{exit}(e, r) \Rightarrow \neg\texttt{in}(e, r) \wedge \neg\texttt{out}(e, r) \wedge \neg\texttt{enter}(e, r) \end{aligned}$$

A region is a subset of states with which *all* transitions labeled with the same event e have exactly the same "entry/exit/unrelated" relation. This relation will become the predecessor/successor relation in the Petri net. The event may either always be an *enter* event for the region (case (I) in the previous definition), or always be an *exit* event (case (II)), or never "cross" the region's boundaries (each transition labeled with e is *internal* or *external* to the region if the antecedents of neither (I) nor (II) hold). The transition corresponding to the event will be predecessor, successor or unrelated with the corresponding place respectively.

Let us consider the TS shown in Fig. 3.10. The set of states $r_3 = \{s_2, s_3, s_6\}$ is a region, since all transitions labeled with a and with b enter r_3, and all transitions labeled with c exit r_3. On the other hand, $\{s_2, s_3\}$ is not a region since transition $s_1 \xrightarrow{b} s_3$ enters this set, while another transition also labeled with b, $s_4 \xrightarrow{b} s_6$, does not. Similar violations of the region conditions exist for two transitions labeled with a. However, there are no violations for c since both transitions labeled with c exit this set of states.

Each TS has two *trivial regions*: the set of all states, S, and the empty set. Further on we will only consider non-trivial regions. The set of non-trivial regions of TS will be denoted by R_{TS}. For each state $s \in S$ we define the set of non-trivial regions containing s, denoted by R_s.

A region r is a *pre-region* of event e if there is a transition labeled with e which exits r. A region r is a *post-region* of event e if there is a transition labeled with e which enters r. The set of all pre-regions and post-regions of e is denoted with ${}^{\circ}e$ and e° respectively. By definition it follows that if $r \in {}^{\circ}e$, then all transitions labeled with e exit r. Similarly, if $r \in e^{\circ}$, then all transitions labeled with e enter r. Let r and r' be regions of a TS. A region r' is said to be a *subregion* of r iff $r' \subset r$. A region r is a *minimal* region if there is no other region r' which is a subregion of r.

There are eight non-trivial regions in the TS from Fig. 3.10:

$$r_1 = \{s_1, s_3, s_5\}; r_2 = \{s_1, s_2, s_4\}; r_3 = \{s_2, s_3, s_6\}; r_4 = \{s_1, s_4, s_5\}$$

$$r_5 = \{s_1, s_2, s_3\}; r_6 = \{s_4, s_5, s_6\}; r_7 = \{s_2, s_4, s_6\}; r_8 = \{s_3, s_5, s_6\}.$$

All of these regions are minimal. Pre-regions and post-regions are defined as follows: ${}^{\circ}a = \{r_1, r_4\}$; ${}^{\circ}b = \{r_2, r_4\}$; ${}^{\circ}c = \{r_3, r_5\}$; $a^{\circ} = \{r_3, r_7\}$; $b^{\circ} = \{r_3, r_8\}$; $c^{\circ} = \{r_4, r_6\}$.

3.5.2 Properties of Regions

The following propositions state a few important properties of regions [82, 78, 80].

Property 3.5.1. If r and r' are two different regions such that r' is a subregion of r, then $r - r'$ is a region.

Property 3.5.2. A set of states r is a region, if and only if its coset $\overline{r} = S - r$ is a region, where S is a set of all states of the TS.

Property 3.5.3. [82, 80] Every region can be represented as a union of disjoint minimal regions.

3.5.3 Excitation Regions

While regions in a TS are related to places in the corresponding PN, the excitation region [64] (ER) for event a is the maximal set of states in which transition a is enabled. Therefore, excitation regions are related to transitions of the PN. Similarly to ERs, we define switching regions as sets of states reached *immediately after* the occurrence of an event.

Definition 3.5.3 (Excitation and switching regions). *A set of states S' is called an* excitation region *for event a, denoted by $ER(a)$, if it is a* maximal *set of states such that for every state $s \in S'$ there is a transition $s \xrightarrow{a}$.*

A set of states S' is called a switching region *for event a, $SR(a)$, if it is a* maximal *set of states such that for every state $s \in S'$ there is a transition $\xrightarrow{a} s$.*

In the TS from Fig. 3.10a, $ER(a) = \{s_1, s_5\}$ and $SR(a) = \{s_2, s_6\}$.

3.5.4 Excitation-Closure

If a TS satisfies the excitation-closure property (as defined below), then deriving an equivalent PN requires only one transition per event of the TS. We first describe the constraints that an excitation-closed TS must satisfy.

Definition 3.5.4 (Excitation-closed TS (ECTS)). *A transition system $TS = (S, E, T, s_0)$ is called* excitation-closed *if it satisfies the following two axioms:*

(A4) Excitation-closure: For each event a: $\bigcap_{r \in {}^\circ a} r = ER(a)$
(A5) Event effectiveness: For each event a: ${}^\circ a \neq \emptyset$

(A4) and (A5) imply that the excitation region for every event a is equal to the intersection of its pre-regions. It is easy to see that the TS shown in Fig. 3.10 is excitation-closed since all axioms (A1) – (A5) are satisfied. For example, there are two pre-regions of event a, $r_1 = \{s_1, s_3, s_5\}$ and $r_4 = \{s_1, s_4, s_5\}$. Since $ER(a) = r_1 \cap r_4 = s_1, s_5$, the excitation-closure condition for event a is satisfied. Similarly, the excitation-closure condition is satisfied for all other events and, therefore, an equivalent PN has one transition per event of the TS.

The TS shown in Fig. 3.11a is a cyclic excitation-closed TS, while Fig. 3.11b shows a TS that is not excitation-closed. The excitation-closure property is violated for events a and b. Let us consider event a. The only minimal pre-region of a is region $\{s_1, s_3, s_5, s_7\}$. However, $ER(a)$ has one state less, $ER(a) = \{s_1, s_3, s_5\}$.

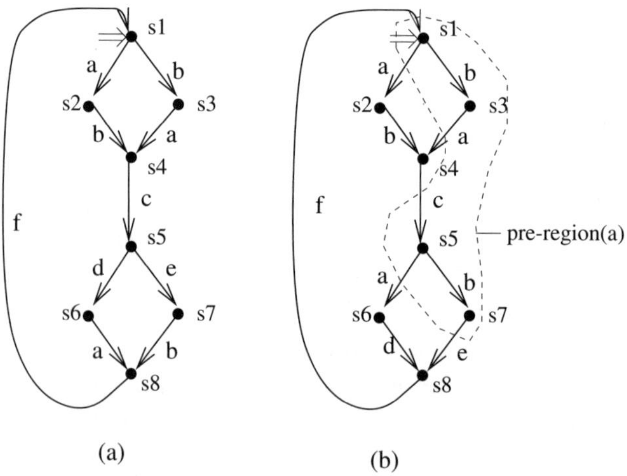

Fig. 3.11. Examples of excitation-closed (**a**) and non-excitation-closed (**b**) TSs

For any ECTS it is possible two derive two equivalent PNs that are called the saturated PN and the minimal saturated PN, using the following algorithm.

Algorithm: (minimal) saturated PN synthesis

- For each event $e \in E$ generate a transition labeled with e in the PN;
- For each (minimal) region $r_i \in R_{TS}$ generate a place r_i;
- Place r_i contains a token in the initial marking iff the corresponding region r_i contains the initial state of the ETS s_0;
- The flow relation is as follows: $e \in r_i\bullet$ iff r_i is a pre-region of e and $e \in \bullet r_i$ iff r_i is a post-region of e, i.e.

$$F_{TS} \stackrel{def}{=} \{(r,e)|r \in R_{TS} \ \wedge \ e \in E \ \wedge \ r \in {}^{\circ}e\} \cup\{(e,r)|r \in R_{TS} \ \wedge \ e \in E \ \wedge \ r \in e^{\circ}\}$$

Intuitively, the net obtained using all regions has as many places as non-trivial regions. Each pre-region (post-region) of an event is a predecessor (successor) place of the corresponding transition. All places that correspond to regions that cover the initial state must be marked in the initial marking.

A PN which is synthesized following this procedure is called a *saturated* net, since all regions are mapped into the corresponding places. Any ECTS has a unique saturated net, however this has a lot of redundancy. As shown in [82], it is enough to consider only minimal regions. The net constructed from all minimal regions is also unique and is called a *minimal saturated net*. We will go two steps further, and will first synthesize a *place-irredundant* PN, and then a *place-minimal* PN, still preserving equivalence between its RG and the ECTS. The above algorithm can be applied to any given set of regions that satisfies the excitation-closure condition, instead of the set of all regions, or the set of all minimal regions.

Fig. 3.12 shows an example of ECTS, the resulting minimal saturated PN, and its RG. Note that the reachability graph generates the same set of traces as the original ECTS, but is not isomorphic to it. In fact it corresponds to a minimized version of the original ECTS. The PN synthesis algorithm indirectly performs TS minimization, similar to the classical state minimization of finite automata. However, the relation with classical state minimization is not straightforward and is analyzed more in detail in [80].

3.5.5 Place-Irredundant and Place-Minimal Petri Nets

A minimal saturated net can be redundant. Many places can still be removed from it while still preserving the equivalence between its RG and the ETS. By analogy with logic minimization, a *saturated net* is like the set of *all implicants* for a Boolean function, while a *minimal saturated net* is like the set of *all prime implicants*. Our goal is to provide a method for constructing an *irredundant net with minimal regions*, which is similar to an *irredundant*

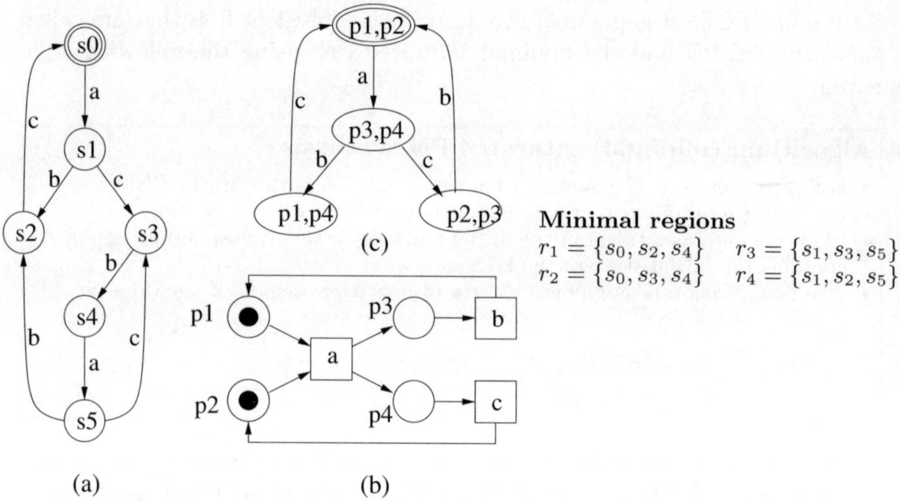

Fig. 3.12. (**a**) An example of ECTS, (**b**) Petri net after synthesis of the ECTS, (**c**) Reachability Graph of a PN equivalent to the original ECTS

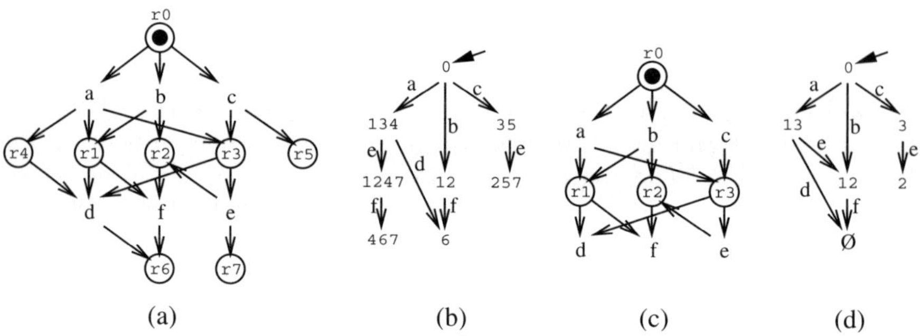

Fig. 3.13. (**a**) Minimal saturated PN, (**b**) RG with each state labeled with the indices of the marked places, (**c**) place-irredundant PN, (**d**) equivalent RG corresponding to the place-irredundant PN

cover of prime implicants [6]. Unfortunately, the analogue of Quine and McCluskey's result does not hold in this case, i.e. there exist ECTSs for which *any minimum* corresponding PN requires using at least one non-minimal region. An example of such a case is given below. Let us describe how to build a place-irredundant and a place-minimal net from an ECTS.

Definition 3.5.5 (Place-irredundant and place-minimal net). *A labeled Petri Net is*

- *place-irredundant if no place can be removed from it without losing the equivalence of the* RG*,*
- *place-minimal if any equivalent* PN *with the same set of transitions contains a greater or equal number of places.*

Definition 3.5.6 (Irredundant and minimal sets of regions).

- *A set of regions R is called* redundant *if there is a region $r \in R$ such that $R - \{r\}$ still satisfies the excitation-closure condition. Otherwise, R is called* irredundant.
- *A set of regions R is called* minimal *if it is irredundant and any other irredundant set of regions contains a greater or equal number of regions.*

Theorem 3.5.1. *Let R be an irredundant or a minimal set of regions of an* ECTS *TS. Let $N_R = (R, E, F_R, s_R)$ be a safe pure* PN *obtained from TS by the* PN *synthesis Algorithm 3.5.4 using R instead of the set of all regions.*

- *If R is an irredundant set of regions, then N_R is place-irredundant.*
- *If R is a minimal set of regions, then N_R is place-minimal.*

Fig. 3.13a shows a minimal saturated PN and Fig. 3.13b shows its RG. Regions $r0$, $r1$, $r2$ and $r3$ are enough to guarantee the excitation-closure of the RG and a place-irredundant PN can be obtained (c) with an equivalent RG (d).

A place-irredundant PN can be always obtained using an irredundant set of *minimal* regions. In that respect minimal regions resemble prime implicants in Boolean minimization. However, a minimal set of regions does not necessarily contain minimal regions only. Fig. 3.14 shows a transition system (on the left) and a place-irredundant net, corresponding to the (unique) irredundant set of minimal regions (in the middle). However, a place-minimal safe pure net (on the right) can be obtained from the place-irredundant net by merging two minimal regions, which are denoted in the TS with dotted lines, into one non-minimal region.

The relationship between irredundant sets of minimal regions and minimal sets of regions is studied in [80]. It is shown there that a place-minimal net can always be obtained from one of the irredundant sets of minimal regions, by merging some disjoint minimal pre-regions into non-minimal pre-regions.

Based on the equivalences for TSs and PNs presented in this section, the framework depicted in Fig. 3.15 and described in the next section can be devised for the synthesis of PNs.

- If the original TS is not excitation-closed, then (as will be shown in the next section) the labels of the TS for some violating event are split to obtain an equivalent ECTS. In this ECTS the same event can be represented by a few different labels. Next, all minimal pre-regions *that are predecessors of some event* are generated to derive a minimal saturated PN. This restriction to

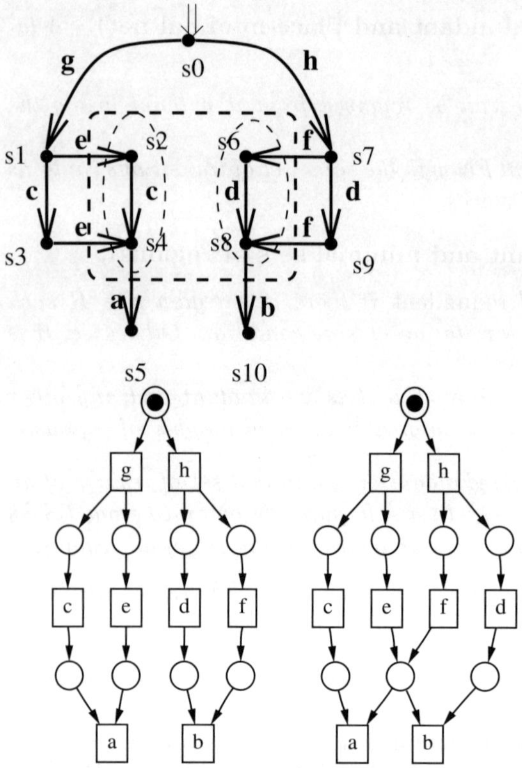

Fig. 3.14. ECTS, its place-irredundant and place-minimal net

event predecessors only is due to our region generation mechanism, and has shown to be sufficient in practice to obtain good results using a reasonable amount of computation time.

- Then, an irredundant subset of pre-regions is calculated and a place-irredundant PN obtained.
- Finally, by merging minimal pre-regions further minimization of regions can be obtained. Exploring all place-irredundant nets can be computationally very expensive. Hence we use only a greedy place merging starting from a place-irredundant net, thus yielding a quasi-place-minimal PN.

3.6 Algorithm for Petri Net Synthesis

The algorithm for synthesis of a PN is shown in Fig. 3.15 and is illustrated by two examples in Fig. 3.14 and Fig. 3.17. The first one illustrates synthesis for an ECTS, the second does it for a TS that does not satisfy the excitation-closure condition. The input of this algorithm is a TS. The output is a PN whose RG is equivalent to the TS.

First the algorithm performs the splitting of labels if the initial TS is not an ECTS. Then all minimal pre-regions for each event are generated, from which an irredundant set of regions is chosen. From this set a place-irredundant net is generated by using Algorithm 3.5.4.

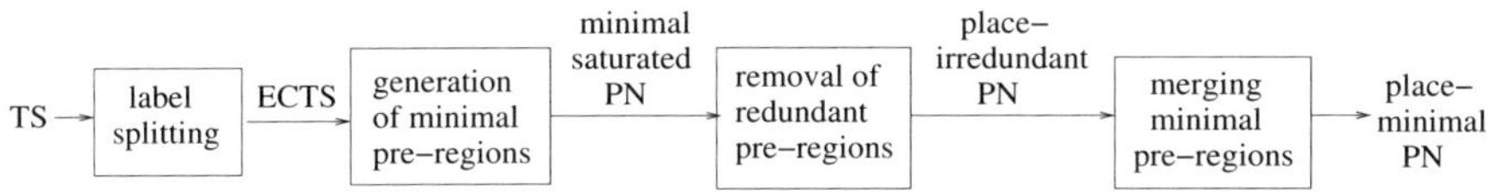

Fig. 3.15. Framework for the synthesis of PNs from TSs

3.6.1 Generation of Minimal Pre-regions

The generation of pre-regions of an event e is based on the fact that any pre-region must cover $ER(e)$. Starting from $ER(e)$, any event with an illegal crossing relation is legalized by adding new states to a set of states containing $ER(e)$ until it becomes a region. An exhaustive search through all the possible legalizations guarantees that all minimal pre-regions that are predecessors of some event will eventually be found.

The restriction to pre-regions is not a severe limitation, because the excitation-closure condition is based only on predecessors of a given event. This means that we can find a PN equivalent to any ECTS by looking only at predecessor regions. We can lose with respect to minimality, but efficient implementation is usually more of a concern than exact minimization.

As illustrated in Fig. 3.17, if $ER(a)$ violates region conditions, then two expansions are possible. The algorithm expands the set of states in both directions, thus implicitly generating a binary exploration tree as shown in Fig. 3.17. The search tree is however reduced in a few ways:

- If the same set of states is generated more than once, then only one branch of the tree is explored.
- If a region is generated during the exploration, then the search along this branch is immediately stopped. This is sound, since the search tree is monotonic in the following sense: each parent vertex of the tree is a subset of the child vertex. Hence, it is not possible to generate minimal regions along the branches starting from another minimal region.
- If the target of the procedure is just to produce an irredundant set of regions, then the excitation-closure for a given event is checked on-the-fly and the search is stopped as soon as it is satisfied. This mechanism can be used to produce a locally optimal solution even for very large TSs.

The complexity of region generation is known to be polynomial in the size of the TS [83], which gives an upper bound on the size of the search tree.

3.6.2 Search for Irredundant Sets of Regions

Let R be a set of regions such that both the excitation-closure and the event effectiveness conditions hold (the set of all minimal pre-regions of an ECTS satisfies these two conditions). Note that if R satisfies the event effectiveness condition, and some regions are removed from R so that the excitation-closure condition still holds, then the event effectiveness condition remains satisfied. Therefore, we need to monitor only the excitation-closure condition, while removing redundant regions.

We will illustrate how an irredundant set of places can be calculated by means of the example of Fig. 3.16. Table 3.1 presents all minimal pre-regions of the TS.

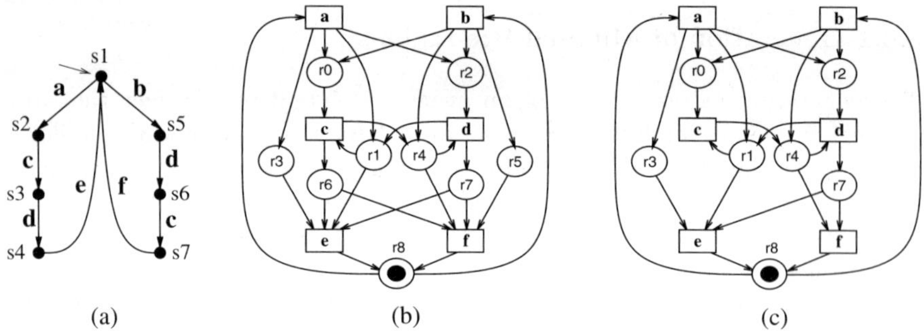

Fig. 3.16. (**a**) Transition system. (**b**) Minimal saturated net. (**c**) Place-irredundant net

Table 3.1. All minimal pre-regions of the transition system depicted in Fig. 3.16a

pre-region	**ev.**	**pre-region**	**ev.**	**pre-region**	**ev.**
$r_0 = \{s_2, s_5, s_6\}$	c	$r_1 = \{s_2, s_4, s_6\}$	c, e	$r_2 = \{s_2, s_3, s_5\}$	d
$r_3 = \{s_2, s_3, s_4\}$	e	$r_4 = \{s_3, s_5, s_7\}$	d, f	$r_5 = \{s_5, s_6, s_7\}$	f
$r_6 = \{s_3, s_4, s_7\}$	e, f	$r_7 = \{s_4, s_6, s_7\}$	e, f	$r_8 = \{s_1\}$	a, b

As a preliminary step, *essential regions* are calculated. A region r is *essential* if there exists a state s and an event e such that $r \in {}^\circ e$, $s \notin r$ and for all $r' \in {}^\circ e$, $r' \neq r$ we have $s \in r'$ (i.e. r is the only region that removes from the intersection of pre-regions a state in which e is not enabled). For example, for event c we have

$$ {}^\circ c = \{r_0, r_1\}; \qquad ER(c) = \{s_2, s_6\} = r_0 \cap r_1 $$

In this case, both r_0 and r_1 are essential, since none of them can be removed from ${}^\circ c$ without violating its excitation-closure. Similarly we can deduce that r_2, r_4 and r_8 are also essential (r_2 and r_4 are essential for d and r_8 for a, b). Thus we have four non-essential regions: r_3, r_5, r_6 and r_7.

Next, for each event with non-essential pre-regions (e and f in the example), all minimal covers are implicitly generated. To reduce the complexity of the problem, essential regions are assumed to be implicitly included in each cover. For event e, we have two minimal covers: $\{r_6\}$ and $\{r_3, r_7\}$. For event f we also have two minimal covers: $\{r_7\}$ and $\{r_5, r_6\}$. Finding a minimum cost cover can be posed as finding a minimum cost solution of a Boolean equation describing the covering conditions [84, 85]. The equation corresponding to the example is as follows:

$$(r_6 + r_3 \cdot r_7) \cdot (r_7 + r_5 \cdot r_6) = 1$$

A cost must be assigned to each region, according to the objective function to be minimized, which depends on the application. For example, if we want to minimize the total number of places and arcs (a heuristic measure of the "simplicity" of the PN), then we can assign to each place p a cost of

$$|\bullet p| + |p \bullet| + 1$$

If we want to minimize only the number of places and obtain a place-minimal PN, then the cost of each place is 1.

In our case,

$$\text{cost}(r_3) = \text{cost}(r_5) = 3; \qquad \text{cost}(r_6) = \text{cost}(r_7) = 4$$

and two minimum-cost covers exist: $\{r_3, r_7\}$ and $\{r_5, r_6\}$ (the former is shown in Fig. 3.16c). There is another possible solution ($\{r_6, r_7\}$), but it has non-minimum cost. The existence of two place-minimal nets for this example gives a negative answer to a question posed in [86]: whether there always exists at most one optimal net which could be considered canonical.

3.6.3 Label Splitting

The set of minimal pre-regions of an event a is calculated by gradually expanding $ER(a)$ to obtain sets of states that do not violate the "entry-exit" relationship. When excitation-closure is not fulfilled (see Definition 3.5.4), i.e.

$$\bigcap_{r \in {}^{\circ}a} r \neq ER(a)$$

some events must be split to make the TS excitation-closed.

The strategy to split events is as follows. During the expansion of $ER(a)$ toward the pre-regions of a, several sets of states are explored. We focus our attention on sets of states S' such that

$$ER(a) \subseteq S' \subset \bigcap_{r \in {}^\circ a} r$$

For each of these sets of states, the number of events that violate the region conditions are calculated. Finally, the set that has the least number of "bad" events is selected. If several sets have the same number of "bad" events, the smallest one is selected.

The selected set of states is then forced to be a region. Informally, this is done by splitting the labels of those events that do not fulfill the region conditions. This strategy guarantees that the new intersection of pre-regions is closer to $ER(a)$.

An example is depicted in Fig. 3.17a and 3.17b for the pre-regions of event c. Initially $ER(c) = \{s_2, s_5\}$ is taken for expansion. Next, two possible legalizations for event b are considered. Further expansions are applied until all branches of the search tree find a region. In this case, regions covering states in $SR(c)$ have been also explored. The example also illustrates how all branches will eventually be pruned, in the worst case, when covering the whole set of states. Let us call r' the intersection of the regions found in the expansion. We have

$$r' = \{s_1, s_2, s_3, s_4, s_5, s_6, s_7\} \cap \{s_2, s_3, s_5, s_6\} = \{s_2, s_3, s_5, s_6\}$$

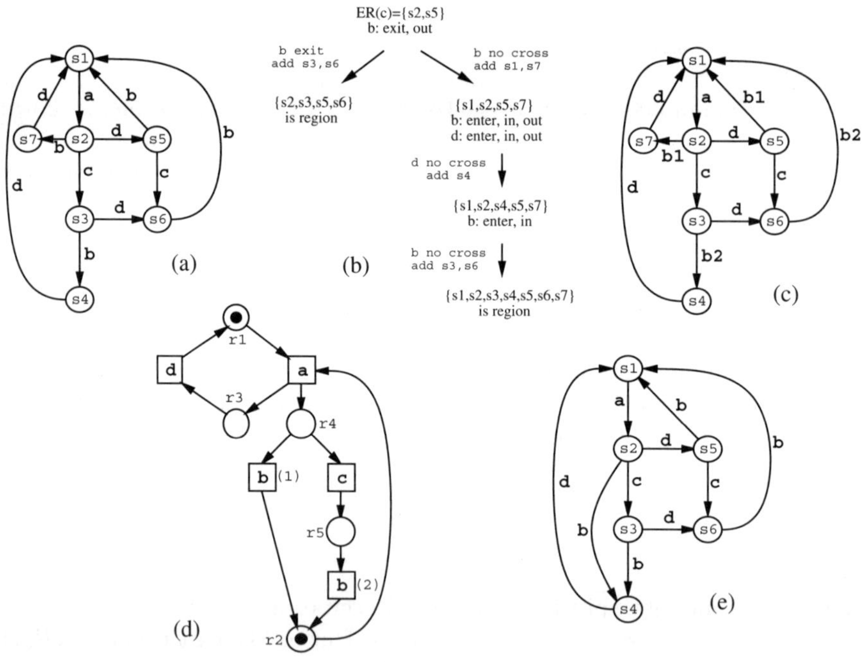

Fig. 3.17. (**a**) TS, (**b**) expansion tree for pre-regions of event c, (**c**) ECTS after label splitting for event b, (**d**) PN, (**e**) RG of the PN

The strategy for label splitting will take all those explored sets r such that

$$\{s_2, s_5\} \subseteq r \subset r'$$

All three states explored before finding regions are good candidates. However the set $\{s_2, s_5\}$ is the best one by the fact that only one event violates the crossing conditions and it makes the intersection of pre-regions smaller (closer to ER). Thus, event b is split into two new events (b_1 and b_2) for $\{s_2, s_5\}$ to become a region. The new TS is equivalent to the original one (if one treats labels b, b_1 and b_2 as belonging to same equivalence class of labels for event b) and is now an ECTS. The corresponding PN is shown in Fig. 3.17d and its RG in Fig. 3.17e. Note that it contains one state less than the original TS, due to the implicit minimization for states $s4$ and $s7$.

3.7 Event Insertion in Transition Systems

Several synthesis steps for asynchronous circuits are based on constrained transformations of a TS derived from a specification in the form of a labeled PN. Two operations form the basis of such transformations:

- event insertion and
- concurrency reduction.

In this section we will discuss the first of them, *insertion of a single event* into a TS. The second will be briefly discussed in Chap. 7 and is studied in more detail in [87, 88].

The goal of event insertion is to refine the specification *while preserving behavioral equivalence*. This is defined here as trace equivalence when the added event is hidden, plus a set of "correctness" properties that are typical of the asynchronous circuit synthesis setting. We will rely on a simple scheme which consists of two steps, while possible alternatives are discussed in [89].

- Choose in the original TS a set of states r in which the new event x will be enabled. r corresponds to the excitation region of event x in the new TS and therefore is denoted as $ER(x)$ in Fig. 3.18.
- Delay all transitions that exit the set of states r until event x fires.

Definition 3.7.1 (Event insertion). *Let $TS = (S, E, T, s_0)$ be a transition system and $x \notin E$ be a new event. Assume that $r \subseteq S$ is an arbitrary subset of states. Let r', $r' \cap S = \emptyset$, be a set of new states such that for each $s \in r$ there is one state $s' \in r'$ and vice versa. The insertion of x in TS by r produces another transition system $TS' = (S', E', T', s'_0)$ defined as follows:*

$$\begin{aligned}
S' &= S \cup r' \\
E' &= E \cup \{x\} \\
T' &= T \cup \{(s \xrightarrow{x} s') | s \in r \ \wedge \ s' \in r'\} \cup \\
&\quad \{(s'_1 \xrightarrow{a} s'_2) | s_1, s_2 \in r \ \wedge \ (s_1 \xrightarrow{a} s_2) \in T\} \cup \\
&\quad \{(s'_1 \xrightarrow{a} s_2) | s_1 \in r \ \wedge \ s_2 \notin r \ \wedge \ (s_1 \xrightarrow{a} s_2) \in T\} - \\
&\quad \{(s_1 \xrightarrow{a} s_2) | s_1 \in r \ \wedge \ s_2 \notin r\}
\end{aligned}$$

The transformation of TS to TS' using Definition 3.7.1 splits each state $s \in r$ in S into two states s and s' in S'. All other states $s \notin r$ have only one state in S'. Fig. 3.18 illustrates how event insertion is performed.

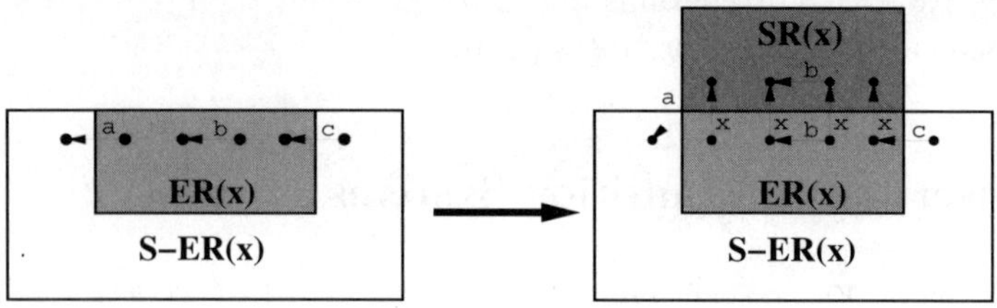

Fig. 3.18. Insertion of event x from $ER(x)$

The following properties must be typically preserved when performing transformations on TSs in the context of asynchronous synthesis: trace equivalence, persistency, commutativity, determinism, and deadlock freedom. Trace equivalence captures the notion of "implementation", in that the specification and the resulting circuit must have the same observable behavior. Persistency and commutativity are related to the notion of speed-independence, i.e. to the property of an asynchronous circuit that its behavior does not depend on the speed of its components (gates). As shown in [79], these two properties ensure that a deterministic TS has a speed-independent implementation. Persistency was discussed in the context of PNs in Sect. 3.1 and states that no event can be disabled by any other event. Commutativity guarantees that the same state of the TS is reached under any order of enabled event firing. Determinism, as in Definition 3.4.2, is an essential property of circuits built from Boolean gates. Finally, deadlock freedom guarantees that the liveness of the initial TS is preserved.

We will now formulate persistency, commutativity, and deadlock freedom of an event within a given set of states for the TSs.

Definition 3.7.2 (Event persistency). *Let $TS = (S, E, T, s_0)$ be a transition system. An event $a \in E$ is said to be* persistent *in $r \subseteq S$ iff:* $\forall s_1 \in r : [s_1 \xrightarrow{a} \ \wedge \ (s_1 \xrightarrow{b} s_2) \in T] \implies s_2 \xrightarrow{a}$

An event $a \in E$ is said to be persistent *if a is persistent in S.*

Definition 3.7.3 (Commutativity). *A transition system TS is called commutative if for any traces ab and ba that are feasible from some state*

$s_1 \in S$ both traces lead to the same state, i.e. if $s_1 \stackrel{a}{\to} s_2$, $s_2 \stackrel{b}{\to} s_4$ and $s_1 \stackrel{b}{\to} s_3$, $s_3 \stackrel{a}{\to} s_5$ then $s_4 = s_5$.

Definition 3.7.4 (Relative deadlock freedom). *Let TS' be a deterministic TS derived by event insertion from a deterministic TS TS. We say that TS' is relatively deadlock free with respect to TS if whenever state s' is a deadlock (i.e. it has no enabled event) in TS' and is reachable from the initial state s'_0 by a feasible trace σ', then also state s of TS, reached from s_0 by a feasible trace $\sigma = \sigma' \downarrow E$ (i.e. σ' projected onto set E), is a deadlock.*

It is easy to show that the insertion of event x by Definition 3.7.1 *always preserves trace equivalence, determinism and deadlock-freedom* [89]. Persistency and commutativity, on the other hand, are not automatically preserved, and need a more careful analysis. We leave that discussion, considering also how a pair of new events for the rising and falling transitions of a new signal can be inserted into a State Graph, to Chap. 5.

4. Logic Synthesis

Chapter 2 has already outlined the main concepts and techniques behind our approach to the design of asynchronous control circuits. The key stage in this approach is logic synthesis from Signal Transition Graphs (STGs), a model which offers important advantages to the asynchronous controller and interface designer. On one hand, STGs are very similar to Timing Diagrams, which can be seen as a conventional pragmatic design notation. On the other hand, they are based on the formally sound theory of Petri nets, with a clearly defined syntax and semantics, and a plethora of algorithms and techniques for model analysis and transformations.

In the previous chapter we have given a prominent role to Petri nets as our main formalism. We have defined the relationship between Petri nets, a model with true concurrency, causality, and conflict relations, and Transition Systems, a model with interleaving semantics of concurrency and a notion of global state. The idea of distributing global states into places of a Petri net, using the theory of regions, is conceptually akin to the way global states are encoded by binary signals in circuits. This relationship lays the formal foundation for the synthesis of control circuits from STGs, the subject of this chapter.

Throughout Chap. 2 we assumed that the STG did not exhibit a behavior in which some output signal is in conflict with, or may be disabled by, some other output or input signal. However, such a conflict may indeed come as an initial specification requirement, for example when describing a circuit involving arbitration. However, it is a well-known fact [90, 35] that it is impossible to implement an arbiter using standard logic gates (in fact any discrete state components). Arbitration is an analog phenomenon, which involves meta-stable or even oscillatory anomalous behavior in discrete latches and cannot be safely resolved in the discrete domain. In order to construct asynchronous systems with output disabling, designers must use special components called Mutex elements [91], which involve analog sub-circuits, specifically aimed to deal with low thresholds and small potential differences. They confine the meta-stable condition within themselves and resolve it to the outside world by simply "waiting long enough".

In our approach, that is based on logic synthesis to a library of Boolean gates, we do not handle such conflicts directly. As we will explain in Chap. 8, we assume that the designer provides an STG in which signals which exhibit conflicts are "factored out" into the environment, implemented using Mutex gates from the library, and declared as inputs to the logical part, which is subject to logic synthesis.

Chapter 2 showed that it is indeed possible to implement STGs with conflicts (or non-persistency) in our design flow, but only involving input signals. This sort of conflicts generally comes into place in order to use non-deterministic choice to *model in an abstract fashion the complex behavior of the external environment*. For example, it is very convenient to abstract the deterministic behavior of a CPU into a non-deterministic sequence of Read and Write cycles, when designing a bus interface.

4.1 Signal Transition Graphs and State Graphs

This section will formally introduce our main specification language, the Signal Transition Graph (STG), and its semantical model, the State Graph (SG), together with the conditions of consistent state encoding.

4.1.1 Signal Transition Graphs

STGs are a particular type of labeled PNs, where transitions are associated with the changes in the values of binary variables. These variables can for example be associated with wires, when modeling interfaces between blocks, or with input, output and internal signals in a control circuit.

Here is the formal definition of an STG.

Definition 4.1.1 (Signal Transition Graphs). *A* Signal Transition Graph *(STG) is a quadruple $G = (N, m_0, X, \lambda)$, where*

- *(N, m_0) is a Petri net based on a net $N = (P, T, F)$,*
- *X is a finite set of binary signals, which generates a finite alphabet $A^X = X \times \{+, -\}$ of signal transitions*
- *$\lambda : T \to A^X$ is a labeling function.*

Labeling λ does not need to be 1-to-1 (some signal transitions may occur several times in the PN), and it may be extended to a partial function, in order to allow some transitions to be "dummy" ones (denoted by ϵ), that is to denote "silent events" that do not change the state of the circuit.

When talking about individual signal transitions, the following meaning will be associated with their labels. A label $x+$ is used to denote the transition of signal x from 0 to 1 (rising edge), while $x-$ is used for a 1 to 0 transition (falling edge). In the following it will often be convenient to associate STG transitions directly with their labels, "bypassing" their Petri net identity. In such cases if the labeling is not 1-to-1 (so called multiple labeling), we will also use a superscript denoting the instance number of the $x+$ or $x-$. Sometimes, when reasoning on a pure event-based level, it will also be convenient to hide the direction of a particular edge and use $x*$ to denote either a $x+$ transition or a $x-$ transition. Putting it all together, the notation x_i^j+ would stand for the j-th rising edge of a signal $x_i \in X$.

An STG inherits the basic operational semantics from the behavior of its underlying Petri net, as defined in Sect. 3.1. In particular, this includes: (i) the rules for transition enabling and firing, (ii) the notions of reachable markings, traces, and (iii) the temporal relations between transitions (precedence, concurrency, choice and conflict). Likewise, STGs also inherit the various structural (marked graph, free-choice, etc.) and behavioral properties (boundedness, liveness, persistency etc.), and the corresponding classification of PNs.

A simple example of STG specification, referred as the xyz example in the following, is shown in Fig. 4.1a. The behavior of this STG can be traced in the reachability graph of its Petri net, shown in Fig. 4.1b. For example, events labeled $y+$ and $z+$ are concurrent, and $x+$ precedes $y+$. There are no conflicts between events, because this Petri net is a marked graph (it contains no choice places). The net is 1-bounded (safe), live and persistent.

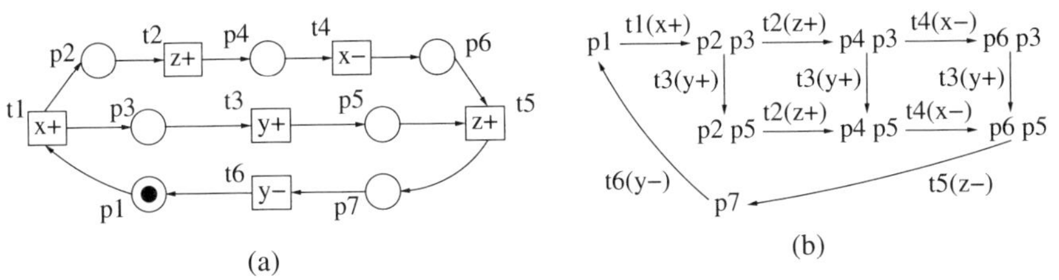

Fig. 4.1. Example xyz: STG (**a**) and its reachability graph (**b**)

The signal transition labeling of an STG may sometimes differentiate between input and non-input signals, thus forming two disjoint subsets, X_I (for inputs) and X_O (for non-inputs, or simply outputs), such that $X = X_I \cup X_O$. An STG is called *autonomous* if it has no input signals (i.e. $X_I = \emptyset$).

Graphically, an STG can either be represented in the standard form of a labeled PN, drawing transitions as bars or boxes and places as circles, or in the so-called STG shorthand form. The latter, as was first shown in Sect. 2.1, designates transitions directly by their labels and omits places that have only one input and one output transition. When multiple transitions have the same label, superscripts are often used to distinguish them. In drawings, indices separated by comma or slash (e.g. $a+,1$ or $a+/1$) are also used.

Examples of STGs, in their shorthand notation, were shown in Fig. 2.3, describing a simple VME bus controller example. It was assumed in them that $X_I = \{dsr, dsw, ldtack\}$ and $X_O = \{lds, dtack, d\}$. The first two STGs, in Fig. 2.3a and 2.3b, are marked graphs (they do not have choice on places). The third one, in Fig. 2.3c, modeling both read and write operation cycles, is not a marked graph because it contains places with multiple input and output transitions. It is not a free-choice net either, because one of its choice places, the input to transitions `lds+/1` and `lds+/2`, is not a free-choice place. The latter is however a unique choice place because whenever one of the above two transitions is enabled the other is not, which is guaranteed by the other choice place, which is a free-choice one. Thus, behaviorally, this net does not lead to dynamic conflicts (arbitration) or confusion, as it is free from any interference between choice and concurrency.

4.1.2 State Graphs

In the same way as an STG is an interpreted PN with transitions labeled with binary signals, a state graph (SG) is the corresponding binary interpretation of a TS in which the events represent signal transitions.

As noted before, the interleaving semantics of an STG can be conveniently expressed by the reachability graph of its underlying PN. In addition to that, bearing in mind the binary signal interpretation of transitions, there is a corresponding binary interpretation of markings or states of the STG. This brings us to the formal notion of a state graph.

Definition 4.1.2 (State Graph). *A* state graph *(SG) is a quadruple* $SG = (TS, X, \lambda_S, \lambda_E)$*, where*

- $TS = (S, E, T, s_0)$ *is a transition system,*
- $X = X_I \cup X_O$ *is a set of binary signals,* $X = \{x_1, x_2, ..., x_n\}$,
- $\lambda_S : S \to \{0,1\}^{|X|}$ *is a* state assignment *function, and*
- $\lambda_E : E \to X \times \{+,-\}$ *is a (possibly partial, in case of dummy events)* event assignment *function.*

When talking about individual states, each state node $s \in S$ in the SG is labeled by λ_S with a *binary vector* $(s(1), s(2), ..., s(n))$, where $s(i) \in \{0,1\}$, in accordance with the order of signals in X, i.e. $s(i)$ denotes the i-th component of s, corresponding to the (binary) value of signal $x_i \in X$. If signals have unique identifiers, we may use the notation $s(x) \in \{0,1\}$ to refer to the value of signal x in state s. When talking about individual transition arcs, notation $(s, x_i*, s') \in T$ stands for $(s, e, s') \in T$ and $\lambda_E(e) = x_i*$.

Returning back to the labeled RG of a particular STG $G = (N, m_0, X, \lambda)$, it is formally an SG $SG = (TS, X, \lambda_S, \lambda_E)$, defined on the $TS = (S, E, T, s_0)$ that is generated by the full reachability analysis of net N starting from the initial marking, m_0. The states $s \in S$ in SG are obtained from the reachable markings of net N. The names of events $e \in E$ in SG, according to the event assignment function, are trivially obtained from the labels of the transitions of N that fire between the corresponding pairs of markings (states). The question of obtaining the binary codes of states $s \in S$ is however slightly less trivial and is closely related to the consistency of the STG as will be shown in Sect. 4.2.2.

Strictly speaking, in some rare cases one marking of a reachability graph can be mapped to a few different binary states of an SG. We will not consider these exotic cases, as they can be reduced to a normal case by state splitting.

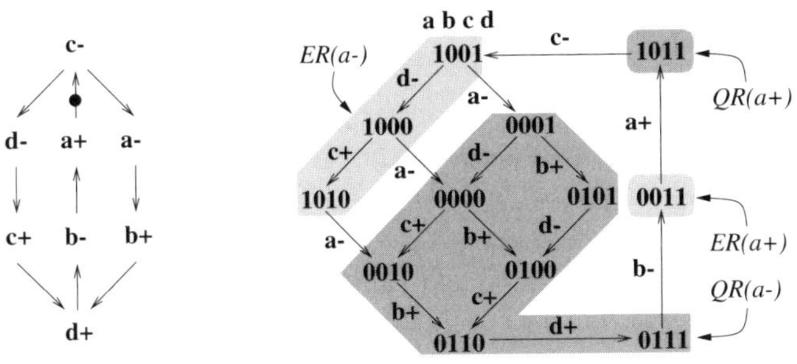

(a) (b)

Fig. 4.2. Example *abcd*: (**a**) Signal Transition Graph, (**b**) State Graph

Fig. 4.2a and 4.2b depict an STG and the corresponding SG. This example will be used along this chapter to illustrate methods for logic synthesis.

4.1.3 Excitation and Quiescent Regions

The following definitions relate signal transitions with states of an SG $SG = (TS, X, \lambda_S, \lambda_E)$, defined on the transition system $TS = (S, E, T, s_0)$.

Definition 4.1.3 (Excitation region). *The positive and negative* excitation regions (ER) *of signal $x \in X$, denoted by* $\mathsf{ER}(x+)$ *and* $\mathsf{ER}(x-)$, *are the sets of states in which $x+$ and $x-$ are enabled, respectively, i.e.*

$$\mathsf{ER}(x+) = \{s \in S \mid s \xrightarrow{x+}\}$$
$$\mathsf{ER}(x-) = \{s \in S \mid s \xrightarrow{x-}\}$$

Definition 4.1.4 (Quiescent region). *The positive and negative* quiescent regions (QR) *of signal $x \in X$, denoted by* $\mathsf{QR}(x+)$ *and* $\mathsf{QR}(x-)$ *are the sets of states in which x has the same value, 1 or 0, and is stable, i.e.*

$$\mathsf{QR}(x+) = \{s \in S \mid s(x) = 1 \ \wedge \ s \notin \mathsf{ER}(x-)\}$$
$$\mathsf{QR}(x-) = \{s \in S \mid s(x) = 0 \ \wedge \ s \notin \mathsf{ER}(x+)\}$$

The excitation and quiescent regions for signal a are shadowed in the SG depicted in Fig. 4.2b.

Whenever we are not specific about the direction of a signal transition, i.e. $x*$, we will use the notation $\mathsf{ER}(x*)$ and $\mathsf{QR}(x*)$ to denote its excitation and quiescent region.

In some cases, we may need to distinguish subsets of states of ERs or QRs corresponding to different instances of the same event, usually coming from different transitions with the same label in the STG. To distinguish such subsets, we will use sub-indices, e.g. $\mathsf{ER}_i(x+)$ or $\mathsf{QR}_j(x-)$.

The triggering relation between events is also important in logic synthesis. It captures the notion of causality at the level of SG and it has a strong relationship with the signals in the support of the Boolean functions that implement the output signals.

Definition 4.1.5 (Trigger signals). *An event $x*$ is called a* trigger *of an event $y*$, denoted as $x* \in \mathsf{Trig}(y*)$, if the firing of $x*$ enters $\mathsf{ER}(y*)$. Formally,*

$$x* \in \mathsf{Trig}(y*) \iff \exists s, s' \in S : s \notin \mathsf{ER}(y*) \ \wedge \ s \xrightarrow{x*} s' \ \wedge \ s' \in \mathsf{ER}(y*).$$

In the example of Fig. 4.2 we can observe several triggering relations. For example, $c-$ triggers $d-$ because of the arc $1011 \xrightarrow{c-} 1001$. However, $a-$ does not trigger $d-$, since there is no arc $s \xrightarrow{a-} s'$ such that $s \notin \mathsf{ER}(d-)$ and $s' \in \mathsf{ER}(d-)$.

4.2 Implementability as a Logic Circuit

This section will formally define important properties that an SG must satisfy in order to be implemented in a hazard-free logic circuit. The same properties can be defined for STGs. Rather than characterizing them at the level of STG, we will simply say that a property holds in an STG if it holds in its underlying SG.

The properties that guarantee the existence of a hazard-free implementation for an SG are the following:

- Boundedness
- Consistency
- Complete state coding
- Output persistency

We next describe the details concerning these properties.

4.2.1 Boundedness

Even though this was not explicitly stated in Definition 4.1.2, we always assumed that the set of states of an SG was finite. This is a necessary property for the implementability as a logic circuit. However, this property is not so evident when the specification is given as an STG. Even though an STG has a finite structure (number of places, transitions and arcs), it may manifest an unbounded behavior.

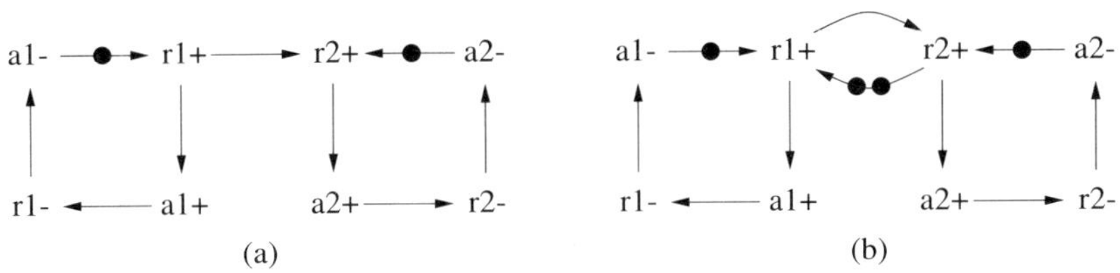

Fig. 4.3. (**a**) Unbounded STG, (**b**) bounded STG

As an example, Fig. 4.3a shows an STG in which the arc $r1+ \rightarrow r2+$ can accumulate an arbitrary number of tokens. From this fact, one can easily deduce that the number of markings of the STG is infinite and so is the set of states of the underlying SG.

When such a situation occurs, the designer may try to change the specification to make it bounded. Fig. 4.3b depicts a similar specification in which an arc $r2+ \rightarrow r1+$ with two tokens has been added. The two arcs between the left and right part of the STG form a place invariant (see Sect. 3.2). As a result of that, the total number of tokens accumulated on those arcs is always constant. This transformation restricts the behavior of the system, but makes it bounded.

4.2.2 Consistency

Each signal x of an SG partitions the set of states S into two subsets, $S(x = 0)$ and $S(x = 1)$, defined as follows:

$$S(x = 0) = \{s \in S \mid s(x) = 0\}$$
$$S(x = 1) = \{s \in S \mid s(x) = 1\}$$

The consistency of an SG refers to the fact that the events $x+$ and $x-$ are the only ones that cross these two subsets according to their meaning: switching from 0 to 1 and from 1 to 0, respectively. This is captured by the definition of consistent SG.

Definition 4.2.1 (Consistent SG).
A state graph (SG) $SG = (TS, X, \lambda_S, \lambda_E)$ *is said to have a* consistent state coding *(we will call such an SG consistent) if for each transition* $s \xrightarrow{e} s'$ *the following conditions hold:*

- *if* $\lambda_E(e) = x+$*, then* $s(x) = 0$ *and* $s'(x) = 1$*;*
- *if* $\lambda_E(e) = x-$*, then* $s(x) = 1$ *and* $s'(x) = 0$*;*
- *in all other cases,* $s(x) = s'(x)$*.*

The SG depicted in Fig. 4.2 is consistent, and so is the STG in the same figure. On the other hand, consider the STG shown in Fig. 4.4a and its reachability graph in Fig. 4.4b. It is easy to observe that there are sequences of transition

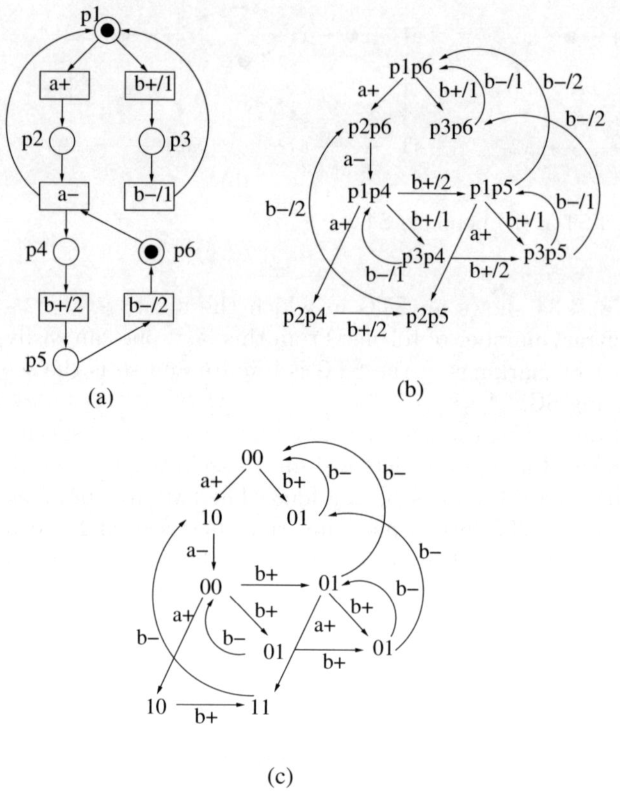

Fig. 4.4. (**a**) Inconsistent STG, (**b**) reachability graph, (**c**) state graph

firings in which an occurrence of $b+$ (corresponding to $b+/1$) is followed by another occurrence of $b+$ (corresponding to $b+/2$) without any intermediate occurrence of $b-$. One may notice that this was caused by the fact that after reaching marking $(p1, p4)$, the STG enables transition $b+/1$ concurrently with $b+/2$, which is clearly wrong.

In an attempt to assign a binary encoding to the states of the SG (see Fig. 4.4c), one immediately realizes that inconsistent arcs, such as $01 \xrightarrow{b+} 01$, are unavoidable.

It is easy to observe the close relationship between the reasons why the STG and the SG violate consistency. This suggests a practical way for a designer to rectify the problem by modifying the STG specification exactly in the position in which this design error occurs. In terms of the SG we must avoid transition enabling which leads to a state with inconsistent encoding. An example of such a successful error correction is shown in Fig. 4.5. The dashed arcs in Fig. 4.5a depict the modification at the STG level.

With Definition 4.2.1 we have identified the first necessary condition on the way from STGs to their circuit implementation. It is clear that only a

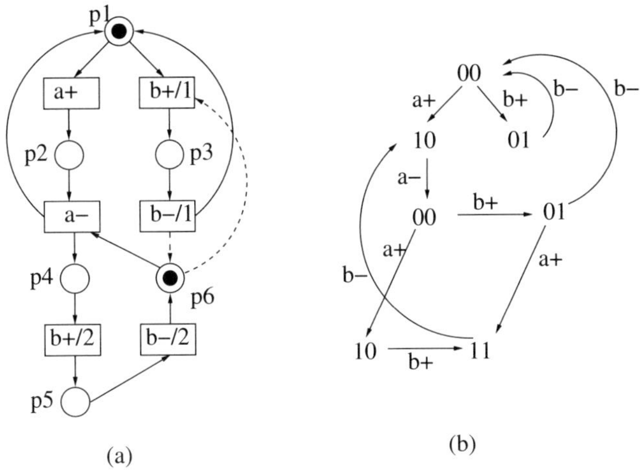

Fig. 4.5. "Fixing" the specification error in the example of Fig. 4.4: (**a**) consistent STG, (**b**) consistent SG

consistent STG can be considered for implementation, because otherwise it does not produce any meaningful behavioral interpretation in terms of binary signals.

4.2.3 Complete State Coding

As was first noted in Sect. 2.4, binary encoded TSs, in particular the SGs generated by STGs, may have two or more *semantically different* (i.e. not equivalent) states encoded with the same binary vectors. Let us examine this situation in more detail.

Definition 4.2.2 (Unique State Coding). *[92] A state graph $SG = (TS, X, \lambda_S, \lambda_E)$, where $TS = (S, E, T, s_0)$, satisfies the* Unique State Coding *(USC) condition if every state in S is assigned a unique binary code, defined on the signal set X. Formally, USC means that the state assignment function, λ_S, is injective.*

As was shown in [92], the USC condition is sufficient for deriving the next state functions (see Chap. 2.3) from the SG without ambiguity, by simply using the binary codes of the states. It is however not a necessary condition because, firstly, some states in the SG specification may be equivalent, and, secondly, the ambiguity may actually happen on input signals, for which the next state functions are not derived.

Definition 4.2.3 (Complete State Coding). *[92] A state graph $SG = (TS, X, \lambda_S, \lambda_E)$, where $TS = (S, E, T, s_0)$, satisfies the* Complete State Coding *(CSC) condition if for every pair of states $s, s' \in S$ having the same binary code, i.e. $\lambda_S(s) = \lambda_S(s')$, the sets of enabled output signals are the same.*

The SG produced by the *abcd* example of Fig. 4.2 satisfies both USC and CSC conditions, whereas the one produced by the example in Fig. 4.5 satisfies neither USC nor CSC, regardless of whether the signals are input or output, but assuming that at least one of them is output. The states that are in coding conflict are, for example, those pairs that are labeled 01 and 10 because they have different patterns of enabled output signals.

Recall the VME bus example. The state graph of the read cycle, shown in Fig. 2.6b, did not satisfy CSC, having two states with different enablings of output signals. Subsequently, this CSC problem was resolved by adding an internal memory signal, as shown in Fig. 2.9.

4.2.4 Output Persistency

While the previous three properties aimed at guaranteeing the existence of a logic implementation of the specified behavior, the property discussed in this section aims at guaranteeing the robustness of the implementation under any delay of the gates of the circuit. This robustness is commonly known as speed-independence. A more detailed discussion of this concept will be presented in Sect. 4.6.

The model for speed-independence is illustrated in Fig. 4.6, in which "circuit" is an acyclic network of zero-delay Boolean gates. The functionality of the circuit is expected to be independent of the delays of the gates that produce the output events. This means that for any delay in the interval $[0, \infty)$, the circuit should behave as specified.

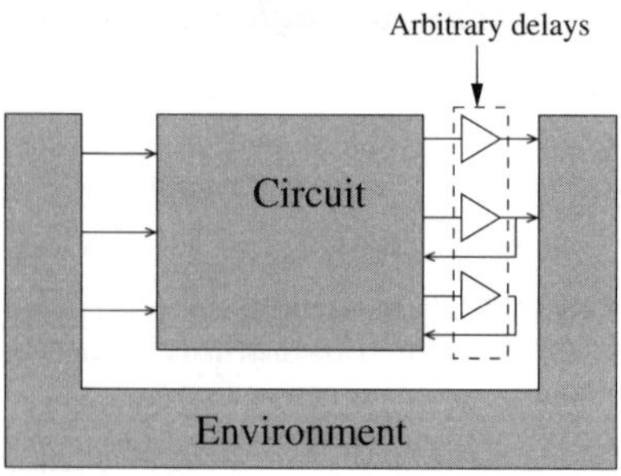

Fig. 4.6. Delay model for speed-independence

Speed-independence is tightly related to the concept of signal persistency. Before giving a formal definition, we will introduce persistency by means of the example in Fig. 4.7. In state 0000, there are two events enabled: $a+$ and $b+$. However, the firing of one of them disables the other. This situation is known as a conflict between $a+$ and $b+$.

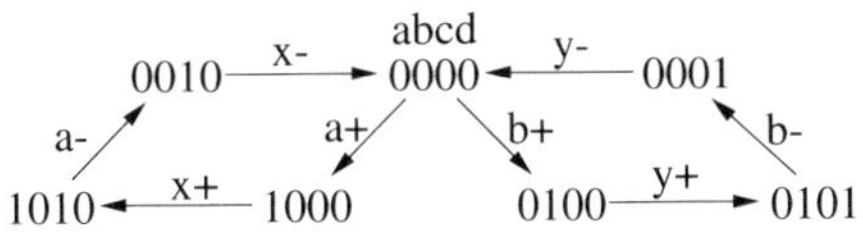

Fig. 4.7. State graph with conflict

Output disables output. Conflicts have a negative effect on the robustness of the circuit to delay variations. Let us assume that a and b are output signals. A possible implementation is shown in Fig. 4.8a. When the circuit is in the state 0010, only the event $x-$ is enabled and no gate is excited. When $x-$ fires, the gates producing a and b are enabled simultaneously and a race for switching one of the outputs starts. The winner of the race will depend on different technological factors and, in particular, on the delays of a and b. If they are similar, a non-deterministic behavior, such as the one depicted in Fig. 4.9a could be produced. This effect is known as metastability [93]. Another possible effect could be the appearance of glitches, as shown in Fig. 4.9b.

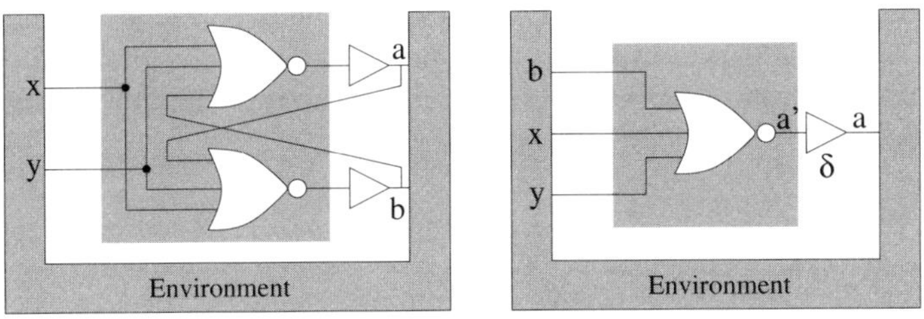

Fig. 4.8. Circuits without output persistency

The impact of these non-deterministic behaviors can be extremely dangerous. Assume that, in a metastable state, the environment "reads" both outputs being at high. That could provoke a malfunction in the environment that could cause, for example, both inputs going high. A similar effect could be produced by glitches.

In synchronous design, glitches are considered as spurious events that are not relevant from the point of view of the functionality of the system, since signals are typically sampled by flip-flops only when clock edges occur. However, asynchronous systems are much more sensitive to signal changes since each transition can be considered as an observable event of the system. For this reason, the delay models used when modeling asynchronous systems are generally conservative and glitches are considered as undesired signal transitions.

Output disables input. Let us assume now that a is the only output of the system, and let us call δ the magnitude of the delay before signal a. A

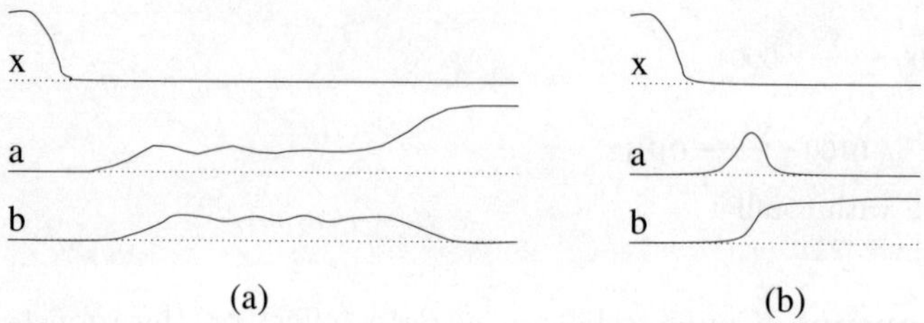

Fig. 4.9. Non-deterministic behaviors produced by a conflict: (**a**) metastability, (**b**) glitch

possible implementation is shown in Fig. 4.8b. Let us also assume that δ is one time unit, the delay of the NOR gate is three time units and the delay of the environment is five time units. The delay of the environment represents the fact that the enabling of an input event and its firing are separated by five time units. With these delays, we could observe a timed trace like this:

$$\overset{x-}{t=0} \cdots \overset{a'+}{t=3} \cdots \overset{a+}{t=4} \cdots \overset{x-}{t=9} \cdots$$

where a' represents the value of the signal before the delay. Event $b+$ would never be able to fire since it requires five time units of enabling time, and $a+$ always fires four time units after $x-$. However, if we assume that δ is four time units, the following behavior could be observed:

$$\overset{x-}{t=0} \cdots \overset{a'+}{t=3} \cdots \overset{b+}{t=5} \cdots \overset{a+}{t=7} \cdots \overset{a'-}{t=8} \cdots$$

where the observable values of a and b are simultaneously at 1, which is an unacceptable behavior according to the specification. This phenomenon occurs due to the fact that $b+$ is enabled after the firing of $x-$. However δ is long enough to hide the firing of $a'+$ to the environment. When $a+$ becomes observable to the environment, $b+$ has already fired, thus producing an unexpected event.

Similarly to the case in which an output disables an output, metastability could also be one of the effects manifested when an input is disabled by an output.

Input disables input. If we assume that a and b are inputs, the conflict between them is internal to the environment and its resolution does not depend on the delays between the circuit and the environment. For this reason, conflicts between input events do not pose any problem for the robustness of the circuit to delay variations.

Formalizing output-persistency. We now formally define output-persistency of an SG.

Definition 4.2.4 (Disabling and conflict state). *An event $x*$ is said to* disable *another event $y*$ if there is a state $s \in \mathsf{ER}(y*)$ and a transition $s \xrightarrow{x*} s'$ such that $s' \notin \mathsf{ER}(y*)$. The state s in which disabling occurs is called a* conflict state.

Definition 4.2.5 (Signal persistency). *A signal x is said to be* persistent *if no event $x*$ of signal x is disabled by any other event $y*$.*

Definition 4.2.6 (Output persistency). *An SG is said to be output persistent if for any pair of signals x and y such that x disables y, both x and y are input signals.*

We can now formulate the condition for the implementability of an STG.

Proposition 4.2.1. *[94] An STG is implementable as a speed-independent circuit iff its SG is finite, consistent, output-persistent and satisfies CSC.*

4.3 Boolean Functions

Before presenting methods to derive Boolean equations for the implementation of asynchronous circuits, we provide some background on Boolean functions and on the realization of functions with gates.

4.3.1 ON, OFF and DC Sets

An incompletely specified (scalar) Boolean function is a functional mapping $F : \mathbb{B}^n \rightarrow \{0, 1, -\}$, where $\mathbb{B} = \{0, 1\}$ and '−' represents the *don't care* (DC) value. The subsets of the domain $\mathbb{B}^n$ in which F has the 0, 1 and DC values are respectively called the OFF-set, ON-set and DC-set. F is *completely specified* if its DC-set is empty. We shall further always assume that F is a completely specified Boolean function unless we explicitly say otherwise.

4.3.2 Support of a Boolean Function

Let $F(x_1, x_2, \ldots, x_n)$ be a Boolean function of n Boolean variables. The set $X = \{x_1, x_2, \ldots, x_n\}$ is called the *support* of the function F. In this section we shall mostly use the notion of *true support*, which is defined as follows. A variable $x \in X$ is *essential* for function F (or F is dependent on x) if there exist at least two elements of $\mathbb{B}^n$, v_1 and v_2, different only in the value of x, such that $F(v_1) \neq F(v_2)$. The set of essential variables for a Boolean function F is called the *true support* of F. It is clear that for an arbitrary Boolean function its support may not be the same as the true support. E.g., for a support $X = \{a, b, c\}$ and a function $F(X) = \overline{b} + c$ the true support of $F(X)$ is $\{b, c\}$, i.e. only a subset of X.

4.3.3 Cofactors and Shannon Expansion

Let $F(X)$ be a Boolean function with support $X = \{x_1, x_2, \ldots, x_n\}$. The *cofactors* of $F(X)$ with respect to x_i and $\overline{x}_i$, respectively, are defined as

$$F_{x_i} = F(x_1, x_2, \ldots, x_i = 1, \ldots, x_n)$$
$$F_{\overline{x}_i} = F(x_1, x_2, \ldots, x_i = 0, \ldots, x_n)$$

The well-known Shannon expansion of a Boolean function $F(X)$ is based on its cofactors:

$$F(X) = x_i F_{x_i} + \overline{x}_i F_{\overline{x}_i}.$$

4.3.4 Existential Abstraction and Boolean Difference

The existential abstraction of a function $F(X)$ with respect to x_i is defined as

$$\exists_{x_i} F = F_{x_i} + F_{\overline{x}_i}.$$

The existential abstraction can be naturally extended to a set of variables. The *Boolean difference*, or *Boolean derivative*, of $F(X)$ with respect to x_i is defined as

$$\delta F / \delta x_i = F_{x_i} \oplus F_{\overline{x}_i}$$

4.3.5 Unate and Binate Functions

A function $F(x_1, x_2, \ldots, x_i, \ldots, x_n)$ is *unate in variable* x_i if either $F_{\overline{x_i}} \leq F_{x_i}$ or $F_{x_i} \leq F_{\overline{x_i}}$ under ordering $0 \leq - \leq 1$, i.e. if $\forall v \in \mathbb{B} F_{\overline{x_i}}(v) \leq F_{x_i}(v)$. In the former case ($F_{\overline{x_i}} \leq F_{x_i}$) it is called *positive unate in* x_i, in the latter case *negative unate in* x_i. A function that is not unate in x_i is called *binate* in x_i. A function is (positive/negative) *unate* if it is (positive/negative) *unate* in all support variables. Otherwise it is binate. For example, the function $F = a + b + \overline{c}$ is positive unate in variable a because $F_a = 1 \geq F_{\overline{a}} = b + \overline{c}$.

4.3.6 Function Implementation

For an incompletely specified function $F(X)$ with a non-empty DC-set, let us define the DC function $F_{DC} : \mathbb{B}^n \to \mathbb{B}$ such that $\mathsf{ON}(F_{DC}) = \mathsf{DC}(F)$. We will say that a function $\widetilde{F}$ is an *implementation of* F if

$$F \cdot \overline{F_{DC}} \leq \widetilde{F} \leq F + F_{DC}.$$

4.3.7 Boolean Relations

A *Boolean relation* is a relation between Boolean spaces [95, 10]. It can be seen as a generalization of a Boolean function, where a point in the domain $\mathbb{B}^n$ can be associated with several points in the co-domain. More formally, a Boolean relation R is $R \subseteq \mathbb{B}^n \times \mathbb{B}^m$. Sometimes, we shall also use the "$-$" symbol as a shorthand in denoting elements in the co-domain vector, e.g. 10 and 00 will be represented as one vector -0. Boolean relations play an important role in multi-level logic synthesis [10]. They will be used in Chap. 6 for logic decomposition.

Consider a set of Boolean functions $\mathcal{H} = \{H_1, H_2, \ldots, H_m\}$. Let $R \subseteq \mathbb{B}^n \times \mathbb{B}^m$ be a Boolean relation with the same domain as functions from $\mathcal{H}$. We will say that $\mathcal{H}$ is *compatible* with R if for every point v in the domain of R the vector of values $(v, H_1(v), H_2(v), \ldots, H_m(v))$ is an element of R.

Fig. 4.10 presents an example of Boolean relation with $n = m = 2$. Each pair of compatible functions shown in Fig. 4.10c, is a subset of the relation in which each point in the domain is related to only one point in the co-domain.

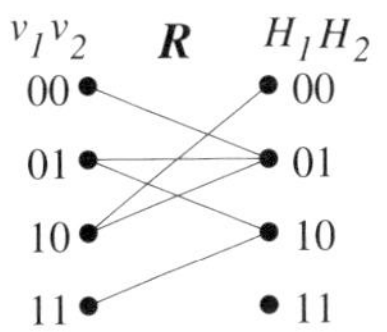

(a)

$v_1\ v_2$	$H_1\ H_2$
00	$\{01\}$
01	$\{01, 10\}$
10	$\{0-\}$
11	$\{10\}$

(b)

$H_1(v) = v_1\ v_2$
$H_2(v) = \overline{v}_1$

$H_1(v) = v_2$
$H_2(v) = \overline{v}_2$

(c)

Fig. 4.10. (**a**) Boolean relation, (**b**) Tabular form, (**c**) Compatible functions

4.4 Gate Netlists

The methods presented in this chapter assume that each Boolean function obtained by logic synthesis can be implemented as an atomic gate without internal hazards [96]. In the most general case, the gates can also be sequential, i.e. the output function may depend on itself. In this section we present three different architectures for the implementation of Boolean functions:

- Complex Gates
- Generalized C-elements
- C-elements with complex gates

To illustrate the implementation with such architectures, we will use a simple example. It corresponds to the implementation of a signal x with the following equation:

$$x = a\ (b + \overline{c} + x)$$

This function is sequential, since the next value of x depends on its current value.

4.4.1 Complex Gates

They are assumed to be implemented as complementary pull-up and pull-down networks in static CMOS. The implementation is shown in Fig. 4.11a.

Recall that, in order for the synthesized circuit to be speed-independent, the atomic complex gate assumption requires that all the internal delays within each gate are negligible and do not produce any spurious behavior observable from the outside. In particular, this requirement includes zero-delay inverters at some gate inputs, which are shown as bubbles in Fig. 4.11.

Complex gates are abstractions of physical implementations that must be adapted to the available technology, possibly based on standard cells or library generators. In Chap. 6 we will discuss how to decompose complex gates to fit a specific standard cell library.

4.4.2 Generalized C-Elements

Here we will assume the availability of a library generator for a family of pseudo-static sequential gates called *generalized C-elements* (gC [40]).

This implementation is based on set, S^x, and reset, R^x, sub-functions of a complex gate function $x = f(X)$, where X is the Boolean vector of input and output signals, including x as well. S^x and R^x are obtained from the Shannon expansion and are often called simply S and R when no confusion arises. The equation of x can be expressed as follows:

$$x = \overline{x} \cdot S + x \cdot \overline{R}$$

where $S = f_{\overline{x}}$ and $\overline{R} = f_x$. If $S \cdot R = 0$, i.e. they are orthogonal, then we have that $S \leq \overline{R}$, thus implying that $f(X)$ is positive unate in x, and the equation can be written as follows:

$$x = S + x \cdot \overline{R}$$

This form of equation is convenient for implementations using gC elements or Set/Reset latches.

For combinational functions, such as AND, NAND, OR, NOR etc, the orthogonality of S and R is guaranteed by definition, because $R = \overline{S}$, where we should take $S = f(X)$. Therefore such functions are commonly implemented as static CMOS transistor networks in which pull-up (for R) and pull-down (S) networks are merely dual.

For sequential gates, i.e. those described by function with feedback, the set and reset functions need not be complementary because the current value of the function does also depend on the state of the feedback signal. A fully

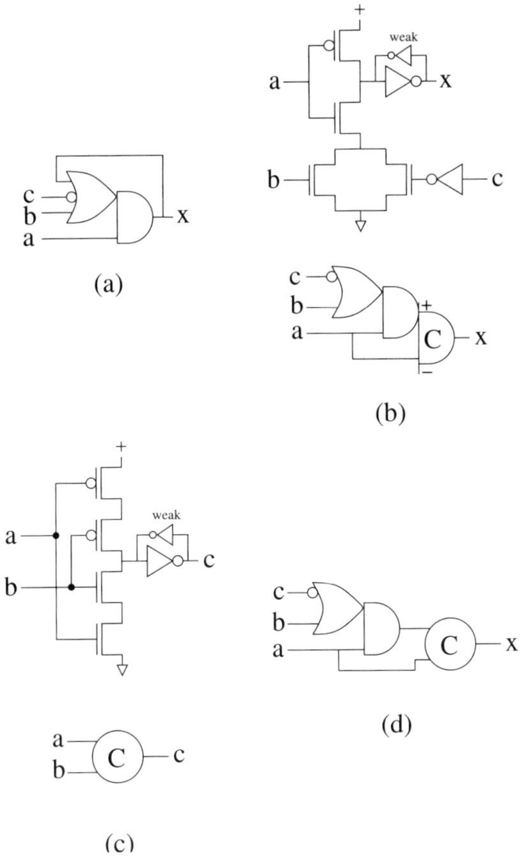

Fig. 4.11. (**a**) Complex gate, (**b**) Generalized C-element, (**c**) C-element, (**d**) Implementation with C-element

static implementation of $f(X)$, with dual pull-up and pull-down networks implementing $\overline{f(X)}$ and $f(X)$ respectively, and hence including feedback signal x, would of course guarantee orthogonality as well. Using such an approach is however often too costly in terms of area and speed, because the transistor interconnection for such fully dual functions may be too complex.

A more efficient implementation can be achieved by cofactoring f with respect to $x = 1$ and $x = 0$, and thus reducing the complexity of S and R. This results in a pseudo-static CMOS implementation with orthogonal S and R. Deriving orthogonal S and R functions can be ensured if all transitions of a signal are separated by another signal, i.e. $x-$ does not trigger $x+$ and $x+$ does not trigger $x-$. In this case, $f(X)$ is positive unate in x, i.e. by changing x from 0 to 1 the value of f cannot change from 1 to 0. This means that for any given input vector v, $f_{\overline{x}}(v) = 1$ implies $f_x(v) = 1$. Thus $f_{\overline{x}}$ covers f_x, and $\overline{f_{\overline{x}}} \cdot f_x = 0$.

In this case, two transistor networks can be used, the pull-down for S, and the pull-up for R. It is also possible to reverse the functionality of R and S if an inverter is added at the output. Due to their orthogonality, it is always guaranteed that the pull-down and pull-up networks are never ON at the same time (in other words, no short circuit between the power and ground is possible). Due to sequentiality, however, it is possible that both pull-up and pull-down networks are OFF, and thus the gate output is neither driven to 1 nor to 0. Since in asynchronous circuits there is no guarantee that the state of such a dynamic gate is refreshed periodically, special circuitry must be used to hold the output value to its previous state. A simple way to achieve this is by means of a weak feedback inverter, that can be overridden by the stronger transistors in the pull-up and pull-down networks.

Going back to our example, the equation for x can be written as

$$x = \underbrace{a\,(b+\overline{c})}_{S} \;+\; x\;\underbrace{\overline{a}}_{\overline{R}}$$

An implementation is shown in Fig. 4.11b, in which S and R have been assigned to the pull-down and pull-up networks, respectively, with an inverter at the output of the gate. The inverter of input c must be considered as part of the "atomic" gate in case a speed-independent implementation is sought. The figure also presents an alternative symbol for the gate, in which S and R are identified by the + and − inputs of the C gate.

4.4.3 C-Elements with Complex Gates

The C-element is the most popular latch in asynchronous design [43]. Its behavior is described by the equation

$$c = a \cdot b + c \cdot (a + b)$$

and is equivalent to the behavior of a gC element with $S = a \cdot b$ and $R = \overline{a} \cdot \overline{b}$. One of the possible implementations and the associated symbol are shown in Fig. 4.11c. However, other implementations have also been proposed and studied [97].

A natural question may arise next: what if we use a C-element as a memory latch, and implement the set and reset functions separately as combinational, yet still complex, gates, as shown in Fig. 4.11d?.

Such a decomposition of a complex gate is quite similar to the standard way of implementing synchronous circuits, where combinational logic generates the excitation functions for memory elements. Using this method for asynchronous circuits could be useful for further logic decomposition and technology mapping, but is unfortunately not as straightforward as for synchronous circuits. The main problem is the violation of the atomicity assumption that was previously used for the complex gate. Using separate logic elements for implementing the S and R inputs of the C-element implies the existence of delays, and hence a possibility of glitches on these signals if the Boolean covers for the S and R functions are not properly chosen. The signal transition order shown in Fig. 4.12a must be properly followed in order to avoid such glitches, and each of these covers must be stable between the rising and falling edges of their respective signals S and R. This leads to the so-called *monotonic cover conditions* for S and R covers.

A more detailed discussion of this topic is postponed until Chap. 6.

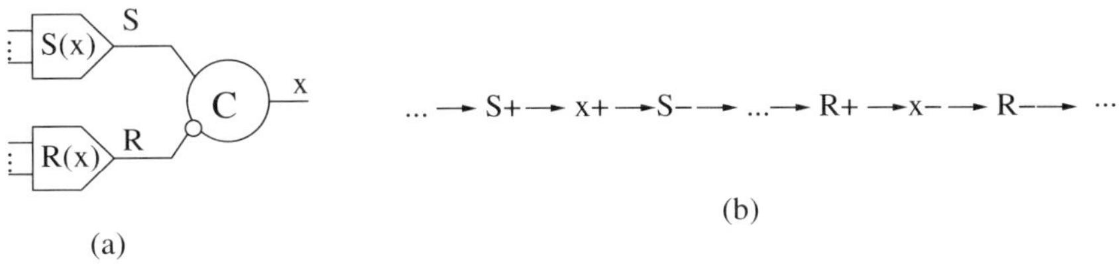

Fig. 4.12. Implementation using C-elements: (**a**) general structure, (**b**) and signal order to guarantee absence of glitches on S and R

4.5 Deriving a Gate Netlist

This section revisits the procedure to derive a logic circuit from an implementable SG. This circuit will be implemented as an interconnection of so-called complex gates, that are gates implementing any Boolean function without restrictions on the complexity, or the fan-in and fan-out. In practice such complex gates can be implemented using generalized C-elements [40]. In case S and R are too complex to be integrated in a single gC element, logic decomposition must be applied (see Chap. 6).

4.5.1 Deriving Functions for Complex Gates

The procedure to derive the Boolean next-state functions for output signals from the SG corresponding to a gate implementable STG specification was described in Chap. 2. It involved finding a minimum cover to the set of binary vectors defined by $\mathsf{ER}(x+) \cup \mathsf{QR}(x+)$, which forms the ON-set of signal x, i.e. $\mathsf{ON}(x)$, using also the DC-set $\mathsf{DC}(x)$.

Given a specification with n signals, the derivation of an incompletely specified function F^x for each output signal x and for each $v \in \mathbb{B}^n$ can be formalized as follows:

$$F^x(v) = \begin{cases} 1 & \text{if } \exists\, s \in \mathsf{ER}(x+) \cup \mathsf{QR}(x+) : \lambda_S(s) = v \\ 0 & \text{if } \exists\, s \in \mathsf{ER}(x-) \cup \mathsf{QR}(x-) : \lambda_S(s) = v \\ - & \text{if } \nexists\, s \in S : \lambda_S(s) = v \end{cases}$$

One might think that the previous definition could be ambiguous when there are two states, s_1 and s_2, for which $\lambda_S(s_1) = \lambda_S(s_2) = v$, $s_1 \in \mathsf{ER}(x+) \cup \mathsf{QR}(x+)$ and $s_2 \in \mathsf{ER}(x-) \cup \mathsf{QR}(x-)$. This ambiguity is precisely what the CSC property avoids, and this is why CSC is a necessary condition for implementability.

Let us apply this procedure to the *abcd* example specified in Fig. 4.2, assuming that signals *a* and *d* are outputs. The incompletely specified functions are represented by the Karnaugh maps in Fig. 4.13.

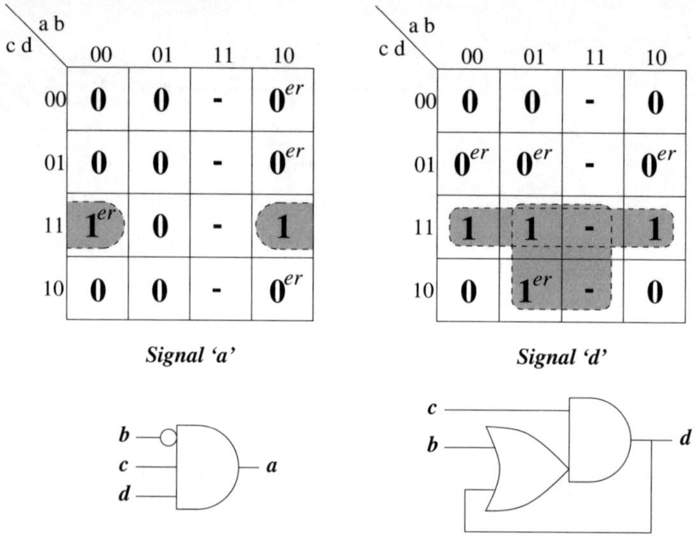

Fig. 4.13. Complex gate implementation for the *abcd* example

The cells containing 0er and 1er correspond to states in the ER of the signal. The cells with – correspond to binary codes with no associated reachable state. The identification of those codes is crucial to define the DC-set of the function for efficient minimization.

The gate implementations in Fig. 4.13 are possible realizations of the incompletely specified functions. Such implementations may have input inverters (bubbles). The most straightforward way to avoid explicit inverters is to "encapsulate" an input bubble into a library gate by appropriate technology mapping. In this case the implementation of the library gate must guarantee that the delay introduced by the bubble is negligible compared to a standard gate's delay. A more conservative approach would argue that delays in input inverters should not be ignored [98]. It requires an appropriate refinement of the STG specification by adding new signals to satisfy the so-called "normalcy" property [98]. Its synthesis then guarantees that the Boolean functions of the implementation gates are monotonic, i.e. either positive unate in all variables or negative unate in all variables.

When analyzing the support of the functions in the implementation, the following property can be proved: if signal x is a trigger of signal y, then x will be in the support of any implementation of y. The proof is based on the fact that the trigger relation implies the existence of two states with codes v_1 and v_2 that only differ in the value of x and for which $F^y(v_1) \neq F^y(v_2)$. Under such condition, x is an essential variable for F^y. Those signals in the support of a function that are not trigger signals are said to be *context* signals.

In our example, b and c are trigger signals for a, whereas d is a context signal. In the case of the implementation of d, b and c are again the trigger signals, whereas d is context.

4.5.2 Deriving Functions for Generalized C-Elements

The implementation with gC elements requires two functions: set (S) and reset (R). As previously discussed, these functions must be orthogonal. In our design framework from STGs, the orthogonality requirement affects only the reachable set of states of the specification. In other words, S and R are allowed to intersect in those input assignments that do not correspond to the binary code of any reachable state.

The definition of the incompletely specified functions for S^x and R^x can be formalized as follows:

$$S^x(v) = \begin{cases} 1 & \text{if } \exists\, s \in \mathsf{ER}(x+) : \lambda_S(s) = v \\ 0 & \text{if } \exists\, s \in \mathsf{ER}(x-) \cup \mathsf{QR}(x-) : \lambda_S(s) = v \\ - & \text{otherwise} \end{cases}$$

$$R^x(v) = \begin{cases} 1 & \text{if } \exists\, s \in \mathsf{ER}(x-) : \lambda_S(s) = v \\ 0 & \text{if } \exists\, s \in \mathsf{ER}(x+) \cup \mathsf{QR}(x+) : \lambda_S(s) = v \\ - & \text{otherwise} \end{cases}$$

According to the previous definition, one may observe that the states in QR($x+$) are in the DC-set of S^x, and the states in QR($x-$) in the DC-set of R^x. These are the states in which the gC element maintains the previous value of the output signal.

Fig. 4.14 depicts the Karnaugh maps for output signals a and d, and the equations obtained for S and R. Rather than using different maps for S and R, we can derive the equations directly from the map of the function by following these rules:

- S^x must cover all cells containing 1er and no cell containing 0 or 0er.
- R^x must cover all cells containing 0er and no cell containing 1 or 1er.

In the example of Fig. 4.14, the following equations are obtained after Boolean minimization:

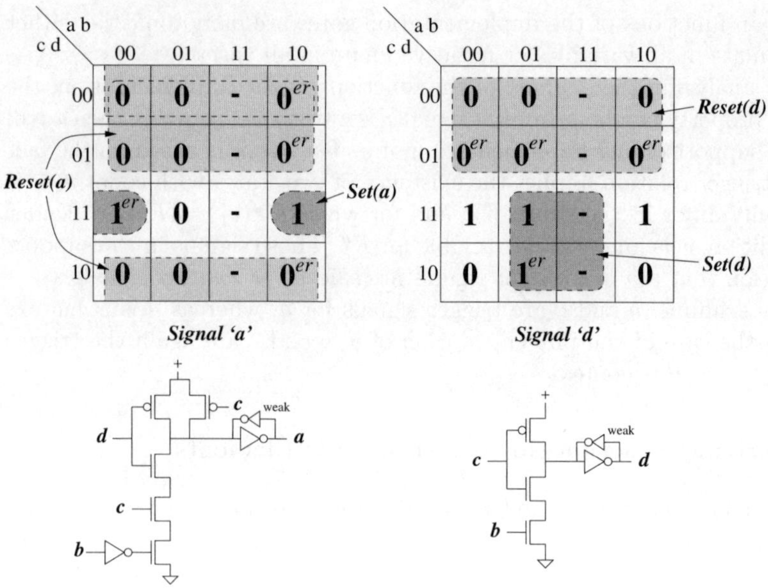

Fig. 4.14. gC implementation for the *abcd* example

$$
\begin{aligned}
S^a &= \overline{b}\,c\,d \\
R^a &= \overline{c} + \overline{d} \\
S^d &= b\,c \\
R^d &= \overline{c}
\end{aligned}
$$

The concept of trigger and context signal is also relevant in this case, and can be individually applied to S and R. As an example, we can identify the following trigger and context signals for a:

b is a trigger signal for $a+$ and thus is essential for S^a,
c and d are context signals for S^a,
c is a trigger signal for $a-$ and thus is essential for R^a, and
d is a context signal for R^a.

It turns out that the covers obtained in Fig. 4.14 also have a monotonic behavior and can be used in an implementation based on C-elements, similar to that in Fig. 4.11d.

4.6 What is Speed-Independence?

In Sect. 4.2, the properties for the implementation of a hazard-free circuit were presented. The absence of hazards in a circuit modeled using unbounded wire delays is intrinsically related to the concept of *speed-independence*. The discussion on speed-independence has been postponed until this section, expecting now from the reader a deeper familiarity with the relationship between specification and implementation and the required properties to derive hazard-free circuits.

Every circuit circuit has two types of objects: gates and wires. A well-accepted intuitive definition of speed-independence is the following:

> *A circuit is speed-independent if its functional behavior does not depend on the delay of its gates.*

A circuit can be speed-independent under a certain environment, but not under another environment. Unless the circuit is autonomous, i.e. it explicitly includes a gate-level model of its environment, the latter must be specified to check the speed-independence of a circuit. To illustrate this fact, we will use the example in Fig. 4.15, that depicts a circuit with two input signals, one observable output signal and one internal signal. The figure also shows three possible environments for the circuit. For the analysis of the behavior, we will assume the following initial values for the signals: $a = 1, b = 0, c = 0, x = 0$.

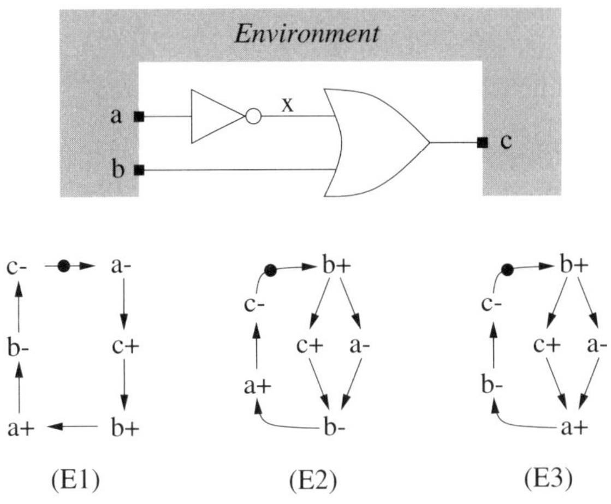

Fig. 4.15. Circuit with different environments

For environment E1, the circuit behaves correctly regardless of the delay of its gates. Its evolution in time can be described by the trace

```
a- x+ c+ b+ a+ x- b- c- ···
```

or the trace

```
a- x+ c+ b+ a+ b- x- c- ···
```

The occurrence of each trace will depend on the relative delay of the inverter with respect to the delay of the environment in producing `b-`. In both cases, the behavior of the circuit is correct and all the gates are persistent, i.e. no enabled gate is disabled without firing.

Let us now consider the environment E2. One might observe the following trace of events from the initial state:

`b+ a- c+ x+ b- a+ x- c- ···`

which is a correct behavior according to the specification. But if the delay of the inverter is long, the following trace can also be observed:

`b+ a- c+ b-` [`c- x+ c+`] `···`

This behavior is unacceptable, since the circuit produces an event `c-` before the environment has produced the event `a+`. Moreover, event `c-` may or may not occur depending on the values of the delays of the gates and of the environment. In other words, gate `c` is not persistent, and hence exhibits hazards. In this case, the functional behavior of the circuit depends on the delay of the inverter and, therefore, the circuit is not speed-independent.

The situation for E3 is a little different. Let us consider the following trace:

`b+ a- c+ a+ b- c- ···`

In this case, the inverter is disabled by event `a+` before `x+` occurs. However, if we disregard the internal signal of the circuit, the observable behavior might still be correct according to the environment. So, is this behavior acceptable?

As shown in Sect. 4.2.4, non-persistency may produce non-deterministic behavior. But how malicious can this non-determinism be? To answer this question properly, we would need to study the analog behavior of the gates. However, the abstract delay models used to characterize gates during logic synthesis are discrete and do not consider such details of gate behavior. Two models have been traditionally considered for this discretization:

- *inertial delay model*, in which gates are assumed not to have "internal memory" and react according to the current value of their inputs signals. In this model, a gate never produces a transition when it is disabled, even if it did not switch when it was enabled.
- *pure delay model*, in which gates have some internal memory that might delay some transition and output it even when a gate is no longer enabled.

Thus, if we consider the pure delay model for the inverter, the following trace could be feasible:

`b+ a- c+ a+ b- c-` [`x+ c+`] `···`

The internally delayed firing of `x+` leads to the disabling of the event `c-` expected by the environment. Therefore, the circuit is not speed-independent. In the pure delay model, persistency is the property that guarantees a hazard-free behavior [99].

Additionally, persistency of input signals with respect to output signals (i.e. no output event can disable an input event) is also required for a circuit to be speed-independent. The reason for that was previously illustrated in Sect. 4.2.4.

4.6.1 Characterization of Speed-Independence

The characterization of speed-independence considered in this book can be stated as follows:

> *A circuit is speed-independent if it is output persistent in all possible behaviors under a given environment.*

The concept of output persistency was introduced in Definition 4.2.6, and it describes the fact that when a signal disables another signal, then both must be input signals. This characterization has some advantages:

- Persistency can be checked locally for each signal of the circuit and the specification (see Definition 4.2.4).
- It guarantees absence of hazards for the pure delay model.

This characterization rejects circuits with correct observable behavior that have internal hazards. Internal hazards can propagate to observable outputs under the pure delay model. They can even propagate under the inertial delay model if a delay element is decomposed as a chain of delay elements. Hence, circuits in which all gates are persistent are more robust to delay models.

The requirement of persistency of internal gates can be waived if more information on the actual delays is known and the inertial delay model can be conservatively used for the analysis of the circuit.

4.6.2 Related Work

Muller [43] presented the first characterization of speed-independence. In his original work, he only considered *autonomous* circuits, i.e. without input signals. Conflict states were defined to characterize non-persistency. A circuit with no conflict states was said to be *semimodular*, according to Muller's definition. However, a non-semimodular circuit could still be called speed-independent according to Muller's definition, if the reachable states belonged to a single *final class* of states. Roughly speaking, non-semimodularity was acceptable if the internal hazards did not cause the circuit to diverge to mutually unreachable behaviors. However, we will not use this broader definition of speed-independence in this book, because it is not robust in presence of a non-inertial delay, e.g. in the presence of non-negligible wire delays, even when a wire does not have any fanout.

Muller's theory was later extended to circuits with external inputs. Semimodularity on output signals (output persistency) was proved to be robust for the pure delay model [99].

Trace theory has also been used to characterize speed-independence [100, 101, 102]. The designer is assumed to specify a design with trace commands, in which the environment controls the transitions at the inputs and the circuit produces the transitions at the outputs. The disabling of a gate is modeled as a transition to a failure state. A failure trace is called *computation interference* [102] or *choking* [100]. Computation interference and choking correspond to violations of semimodularity in Muller's model.

The concept of speed-independence has also been extended to high-level designs with data types and non-determinism [103]. In such framework, verification can be done mechanically with the assistance of theorem provers [104].

Quasi-delay-insensitivity is another concept very close to speed-independence that was originally proposed by Martin [105]. A circuit is said to be quasi-delay-insensitive if its behavior does not depend on the wire delays, with the exception of some isochronic wire forks [46]. A speed-independent circuit can also be considered quasi-delay-insensitive once the forks that require isochronicity have been identified. Even though the class of circuits is the same, the design flows associated to each concept are significantly different. Speed-independence has mostly been related to logic synthesis methods, whereas quasi-delay-insensitivity has been related to the synthesis of asynchronous circuits from high-level specifications using syntax-directed translation methods [105, 106].

4.7 Summary

This section concludes the chapter on synthesis of speed-independent circuits from STGs using complex gates. We have defined STGs, our basic specification model for synthesis, and the conditions for their implementability in logic circuits. These conditions fall into two categories from the point of view of the existence of the binary interpretation of their reachability space. The first category, consistency and boundedness, guarantees that the STG generates a correctly encoded SG that may be seen as a candidate for circuit implementation, perhaps requiring some transformations to ensure its logic circuit implementability. The second category, CSC and output-persistency, are necessary and sufficient to generate a set of hazard-free boolean functions from a consistent and bounded STG, that in turn can be used for constructing a logic circuit implementation using complex gates.

Another form of categorization of properties is by their treatment in the design flow, namely whether our approach rejects an STG not satisfying the property, or it applies some appropriate resolution technique. As discussed in Chap. 5, CSC violations are considered as automatically resolvable (i.e. the specification is consistent but incomplete), while violations of all other properties must be resolved by hand.

Two aspects pertaining to the synthesis of practically implementable circuits have been left out so far. The problem of logic decomposition of complex gates is covered in detail in Chap. 6. The problem of implementing specifications with intended internal conflicts (non-persistency with respect to output signals) is covered from a practical viewpoint in Chap. 8, and more theoretically in [107].

5. State Encoding

As we discussed in the previous chapter, some violations of STG and SG implementability conditions, such as output persistency and consistency, are considered to be essential for implementability, and thus if violated they must be fixed by the designer. Complete State Coding (CSC) is also required for implementability, however a specification that does not satisfy it can be transformed in such a way that its behavior remains externally equivalent to the original specification, yet the new SG satisfies CSC. One of the important conclusions of this chapter is that under reasonable assumptions about the class of specifications, the suggested methods to ensure CSC are effective and always converge.

The CSC property states that the binary encoding of TS states is unambiguous with respect to the internal behavior of a system, i.e. every pair of different states that are assigned the same binary code enables exactly the same set of output signals. This requirement is well-known in the synthesis of asynchronous FSMs, where it was also called race-free encoding [5]. We will first illustrate the similarities between resolving CSC conflicts in SGs and encoding states in FSMs, and then discuss why the former is somewhat more difficult than the latter.

Example 5.0.1. A Reflex Analyzer The verbal specification of this simple device to check human reflexes and accurate eye is as follows. It consists of a monitor and two buttons V and H to control the vertical and horizontal movements of a cursor. A person under test needs to draw a perfect rectangle by pushing the V and H buttons. The first push of button V moves the cursor up. The first push of button H changes the direction of the cursor and moves it right. The next push of V moves the cursor down, and the next push of H moves it left. The results of the test for persons with good and poor reflexes are shown in Fig. 5.1.

The formal specification of the reflex analyzer as an FSM is shown in Fig. 5.2a.

The FSM is specified using symbolic (abstract) states and inputs. They must be binary encoded for implementation in Boolean logic. If the FSM is implemented using synchronous logic, then any unique encoding is correct (even though different encodings lead to different complexity, delay and power

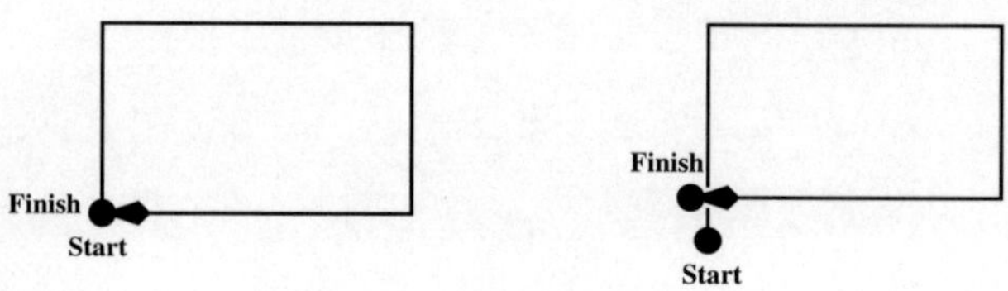

Fig. 5.1. Checking reflexes

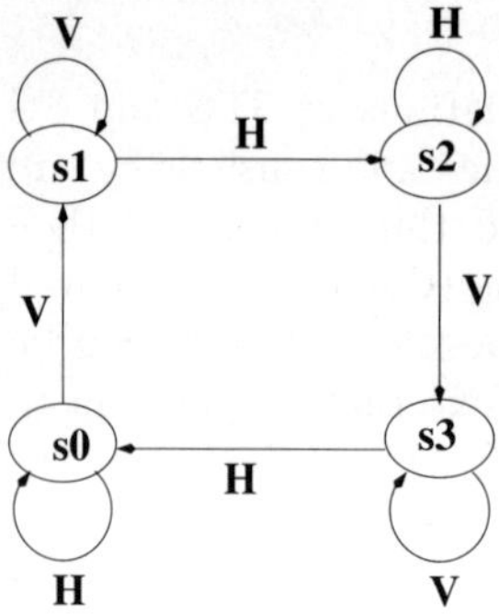

(a)

Encoding table

Inputs		States	
abstract	binary x	abstract	binary y1,y2
V	1	s0	00
H	0	s1	01
		s2	10
		s3	11

(b)

Fig. 5.2. Reflex Analyzer specification diagram (**a**) and table (**b**)

consumption measures [10]). This is due to the fact that for a synchronous implementation only the settled values of state signals matter.

In an asynchronous implementation, though, every state (including transient states) can potentially propagate through the feedback paths back to the FSM present state inputs. This means that encoding may also affect its *functionality*, not only its cost.

Let us assume, for example, that the encoding of inputs and states of the reflex analyzer corresponds to the table in Fig. 5.2b. Some state transitions (e.g. $s1 \rightarrow s2$ or $s3 \rightarrow s0$) involve non-adjacent binary codes. All possible trajectories between states must be considered in an asynchronous implementation, due to unknown gate delays. The expanded binary state diagram using the encoding scheme in Fig. 5.2b is shown in Fig. 5.3, where transient states are drawn with dashed lines.

Let us have a closer look at stable state 00_1 and transient state 00_2 (the latter is potentially reachable during transition $01 \rightarrow 10$). Both have the same binary code for present state (00) and input (0) signals. However, the next states for 00_1 and 00_2 under input 0 are different: 00_1 should perform a self-loop transition to itself, while 00_2 should go to 10. This makes it impossible to implement in logic the behavior of state signal $y2$: under the very same values of state and input signals $y2$ should keep its value for 00_1 and change it for 00_2.

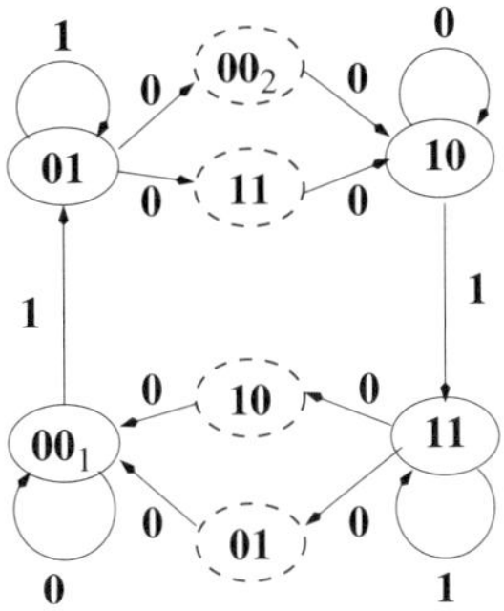

Fig. 5.3. State diagram of encoded reflex analyzer

States like 00_1 and 00_2 define different behaviors of a system, therefore they must be distinguished explicitly to resolve the ambiguity. Such states are called *CSC conflicts*. CSC conflicts must be avoided by a careful choice of encoding.

Fig. 5.4 shows two ways to approach the conflict-free encoding of asynchronous FSMs. The first way is to refine the encoding incrementally, i.e. to keep the value of the existing signals in each state code unchanged, and to add auxiliary state signals for disambiguating CSC conflicts. For the FSM of the reflex analyzer one auxiliary state signal ($y3$) is enough to resolve all CSC conflicts. The encoding table and the resulting state diagram are shown in Fig. 5.4. Note that even though some transient states are encoded by the same code of state signals (see state 10_1 in transitions 10_0 $\rightarrow$ 11_1 and 11_1 $\rightarrow$ 00_1) these states are not in CSC conflict, because they are distinguished by different values of the input signal: 1 for 10_0 $\rightarrow$ 11_1 and 0 for 11_1 $\rightarrow$ 00_1.

If the original encoding of FSM can be changed, then one can improve the quality of the solution by properly solving all CSC conflicts from the beginning. The result of this technique (called global encoding) is shown in the lower right corner of Fig. 5.4. Clearly the resulting FSM is much simpler.

Usually a designer has a full freedom in choosing a particular encoding of state signals of an FSM implementation. This is due to the fact that the interface between the FSM and the environment is represented by its input and output signals, and state variables are invisible to the environment. Thus in FSMs we have a clean separation between observable interface signals and non-observable "memory" signals that keep track of system states.

The history of research on universal methods for race-free encoding of asynchronous FSM goes back to the early 60's. All of them fall into the category of global encoding methods. The works of [108, 109] resulted in an upper bound on the number of state signals for race-free encoding of an arbitrary FSM with a given number of states. Even though the encoding used in the construction of these upper bounds is inefficient (the size of the

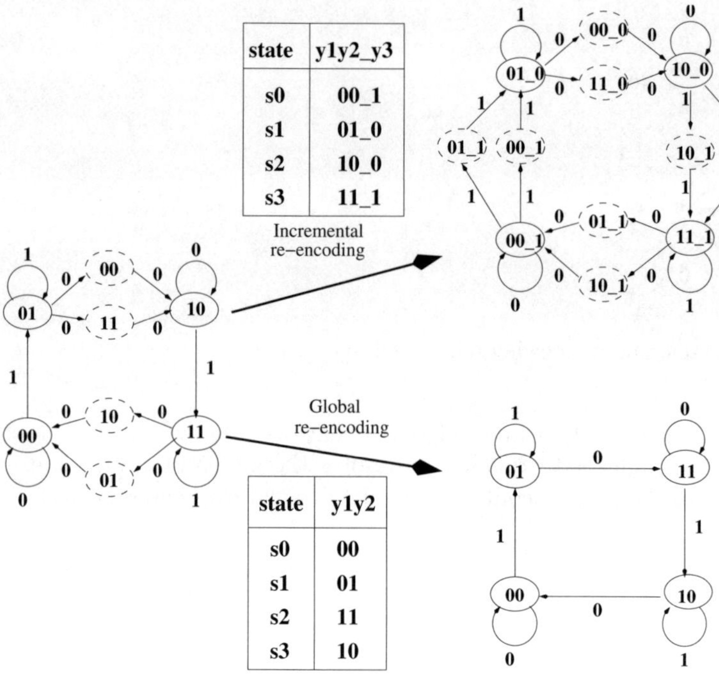

state	y1y2_y3
s0	00_1
s1	01_0
s2	10_0
s3	11_1

state	y1y2
s0	00
s1	01
s2	11
s3	10

Fig. 5.4. Reduction to race-free encoding

obtained code even exceeds that of one-hot encoding), the result is of theoretical importance. It shows that the global encoding step is always effective.

An encoded TS is a specification at a lower abstraction level than an FSM with symbolic inputs, outputs and states, because its events already have a binary interpretation, and encoding is part of the original specification. Moreover, there are no specially designated state signals, since the values at inputs and outputs themselves are used to define the state of a system. Adding auxiliary signals to a specification is allowed, but the encoding of the original inputs and outputs must be preserved. This does not allow one to use the global encoding method for eliminating CSC conflicts, and gives a pivotal role to incremental re-encoding. The key result shown in this chapter is that, under reasonable restrictions on the class of specifications, one can always guarantee that CSC conflicts will be eliminated by adding auxiliary signals. This puts incremental re-encoding for binary TSs in the same position as the former known result about the effectiveness of race-free global encoding for asynchronous FSMs.

5.1 Methods for Complete State Coding

Let us first informally explain methods for complete state coding in binary TSs by considering two intuitive examples. In both cases, we assume that originally the specification is an STG, and that a binary TS (namely an SG) is derived from the STG through reachability analysis.

Example 5.1.1. Frequency divider The purpose of this device is to divide the input transition frequency by two (see Fig. 5.5a). The behavior of the divider with input signal a and output signal b is given by the STG and the corresponding SG in Fig. 5.5b. Starting from the initial state a=b=0 (0*0 in the SG), the divider waits for two transitions at its input and replies by a transition at the output.

Unfortunately, describing the behavior solely in terms of input-output signals is ambiguous, because after input a changes twice, the system returns to the very same values of input and output signals (a=b=0, that is state labeled 00* in the SG) but needs to behave differently than in the initial state. In the initial state 0*0 the system must wait for input changes, while in 00* it needs to produce an output transition. A similar conflict is observed between SG states 0*1 and 01*. To implement the frequency divider one must keep track of the pre-history for states that are involved in a CSC conflict. This in turn requires to disambiguate the conflict states by adding auxiliary (state) signals. The SG in Fig. 5.5b shows that CSC conflicts can be solved by separating states into two groups (indicated by shadowed regions), by using auxiliary signal that have different values in the two groups, and make transitions in between. This leads us to the STG and SG in Fig. 5.5c that satisfy CSC requirements and can be implemented by the logic circuit in Fig. 5.5d. An experienced designer would anticipate this result from the beginning, because it corresponds to the conventional way of implementing a toggle flip-flop as a master-slave. The value of the above example is in showing that formal methods to ensure implementability make sense, as they can produce the very same result that an experienced designer would expect.

Example 5.1.2. The Mad Hatter's tea party Let us take a second example from the never fading "Alice's Adventures in Wonderland". We will analyze one of the best episodes of the tale (the Mad Hatter's tea party). Simplifying the story, let us restrict the list of party participants to the March Hare and Alice herself (our sincere apologies to the Hatter and Dormouse).

A paraphrase of Lewis Carrol's story looks as follows. There is a round table with two cups and a teapot. March Hare and Alice make clockwise moves around the table (see Fig. 5.6a), fill their cups with tea (events $a+$ and $mh+$), and then drink it (events $a-$ and $mh-$). As time has been frozen at six o'clock, it is always tea-time for party participants, and they are repeating this loop infinitely.

Note that Alice and March Hare have some relative freedom in choosing the moment for pouring and drinking their cups of tea. However, they must

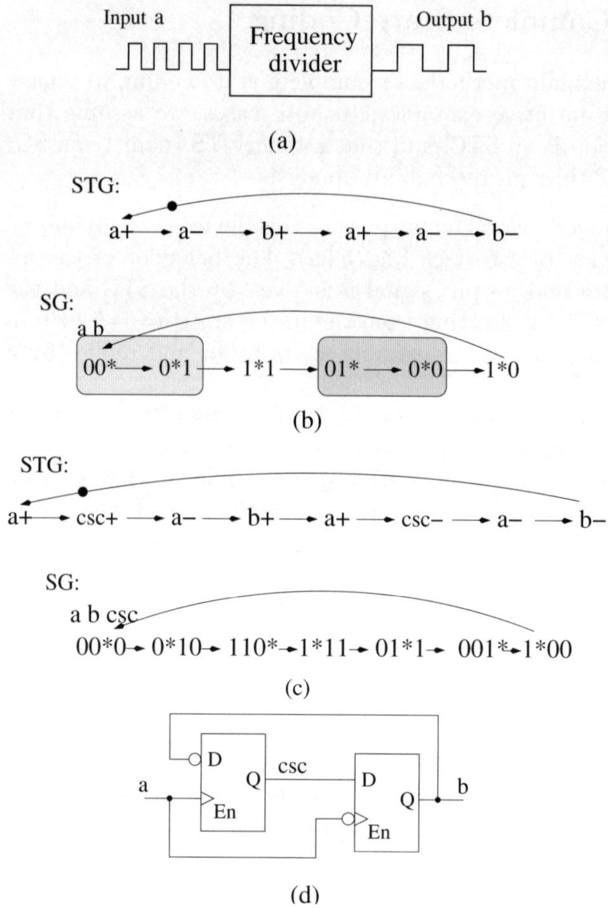

Fig. 5.5a–d. Implementation of the frequency divider

synchronize after drinking, in order to be able to make simultaneous moves around the table. Their behavior can be described by the STG in Fig. 5.6b, where λ is a dummy event that serves the purpose of synchronization after $a-$ and $hm-$. Observing the corresponding SG (Fig. 5.6c) one can find plenty of CSC conflicts which could be separated into three blocks of states (see shadowed regions in Fig. 5.6c). At least two auxiliary signals are needed to disambiguate these conflicts, because all three groups should be encoded differently (to distinguish three states with the same binary code 00). Insertion of signals resolving CSC conflicts is non-trivial for this example, and we will postpone it to a later section, in order to keep the explanation simpler.

However, there is a new option to ensure CSC in STGs, with respect to the FSM approach. Note that the STG description provides an upper bound to the set of potential signal values reached during system operation. In any

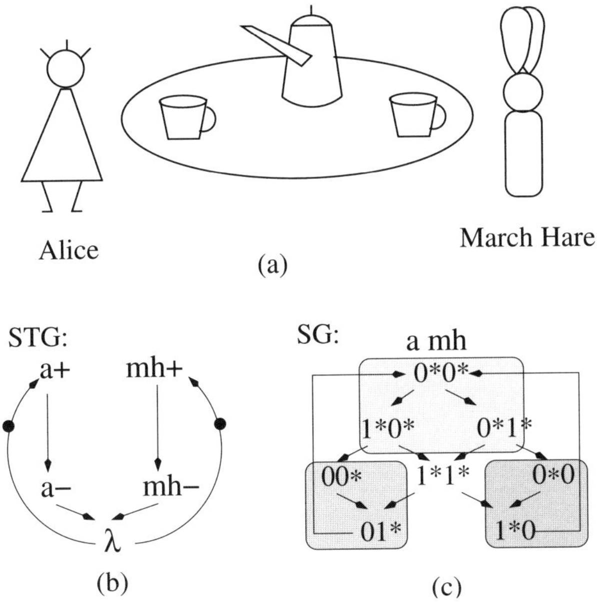

Fig. 5.6a–c. The Mad Hatter's tea party model

implementation, due to fixed (or slowly varying) delays of the gates, the system may follow only a few trajectories out of the several allowed by the speed-independent STG. Indeed the STG in Fig. 5.6b tells, e.g., that Alice could finish her cup even before March Hare has poured tea (events $a-$ and $mh+$ are concurrent in STG). In real life, drinking hot tea usually takes much longer than filling a cup. Such physical features of implementation make many potential sequences of STG events infeasible (through concurrency reduction between events). Keeping this in mind and exploiting the idea of concurrency reduction at all stages of synthesis gives powerful means for design optimization. In particular, based on assumptions about the time required to pour and drink tea, one could *require* that both Alice and March Hare finish filling their cups before either of them finishes drinking hot tea. This introduces an additional synchronization after events $a+$ and $mh+$, which is shown in STG of Fig. 5.7a by the dummy event τ. The SG in Fig. 5.7b, extracted from the STG with reduced concurrency, shows fewer CSC conflicts, which are easily disambiguated by a single auxiliary signal. The corresponding circuit implementation is shown in Fig. 5.7c. The circuit synchronizes two independent processes (inverter a and inverter mh) with the help of a C-element. The latter exactly corresponds to the behavior described by the STG of Fig. 5.7a.

Pushing the idea of concurrency reduction even further, results in a fully sequential process, where Alice and March Hare do everything in turns (see Fig. 5.7d). This process has no CSC conflicts at all, and can be implemented by a simple circuit composed of an inverter and a buffer (see Fig. 5.7e). This

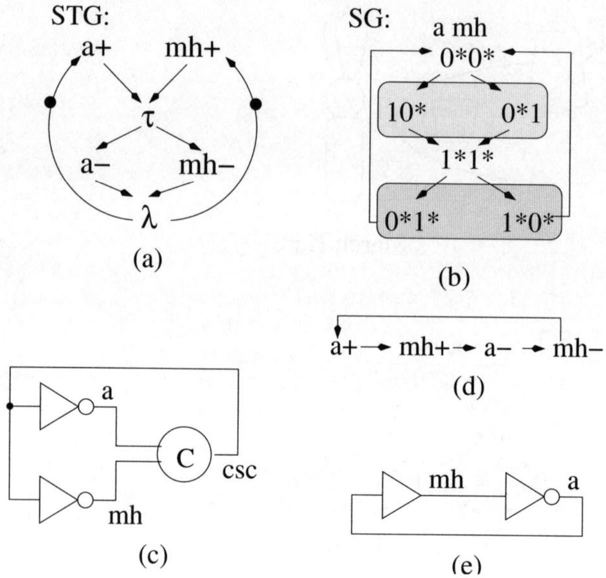

Fig. 5.7a–e. Constrained models for the Mad Hatter's tea party

may or may not be acceptable in practice, depending on performance requirements, because implementation complexity is traded for process performance.

The above examples allow us to draw several conclusions:

- CSC conflicts can be efficiently resolved by adding auxiliary signals, and
- some CSC conflicts could be eliminated by reducing concurrency between events in the specification.

In the rest of the chapter we will concentrate on elimination of CSC conflicts via adding auxiliary signals. The ideas on concurrency reduction will be considered later, in a broader scope of design optimization, in Chap. 7.

5.2 Constrained Signal Transition Event Insertion

In order to eliminate CSC conflicts by adding auxiliary signals, one must answer two key questions:

1. **How** to insert new signal transition events?
2. **Where** in the specification to insert new signal transition events?

The answer to the second question affects the quality of the final implementation, while the answer to the first one affects its correctness. Since it makes little sense to talk about the quality of an incorrect implementation, we will start by investigating solutions to the first question.

5.2.1 Speed-Independence Preserving Insertion

Our final goal is to get an implementation which behaves correctly independent of gate delays. We first assume that a TS is used as the initial specification, and then consider its binary encoded version, the SG. For any transformation on the TS that strives to ensure CSC, we would like to get closer to the final goal, and not introduce any new violation of the speed-independence conditions. The insertion of a single event into an appropriately chosen "area" of a speed-independent TS is considered as the basic operation in these transformations. We will later discuss some heuristics about how to choose such "areas", by considering both correctness and cost.

Formally, this task is stated as follows:

given: 1) a speed-independent TS $A = (S, E, T)$, 2) an event $x \notin E$ and 3) an arbitrary subset of states $r \subseteq S$ that defines the area where x must be inserted into A,

produce a new equivalent TS $A = (S \cup r', E \cup x, T \cup T')$ which satisfies the speed-independence conditions.

It was shown in Sect. 3.7 that the simple insertion scheme from Definition 3.7.1 can be a good match for our purposes, because it *always preserves trace equivalence, determinism and deadlock-freedom.* The rest of the chapter discusses the constraints imposed on this insertion scheme by the speed-independence requirements.

Let us first recall how a new event is inserted in a TS, and introduce a bit of notation to reason about the initial TS and the new one after insertion. Fig. 5.8 shows the way to transform TS A into TS A' through the insertion of x by a subset of states $r \in S$. When the need arises, we will refer to objects in the original TS A and the transformed TS A' by using the name of the TS as a subscript. For example, the sets of states S of TSs A and A' will be denoted as S_A and $S_{A'}$ respectively. The subscripts will be omitted in non-ambiguous contexts.

The insertion of a new event x defines a surjective mapping between the sets of states S_A and $S_{A'}$. This mapping is illustrated in Fig. 5.8, where

- each state s_A outside r is mapped into a single corresponding state $s_{A'}$ of A' ($s_{A'}$ is called the image of s_A), and
- each state $s_A \in r$ is mapped into a pair $s_{A'}, s'_{A'} : s_{A'} \xrightarrow{x} s'_{A'}$ (the pair $s_{A'}, s'_{A'}$ is called the *image* of s_A).

Informally, the insertion simply splits every state $s_A \in r$ into two states $s_{A'} \xrightarrow{x} s'_{A'}$, while leaving the rest of the TS unchanged.

Sect. 3.7 showed that the speed-independence conditions for a deterministic TS are refined to *persistency* and *commutativity* requirements. These requirements can be formulated through efficient local checks on states of a TS and their neighborhood, as shown in Fig. 5.9. When constraining the insertion procedure to preserve speed-independence, we will bear Fig. 5.9a and 5.9b in mind.

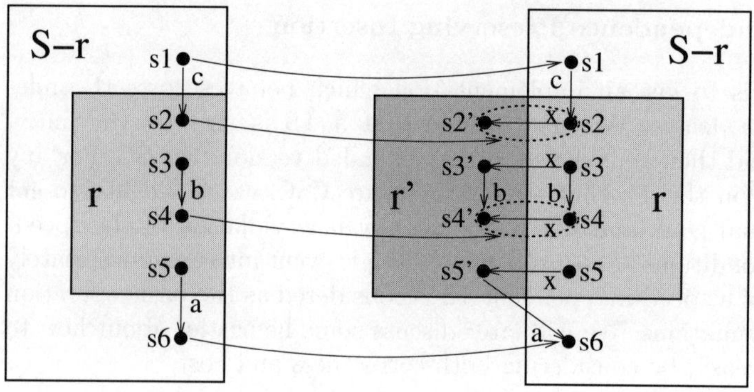

Fig. 5.8. Insertion of a new event into a TS

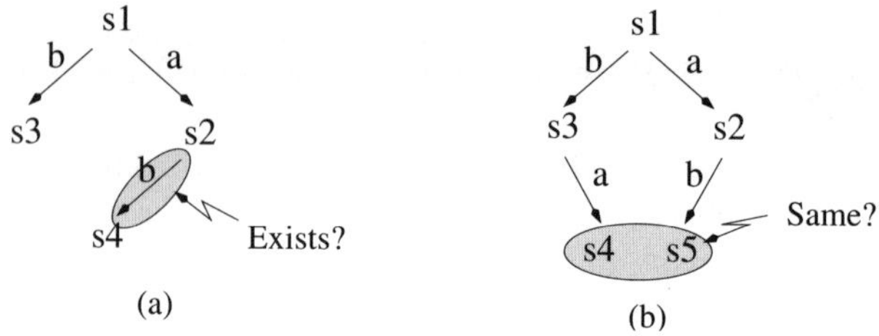

Fig. 5.9. Local persistency (**a**) and commutativity (**b**) analysis

Let us first prove that the insertion procedure shown in Fig. 5.8 is correct when considering the behavior of the newly inserted event x (we will consider preserving speed-independence of other events later).

Property 5.2.1. Let $A = (S, E, T)$ be a transition system, $x \notin E$ be a new event and $r \subseteq S$. Let $A' = (S', E', T')$ be the transition system obtained after inserting x by r as illustrated in Fig. 5.8. Then,

1. x is persistent in A', and
2. x is commutative in A'.

Proof.

1. (by contradiction) Assume that in TS A': $s1 \xrightarrow{x} s3$, $s1 \xrightarrow{a} s2$ and x is not enabled in $s2$ (see Fig. 5.9a with $x = b$). Therefore in TS A, $s1 \in r$ and $s2 \notin r$. By the definition of event insertion for a state at the border of r (state $s2$ in Fig. 5.9a and state $s5$ in Fig. 5.8), first a transition labeled with x should take place, and then r is left via some other event a. Hence no transition $s1 \xrightarrow{x} s3$ would be possible in TS A', which contradicts the initial assumption.

2. Assume that traces ax and xa start from state $s1$ (see Fig. 5.9b with $x = b$). Then $s1, s2 \in r$ in A, since x is enabled in both of them. For internal states of r, the enabling of the original events (e.g. a) is preserved, and hence $s3 \xrightarrow{a} s5$. Since A' is deterministic, then $s4 = s5$.

According to Property 5.2.1, the inserted event x itself does not cause speed-independence violations. However, other events that previously behaved correctly can now suffer from the insertion of x. Once the insertion scheme of Fig. 5.8 has been selected, the only freedom of choice left is the area in the TS where to insert event x, i.e. the set of states r. The correct choice for r is formalized in the following definition.

Definition 5.2.1 (SIP-set). *Let $A = (S, E, T)$ be a transition system, $x \notin E$ be a new event and $r \subseteq S$. Let $A' = (S', E', T')$ be the transition system obtained after inserting x by r as in Fig. 5.8. r is said to be a* speed-independence preserving set *(*SIP-set*) iff:*

1. $\forall a \in E : a$ is persistent in $A \implies a$ is persistent in A'
2. A is commutative $\implies A'$ is commutative

If r satisfies only condition 1, then r is a persistency preserving *set.*

The following theorems determine the conditions for preserving persistency and commutativity during the insertion of a new event. They are formulated in terms of local properties of small fragments of the TS, such as those shown in Fig. 5.9. The check for a subset r to be a SIP-set is performed in two steps: first we check r to be a persistency preserving set (Theorem 5.2.1), then impose additional constraints for such a persistency preserving set to satisfy the commutativity conditions as well (Theorem 5.2.2).

Theorem 5.2.1. *Let $A = (S, E, T)$ be a TS and $r \subseteq S$ be a subset of states. r is a persistency-preserving set iff for any TS states $s1, s2, s3, s4$ (see Fig. 5.9) the following Condition 1 is satisfied:*

$$\forall a \forall b \in E : [s1 \xrightarrow{b} s3,\ s1 \xrightarrow{a} s2 \xrightarrow{b} s4 \in T \ \wedge\ s2 \in r, s4 \notin r] \implies s1 \in r \ \wedge\ s3 \notin r \qquad (1)$$

Proof. $\Longleftarrow$. There are two different cases that violate Condition (1):

1. $s2 \in r$, $s4 \notin r$, but $s1 \notin r$ (see Fig. 5.10a).
2. $s2 \in r$, $s4 \notin r$, but $s1, s3 \in r$ (see Fig. 5.10b).

For Cases 1 and 2, Fig. 5.10 shows the fragments of TS A' obtained by the insertion of x by r. One can see that in both cases persistency is violated for state $s2$, because event b, that was enabled in $s1$, becomes disabled in $s2$ without firing. Thus, Condition (1) is necessary for preserving event persistency.

$\Longrightarrow$. Let us consider a pair of events a, b for which persistency of A' is violated in the quadruple of states shown in Fig. 5.9a. Neither a nor b could

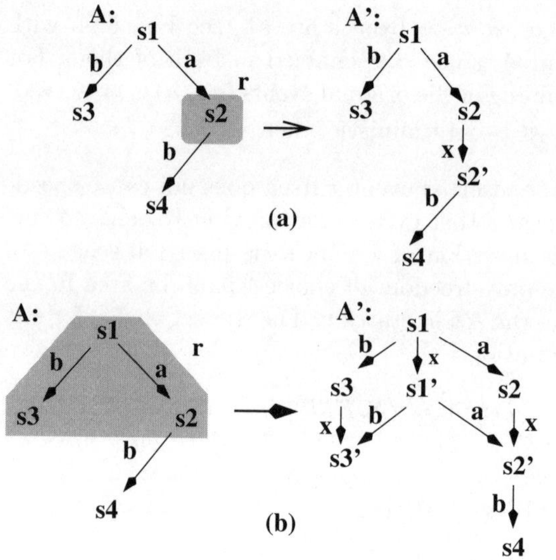

Fig. 5.10a,b. Sets of states not preserving persistency

coincide with the newly inserted event x. Indeed the case when the firing of some event disables x contradicts Property 5.2.1 about the persistency of x in A', while the case when x disables some other events does not match the rules of insertion (see Fig. 5.8). Therefore states $s1_{A'}$ and $s2_{A'}$ are images of states $s1, s2 \in A$ for which $s1 \neq s2$. Four cases are possible:

- $s1, s2 \notin r$ (Fig. 5.11a).
  By the rules of transformation enabled events are the same in any state $s \in A$, $s \notin r$ and its image $s_{A'} \in A'$. Therefore an event b that is enabled in $s2$ remains enabled in $s2_{A'}$ and no violation of persistency by b is possible in state $s1_{A'}$.
- $s1 \in r, s2 \notin r$ (Fig. 5.11b).
  Disabling of b by x cannot take place (no persistency violations are caused by x). Disabling of b by a in the pair of states $s1'$, $s2$ is reduced to consideration of the previous case.
- $s1 \notin r, s2 \in r$. Transformation of A into A' is shown in Fig. 5.11c.
  Since $s2 \in r$ and $s1 \notin r$, then to satisfy Condition (1) state $s4$ has to belong to r. When both states belong to r, the enabling of the original events is preserved during the transformation (see Fig. 5.8). Hence in A' event b must be enabled in both $s1_{A'}$ and $s2_{A'}$ and its persistency is preserved.
- $s1, s2 \in r$. There are two different sub-cases: $s4 \notin r$ (Fig. 5.11,d) and $s4 \in r$ (Fig. 5.11e).
  If $s4 \notin r$ then by Condition (1) $s3$ also has to be out of r. Persistency of b with respect to states $s1'_{A'}$ and $s2'_{A'}$ follows from the fact that in A' $s1'$

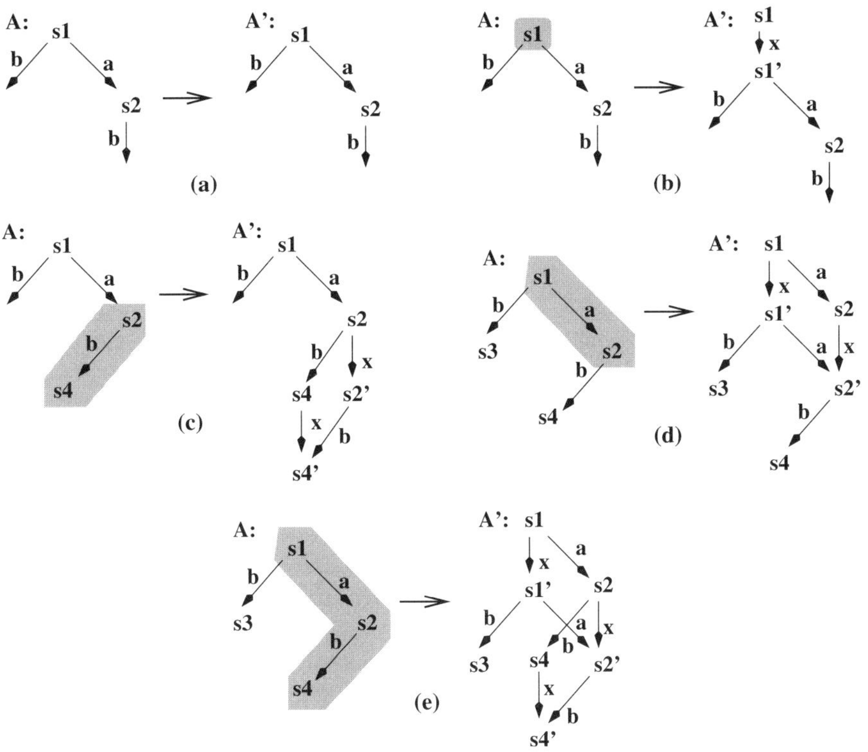

Fig. 5.11a–e. Adding event x to transition system. Cases 1-4

and $s2'$ first make transitions by x and then their successors must perform transitions by b.
If $s4 \in r$ (Fig. 5.11e), then, independent of whether $s3 \in r$ or $s3 \notin r$, transition $s2 \stackrel{b}{\rightarrow} s4$ is internal for r, and b will be enabled in both $s2_{A'}$ and $s2'_{A'}$. Again this excludes the possibility of non-persistency due to b for states $s1_{A'}$ and $s2_{A'}$ and for states $s1'_{A'}$ and $s2'_{A'}$.

Therefore, in none of Cases 1-4 event persistency can be violated. Hence, Condition (1) is sufficient for preserving persistency.

Theorem 5.2.1 suggests simple conditions to check whether a chosen subset of states r is persistency preserving. The next theorem refines these conditions to capture commutativity properties as well.

Theorem 5.2.2. *Let $A = (S, E, T)$ be a commutative transition system and $r \subset S$ be a persistency preserving set. Then r is a SIP-set iff for any* TS *states $s1, s2, s3, s4$ (see Fig. 5.12) the following Condition 2 is satisfied:*

$$\forall a \forall b \in E : s1 \stackrel{a}{\rightarrow} s2 \stackrel{b}{\rightarrow} s4,\ s1 \stackrel{b}{\rightarrow} s3 \stackrel{a}{\rightarrow} s4 \in T \wedge (s1, s2 \in r \wedge s3 \notin r) \Longrightarrow s4 \notin r \qquad (2)$$

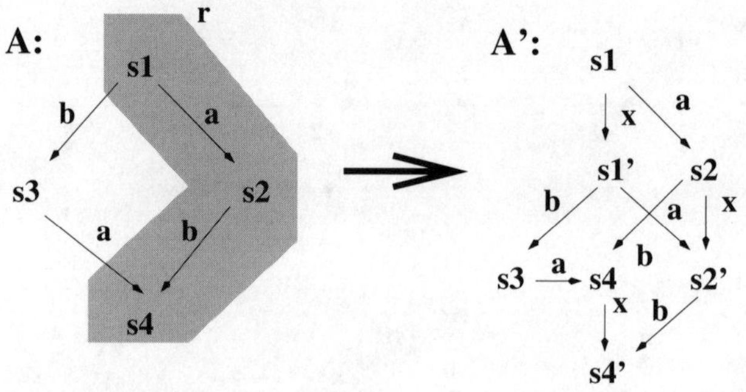

Fig. 5.12. Commutativity violation after event insertion

Proof. $\Longleftarrow$. Suppose Condition (2) is violated. A transformation from A to A' in such case is illustrated by Fig. 5.12. Clearly, commutativity is violated for state $s1'$ because $s1' \overset{ab}{\to} s4'$ but $s1' \overset{ba}{\to} s4$, where $s4 \neq s4'$.

$\Longrightarrow$. Suppose Conditions (1) and (2) are satisfied for a set of states $r \subset S$. By Property 5.2.1 the new event x is commutative in A' and therefore, we need to consider commutativity only between original events of A. Two different orderings of events a and b, each of which may fire in the same state in A, give rise to the quadruple of states $s1, s2, s3, s4$ that we will call a *state diamond* (see Fig. 5.13a). The major cases of intersections of a diamond in A with a set of states r, that are legal according to Conditions (1) and (2), are shown in Fig. 5.13.

Cases $s1 \in r,\ s2, s3, s4 \notin r$ and $s4 \in r,\ s1, s2, s3 \notin r$ are covered by Fig. 5.13b. Cases $s1, s3 \in r,\ s2, s4 \notin r$ and $s2, s4 \in r,\ s1, s3 \notin r$ are symmetrical to Fig. 5.13c and 5.13d respectively. All other cases of intersections are forbidden by Conditions (1) and (2).

It is easy to check that by applying the transformation rules to these five major cases (as it was done in the proof of Theorem 5.2.1) we never get violations of commutativity.

Theorems 5.2.1 and 5.2.2 give a simple and efficient way for checking whether a chosen subset of states r is a SIP-set or not. This check is reduced to the analysis of intersections of r with state diamonds in a TS, where all legal intersections are presented by Fig. 5.13.

This approach to check a set r for being SIP, based on "legal pattern" recognition, is illustrated in Fig. 5.14. The subset of states r, shadowed in Fig. 5.14a, intersects with four diamonds of states: $d1 = \{s1, s2, s3, s4\}$, $d2 = \{s3, s4, s5, s7\}$, $d3 = \{s2, s4, s6, s8\}$, and $d4 = \{s4, s7, s8, s1\}$. Intersections with $d1$ and $d4$ are legal (they correspond to the case of Fig. 5.13b), while intersections with $d2$ and $d3$ are illegal (they correspond to persistency violation as in Fig. 5.10a). The latter means that the subset of states r in Fig. 5.14a is not a SIP-set. Similarly for Fig. 5.14b, one can conclude that r is not a SIP-set because its intersection with state diamond $\{s1, s2, s3, s4\}$

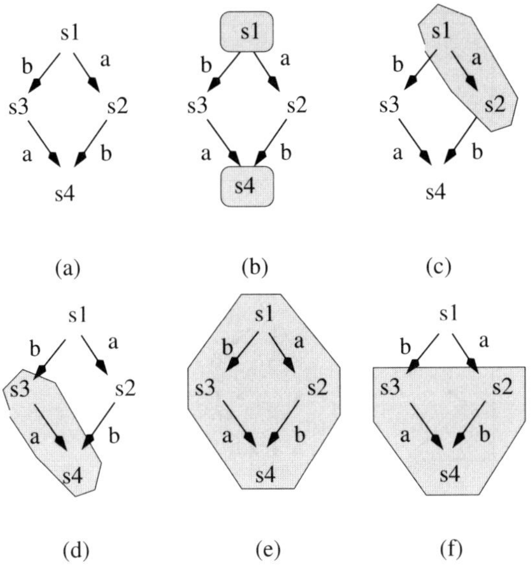

Fig. 5.13a–f. SIP-sets for a state diamond

is illegal. For the set r shadowed in Fig. 5.14c, the reader can check that all intersections with state diamonds are covered by legal cases c, e and f of Fig. 5.13 and hence r is a SIP-set. The insertion of a new event x based on this set is shown in Fig. 5.14d. It is easy to see that the obtained TS is speed-independent. This result is what one can indeed expect from an insertion based on a SIP-set.

5.3 Selecting SIP-Sets

This section presents several basic properties which allow us to formulate heuristic strategies for the selection of SIP-sets. We will show below that manipulating regions (Definition 3.5.2) helps finding valid SIP-sets *automatically*, rather than enumerating all possible insertions and checking for SIP *a posteriori*, which is considerably less efficient.

Property 5.3.1. If r is a region in a commutative excitation-closed transition system, then r is a SIP-set.

The proof of this property is trivial. At the PN level this property corresponds to the following structural transformation: place r is substituted by two places r and r' with a new intermediate transition labeled with x. Place r has only one output transition, x, and all transitions which belong to $r\bullet$ in the initial PN belong to $r'\bullet$ in the new PN. Obviously, such transformations cannot violate persistency or commutativity for any event.

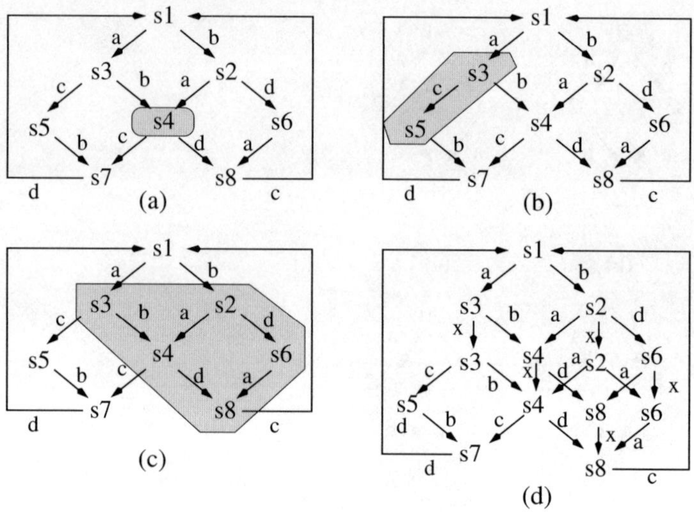

Fig. 5.14a–d. Checking SIP-sets by state diamonds

Property 5.3.2. If r is the excitation region of event c in a commutative transition system A and c is persistent in r, then r is a SIP-set.

Proof.

1. Let us consider the possible cases of persistency violation (see Fig. 5.10a and b). In both cases $s2 \in r$, $s4 \notin r$ and $s2 \xrightarrow{b} s4$. If $b \neq c$, then the firing of b disables c and contradicts the assumption of event persistency for c. If $b = c$, then $s1$ belongs to $r = ER(c)$ (a contradiction with Fig. 5.10a) and $s3$ must be out of $r = ER(c)$ (a contradiction with Fig. 5.10b).
2. Let us consider commutativity violations (see Fig. 5.12). If $b \neq c$, then it follows from $s1 \in r$ and $s3 \notin r$ that the firing of b in state $s1$ disables c. This is a contradiction with persistency of c. If $b = c$, then $s4$ must be out of $ER(c)$, and this contradicts Fig. 5.12.

Intuitively, this property can be stated as follows: delaying a persistent event cannot create violations of persistency or commutativity. At the PN level this means that substituting a persistent transition with a sequential composition of two transitions preserves persistency and commutativity. At the circuit level this property corresponds to a well-known fact: inserting a delay at a gate output before its fanout does not violate speed-independence of the circuit [43].

The candidates to generate SIP-sets considered so far have a clear interpretation in PN terms: they correspond to places and transitions. To achieve a finer granularity in the construction of SIP-sets (though at the expense of losing the clear link with PN objects), let us consider the so called **input** and

exit borders for a subset of states. Informally, an input (exit) border for a subset of states r is the set of states by which one can enter (exit) r.

Definition 5.3.1 (Exit and input border). *Let $A = (S, E, T)$ be a transition system. Given a subset of states $r \subseteq S$, the* exit border *of r (denoted as $EB(r)$) and the* input border *of r (denoted as $IB(r)$) are defined as follows:*

$$EB(r) = \{s \in r | \exists a \in E, s' \in S : s \xrightarrow{a} s' \in T \ \wedge \ s' \notin r\}$$
$$IB(r) = \{s \in r | \exists a \in E, s' \in S : s' \xrightarrow{a} s \in T \ \wedge \ s' \notin r\}$$

Exit borders of regions can be safely used as SIP-sets under the following conditions.

Property 5.3.3. Let $A = (S, E, T)$ be a commutative excitation-closed transition system and let r be a region in A. If all the exit events of r are persistent, then $EB(r)$ is a SIP-set.

Proof.

1. Let us consider the possible cases of persistency violations (see Fig. 5.10a and 5.10b). In both cases $s2 \in EB(r)$ and $s4 \notin EB(r)$. Hence, there exists event c such that $s2 \xrightarrow{c} s5$, $s5 \notin r$. Clearly c is an exit event for r and from the properties of a region any state in which c is enabled belongs to r. If $c \neq b$, then it follows from $s4 \notin EB(r)$ that event c, which is enabled in $s2$, becomes disabled in $s4$. This contradicts the assumption that all exit events of r are persistent.
   If $c = b$ then $s1$ belongs to $EB(r)$ (contradiction with Fig. 5.10a) and $s3$ must be out of r (contradiction with Fig. 5.10b).
2. Commutativity violations (Fig. 5.12) are not possible, because by the same consideration if $c \neq b$, then c becomes disabled in $s3$. If $c = b$, then $s4$ cannot be in $EB(r)$.

One more option in the construction of SIP-sets through regions is given by pre-regions of the same events. It can be shown that under certain conditions [110] intersections of these pre-regions give SIP-sets. An important consequence of the above properties is that good candidates for an insertion can be built on the basis of regions and their intersections, since they preserve equivalence and speed-independence. One may also conclude that SIP-sets for an event insertion can be built more efficiently from regions rather than states.

5.4 Transformation of State Graphs

If a TS is binary encoded, i.e. it is an SG, then some additional constraints for inserting new events are required. Each inserted event has to be interpreted as a signal transition, therefore consistency of state assignment must be preserved. Any event insertion scheme which preserves trace equivalence (like those in Definition 3.7.1) obviously preserves consistency for the original signals. Special care must be taken to ensure consistency of the new signals (that are usually called *state* signals).

A specific class of SG transformations can be defined as follows:

1. Insertion is made by signals, not by events. Therefore, instead of inserting a single event, two signal transitions of a new signal are inserted at each step: $x+$ and $x-$. Two sets of states for insertion, $\mathsf{ER}(x+)$ and $\mathsf{ER}(x-)$, are defined simultaneously such that $\mathsf{ER}(x+) \cap \mathsf{ER}(x-) = \emptyset$.
2. Similar to TS transformations, both sets for insertion, $\mathsf{ER}(x+)$ and $\mathsf{ER}(x-)$, must be SIP-sets. In addition, consistency of state assignment for signal x is required.

Given an SG with a set of binary states S, a partition for the insertion of signal x, called *I-partition*, is a partition of S into four blocks ([111]): S^0, S^1, S^+ and S^-. $S^0$and S^1define the states in which x will have the stable values 0 and 1 respectively. $S^+$and S^-define $\mathsf{ER}(x+)$ and $\mathsf{ER}(x-)$ respectively.

Property 5.4.1. Let A be a consistent SG and let $I = < S^0, S^1, S^+ S^- >$ be an I-partition of A. SG A' obtained by inserting signal x by partition I is consistent iff the only allowed arcs crossing boundaries of the partition blocks are the following: $S^0 \rightarrow S^+ \rightarrow S^1 \rightarrow S^- \rightarrow S^0$, $S^+ \rightarrow S^-$ and $S^- \rightarrow S^+$, $S^+ \rightarrow S^0$ and $S^- \rightarrow S^1$.

Arcs like those shown in the Fig. 5.15 are forbidden by Property 5.4.1. The proof of this Property directly follows from the rules of insertion for events $x+$ and $x-$ (see Definition 3.7.1).

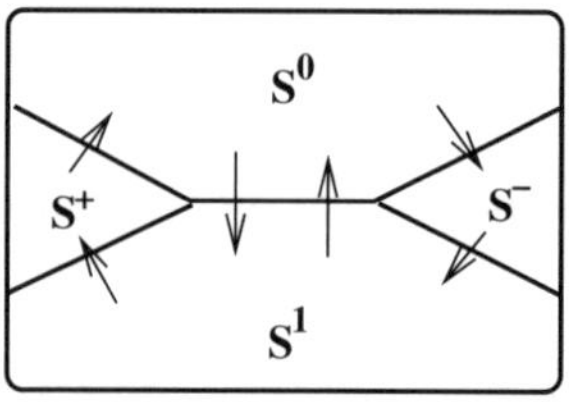

Fig. 5.15. Illegal transitions in an I-partition

An I-partition can be found in two steps:

- Find a bipartition $\{b, \overline{b}\}$, ($\overline{b} = S - b$) of the set of states S. The value of signal x is constant inside blocks b and $\overline{b}$.
- Choose $\mathsf{ER}(x+)$ and $\mathsf{ER}(x-)$ at the boundaries of blocks b and $\overline{b}$ respectively.

The boundaries may be defined in two ways: as exit borders or as input borders. Fig. 5.16a shows insertion by exit borders: given a bipartition $\{b, \overline{b}\}$, $\mathsf{ER}(x+) = EB(b)$ and $\mathsf{ER}(x-) = EB(\overline{b})$ (or vice versa). Fig. 5.16b illustrates insertion by input borders. In this case $\mathsf{ER}(x+) = IB(b)$ and $\mathsf{ER}(x-) = IB(\overline{b})$ (or vice versa).

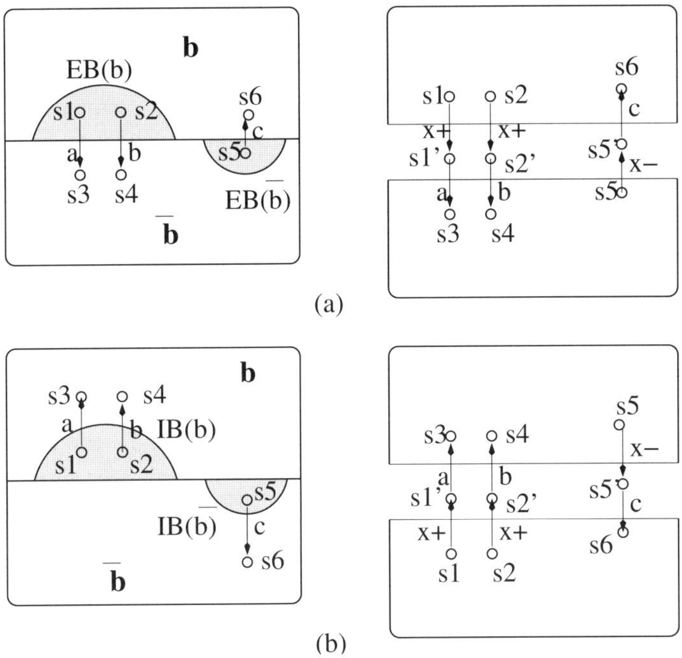

Fig. 5.16. Examples of signal insertion by exit (**a**) and input (**b**) borders

In general, using exit and input borders as insertion sets for new signal transitions does not always guarantee consistency for a new signal. It may be necessary to enlarge an exit border $EB(b)$ with those states of block b which are directly reachable from $EB(b)$. Similarly, for input border $IB(b)$ an enlargement may be required with those states of b from which $IB(b)$ can be entered. Such an enlargement is not necessary if a border is *well-formed*. Informally the notion of well-formed borders can be illustrated by the behavior of a determined traveler: once the traveler decides to leave a country (block b) and enters the custom area (exit border of b or input border of $\overline{b}$) he or she either moves within the custom area (exit or input border) or enters another country (block $\overline{b}$), but never returns to b without visiting $\overline{b}$.

Definition 5.4.1. *Let $\{b, \overline{b}\}$ be a bipartition of a* SG *states.*

1. *The exit border $EB(b)$ is called well-formed iff $\forall s \in EB(b)$: $[\forall s \xrightarrow{a} s' : s' \in \overline{b} \cup EB(b)]$ (similarly for $EB(\overline{b})$);*

2. *The input border $IB(b)$ is called well-formed iff* $\forall s \in IB(b)$*:*
$[\forall s' \xrightarrow{a} s : s' \in \overline{b} \cup IB(b)]$ *(similarly for $IB(\overline{b})$)* .

Let us refer to Fig. 5.15. If well-formed exit borders are chosen for inserting new signal transitions, then the I-partition is defined as follows: $S^0 = b - EB(b)$, $S^+ = EB(b)$, $S^1 = \overline{b} - EB(\overline{b})$, $S^- = EB(\overline{b})$. Since a transition can exit b only through the $EB(b)$, no arcs $S^0 \to S^1$ and $S^0 \to S^-$ in Fig. 5.15 are possible. Due to the well-formedness of $EB(b)$ it is not possible to return from $EB(b)$ to $b - EB(b)$, hence arcs $S^+ \to S^0$ are not possible either. A similar reasoning holds for $EB(\overline{b})$, hence none of the illegal transitions from Fig. 5.15 can occur. This shows the validity of inserting new signals by well-formed input or exit borders.

Property 5.4.2. Let A be a consistent SG, and let $\{b, \overline{b}\}$ be a partition of its set of states S. The SG A' obtained by inserting signal x by exit borders of $\{b, \overline{b}\}$ is consistent iff these borders are well-formed. The same holds for input borders.

If the borders of a given partition $\{b, \overline{b}\}$ of S are not well-formed, a *larger* sets of states must be considered to guarantee consistency. Namely, given $\{b, \overline{b}\}$, we can define the minimal well-formed *extended* EB and IB (denoted MWFEB(b) and MWFIB(b)) as the minimal well-formed enlargements of exit and input borders respectively. MWFEB(b) can be calculated as the least fix point of the following recursion (and is hence unique, due to the determinism of the procedure):

1. $\mathsf{MWFEB}(b) = EB(b)$
2. $[s \in \mathsf{MWFEB}(b) \ \wedge \ s' \in b \ \wedge \ s \to s'] \ \Rightarrow \ s' \in \mathsf{MWFEB}(b)$

A similar recursion can be applied for calculating MWFEB($\overline{b}$), MWFIB(b), and MWFIB($\overline{b}$). Minimal well-formed extended borders hence are minimal sets of states for signal transition insertion which guarantee consistency.

Fig. 5.17 illustrates the process of calculating MWFEB(b) for a block b shadowed in an SG A. Initially the exit border consists of two states 01*10* and 0001* (MWFEB(b) = {01*10*, 0001*}), because these are the only states through which one can leave b. MWFEB(b) is a SIP-set since all state diamonds are intersected with it legally. It is however not well-formed, because once it is entered through the state 01*10* it is possible to retreat back to b by moving into state 0010∗. The first iteration in the least fix point calculation expands MWFEB(b) up to {01*10*, 0001*, 0010*}. This is still not sufficient to make it well-formed and at the next iteration MWFEB(b) is expanded into {01*10*, 0001*, 0010*, 001*1}, reaching the fixed point.

This process is always effective, because it monotonically increases the size of MWFEB(b). The existence of an upper bound follows from the fact that b itself is a well-formed exit border for b. Note, however, that during the expansion some other correctness properties can be violated. This is the case of Fig. 5.17, where the SIP not well-formed set $\mathsf{MWFEB}(b) = \{01{*}10{*}, 0001{*}\}$ is expanded into a well-formed not SIP-set $\mathsf{MWFEB}(b) = \{01{*}10{*}, 0001{*}, 0010{*}, 001{*}1\}$ (it has an illegal intersection with a state diamond $\{01{*}10{*}, 0010{*}, 01{*}11, 001{*}1\}$). In fact for the bipartition $\{b, \bar{b}\}$ given in Fig. 5.17, any insertion of the new event x will violate either consistency or speed-independence. The problem is not how x is inserted (e.g. by exit or input borders) but in the choice of the partition itself. The search for "good" partitions will be discussed later.

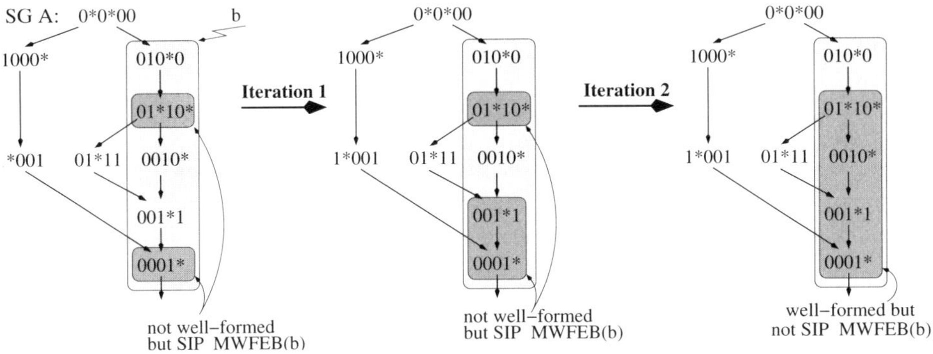

Fig. 5.17. Derivation of $\mathsf{MWFEB}(b)$

5.5 Completeness of the Method

This section establishes a theoretically important result about the effectiveness of the suggested approach for resolving CSC conflicts. This result is technically difficult to prove. Many readers will be satisfied with the following brief statements:

- For excitation closed TSs the method of CSC conflict elimination by inserting new signals based on bipartitions is complete (i.e. all conflicts can be eliminated).
- For non-excitation closed TSs the completeness of the method is an open problem, even though the authors have not found any practical example for which the insertion was not effective.

The former statement is a proven theoretical result that is valid for fairly large class of TSs, while the latter is a useful practical observation about the universal nature (though not proven theoretically) of the suggested approach.

Theorem 5.5.1 links together a choice of bipartition $\{b, \bar{b}\}$ in the original SG and a set of CSC conflicts that can be distinguished by insertion of state signals according to $\{b, \bar{b}\}$. It gives the basis for the iterative transformation of an SG to an equivalent SG satisfying CSC. Different bipartitions are tried for insertions of new signals. The procedure is illustrated in Fig. 5.18. CSC conflicts in Fig. 5.18 are visualized graphically by a conflict graph. An edge connects a pair of states that are in conflict. Every set of conflicting states is represented by a single strongly connected component of the conflict graph. By inserting a new signal according to a chosen bipartition some of the conflicts are removed. This results in the removal of edges from CSC conflict graphs. Only the edges that cross a bipartition boundary can be removed. Note, however, that due to reasons that will be clarified later, not all edges crossing a bipartition boundary are removed.

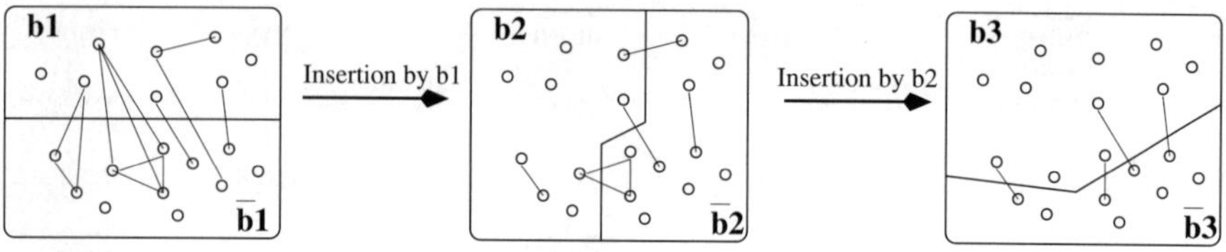

Fig. 5.18. An iterative procedure of reducing CSC conflicts

Then a new bipartition can be chosen to remove some of the remaining conflicts, and the process repeats until all conflicts are eliminated.

Theorem 5.5.2 shows how to eliminate CSC conflicts through iterative insertion of new signals. It suggests a constructive technique to satisfy the CSC requirement for excitation closed TSs. Even though it is impractical (it requires too many signals), it provides a guarantee that every CSC conflict will be solved at some step of the insertion method. From this follows that for any pair of CSC conflicting states $\{s1, s2\}$ the following two-step procedure is possible:

1. find a bipartition $\{b, \overline{b}\}$ separating $s1$ and $s2$,
2. insert new signals based on $\{b, \overline{b}\}$ in such a way that they distinguish states $s1$ and $s2$.

The rationale behind this "monotonic improvement" strategy to solve CSC by inserting state signals according to a chosen bipartition (see Theorem 5.5.1) is given by Fig. 5.19.

Let us assume that $\{b, \overline{b}\}$ is a bipartition of the states of SG A. This bipartition uniquely defines a set C of state pairs that are in CSC conflict and are separated by $\{b, \overline{b}\}$. Elimination of all conflicts in C is the maximum that one could hope to obtain from insertion based on $\{b, \overline{b}\}$. Let us for the moment consider only CSC conflicts in C, and ignore those outside C. A comparison between SGs A and A' (before and after insertion of a new signal) can use the inverse mapping of states from A' to states of A, together with information about their CSC conflicts. If for conflicting states $s1$ and $s2$ of A their images in A' are also in a CSC conflict between each other, then the conflict is not solved by the new signal. The target of inserting new signals is to distinguish as many conflicts in C as possible.

When a new signal x is inserted by exit borders (see Fig. 5.19b) in the new SG A', we can have the following cases:

1. $s1 \in b \setminus EB(b), s2 \in \overline{b} \setminus EB(\overline{b})$
2. $s1 \in EB(b), s2 \in \overline{b} \setminus EB(\overline{b})$
3. $s1 \in b \setminus EB(b), s2 \in EB(\overline{b})$
4. $s1 \in EB(b), s2 \in EB(\overline{b})$

Consider the first case. States from $b - EB(b)$ and $\overline{b} - EB(\overline{b})$ cannot be in mutual CSC conflicts because they have different values of signal x.

Unfortunately, this is not the case for the states from $EB(b)$ and $EB(\overline{b})$. Each state $s \in EB(b)$ ($s \in EB(\overline{b})$) is mapped into two states s and s' in A', such that $s \stackrel{x*}{\to} s'$. Thus every CSC conflict involving states that belong to exit borders is still present in the new SG A' after the insertion of x.

However, adding x simplifies further elimination of the other CSC conflicts, because in the original SG A, CSC conflicts in C could be distributed anywhere in the state space (see Fig. 5.19a), while in A' they are localized in the former exit borders (Fig. 5.19b).

One more signal (call it y) must be inserted to resolve the remaining conflicts in A'. The insertion of y by input borders of bipartition $\{b, \overline{b}\}$ gives the result shown in Fig. 5.19: sets of states b–$EB(b)$–$IB(b)$ and $\overline{b}$–$EB(\overline{b})$–$IB(\overline{b})$ have no state conflicts with respect to set C (Fig. 5.19c). The remaining conflicts in C, those belonging to the input and exit borders of $\{b, \overline{b}\}$, could be eliminated by choosing a new bipartition in such a way that its input and exit borders lie within conflict free sets of states, e.g. within b–$EB(b)$–$IB(b)$ and $\overline{b}$–$EB(\overline{b})$–$IB(\overline{b})$.

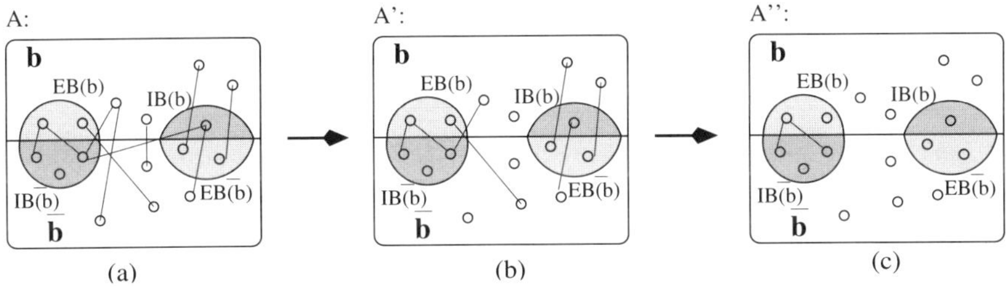

Fig. 5.19. Signal insertion using exit and input borders: original state conflicts (**a**), insertion by exit borders and inverse mapping of conflicts (**b**), insertion by input borders and inverse mapping of conflicts (**c**)

The formal justification of the above explanation is given by Definition 5.5.1 and Theorem 5.5.1.

Definition 5.5.1. *Let $\{b, \bar{b}\}$ be a bipartition of SG states. Let $s1$ and $s2$ have the same binary code, $s1 \in b$, and $s2 \in \bar{b}$. States $\{s1, s2\}$ are said to be* distinguishable *by partition $\{b, \bar{b}\}$ if the following condition does not hold:* $(s1 \in \mathsf{MWFEB}(b) \wedge s2 \in \mathsf{MWFIB}(\bar{b})) \vee (s1 \in \mathsf{MWFIB}(b) \wedge s2 \in \mathsf{MWFEB}(\bar{b}))$

Theorem 5.5.1. *Let $\{b, \bar{b}\}$ be a bipartition of a SG states. All distinguishable pairs of states $\{s1, s2\}$ ($s1 \in b, s2 \in \bar{b}$) will obtain different binary codes after inserting two state signals by* MWFEB *and* MWFIB.

Proof. Consider two auxiliary signals x and y that have been inserted by exit and input borders of partition $\{b, \bar{b}\}$ (see Fig. 5.20b and 5.20c respectively). According to the values of state signals x and y we can separate all states of A'' into four sets $S^{00}, S^{10}, S^{01}, S^{11}$. Let us examine the values of x and y (shown in brackets near states) that are assigned to the states of the original SG A during the insertion. The following cases are possible for each $s \in b$:

1. $s \notin (\mathsf{MWFIB}(b) \cup \mathsf{MWFEB}(b))$. It has only one image $s_{A''} : s_{A''} \in S^{00}$ (see state $s1$ in Fig. 5.20a)
2. $s \in \mathsf{MWFEB}(b)$, $s \notin \mathsf{MWFIB}(b)$. It has images $s_{A''} \in S^{00}$ and $s'_{A''} \in S^{10}$ (see state $s2$ in Fig. 5.20a)
3. $s \in \mathsf{MWFIB}(b)$, $s \notin \mathsf{MWFEB}(b)$. It has images $s_{A''} \in S^{00}$ and $s''_{A''} \in S^{01}$ (see state $s8$ in Fig. 5.20a)
4. $s \in \mathsf{MWFEB}(b) \cap \mathsf{MWFIB}(b)$. It has images $s_{A''} \in S^{01}$, $s''_{A''} \in S^{00}$ and $s'_{A''} \in S^{10}$ (see state $s6$ in Fig. 5.20a)

Similar considerations could be made about states in $\bar{b}$, based on the duality relation, i.e. by replacing b with $\bar{b}$ and replacing all rising transitions of inserted signals x and y with their falling counterparts.

1. $s \notin (\mathsf{MWFIB}(\bar{b}) \cup \mathsf{MWFEB}(\bar{b}))$ has the image $s_{A''} : s_{A''} \in S^{11}$
2. $s \in \mathsf{MWFEB}(\bar{b})$, $s \notin \mathsf{MWFIB}(\bar{b})$ has images $s_{A''} \in S^{11}$ and $s'_{A''} \in S^{01}$
3. $s \in \mathsf{MWFIB}(\bar{b})$, $s \notin \mathsf{MWFEB}(\bar{b})$ has images $s_{A''} \in S^{11}$ and $s''_{A''} \in S^{10}$
4. $s \in \mathsf{MWFEB}(\bar{b}) \cap \mathsf{MWFIB}(\bar{b})$ has images $s_{A''} \in S^{10}$, $s''_{A''} \in S^{11}$ and $s'_{A''} \in S^{01}$

Observing the values of signals x and y for CSC conflicting states of b and $\bar{b}$, one can see that x and y cannot resolve the conflict between a pair of states $s1$ and $s2$ from different partition blocks only when one of them belongs to the exit border of one block while another to the input border of the other block. The latter however contradicts the condition of distinguishability of states $s1$ and $s2$ by the partition.

It follows from the proof of Theorem 5.5.1 that if $s1$ and $s2$ are distinguishable by partition $\{b, \bar{b}\}$ in SG A, then the corresponding states in A'', obtained after insertion of x and y, are distinguishable by partition $\{b'', \overline{b''}\}$, where b'' and $\overline{b''}$ are "images" of b and $\bar{b}$ in the new SG A''. Hence, if all CSC conflicts can be distinguished by k bipartitions, then no more than $2k$ state signals are needed for solving all CSC conflicts in a SG.

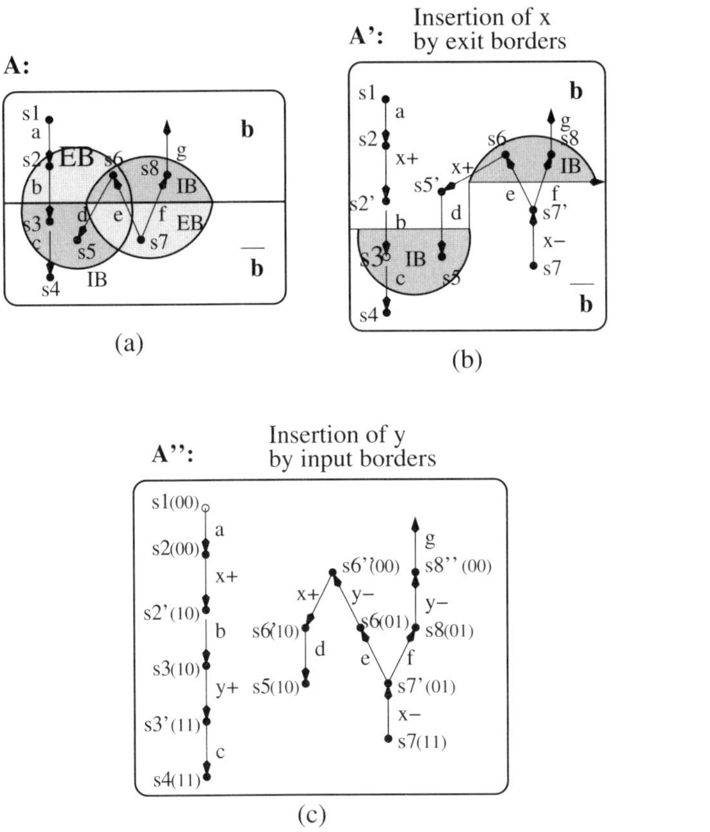

Fig. 5.20a–c. Insertion of state signals by Exit and Input Borders

This shows that our procedure for solving CSC conflicts is monotonous and every new SG is better than the previous one in terms of CSC conflicts. Given an SG, the minimal number k of partitions which solve all CSC conflicts can be calculated. It gives an upper bound on the number of state signals which are necessary for complete state coding.

The following corollary of Theorem 5.5.1 states the conditions for implementability of an SG as a speed-independent circuit.

Corollary 5.5.1. *Let A be a deterministic, consistent, commutative and output-persistent SG. Assume that for every pair of states $s1, s2$ with the same binary code there exists a partition $\{b, \overline{b}\}$ distinguishing $s1$ and $s2$ such that minimal well-formed extended exit and input borders of b and $\overline{b}$ are SIP-sets. Then there exists a finite sequence of SIP insertions that yields an SG A' such that: (1) A' is trace equivalent to A, (2) A' is deterministic, consistent, commutative and output-persistent, and (3) A' satisfies CSC.*

Indeed, if every pair of states with the same binary code is distinguishable by some partition, then according to Theorem 5.5.1 all CSC conflicts in this SG can be solved by inserting state signals. Consistency is preserved since signals are inserted by well-formed borders. These borders are SIP-sets (by the condition of Corollary 5.5.1), hence commutativity and persistency are also preserved.

We will now show that for any pair of CSC conflict states there is a bipartition that distinguishes them and provides SIP-sets for a signal insertion.

This fact is proven (Theorem 5.5.2) for the class of excitation closed TSs in an implicit way. Theorem 5.5.2 gives a constructive method for eliminating CSC conflicts in a TS. The method is based on two important notions:

1. The correspondence between excitation closed TSs and safe PNs (see Chap. 3).
2. The correspondence between global states and place markings in a PN.

To eliminate CSC conflicts Theorem 5.5.2 first suggests to construct an equivalent PN from an excitation closed TS and then to insert additional signals in that PN (one signal for each place) to emulate the process of marking transformations. Since in PNs all markings are different, then if the new signals emulate the value of place markings properly they should distinguish all states of the corresponding SG.

Theorem 5.5.2. *Let A be a deterministic, consistent, commutative, output-persistent, and* excitation-closed *SG. Then there exists an SG A' that is trace equivalent to A, has no CSC conflicts, and is deterministic, consistent, commutative and persistent with respect to output signals.*

Proof. As discussed in Sect. 3.5, any excitation-closed SG A has a corresponding safe Petri Net without self-loops with a reachability graph that is bisimilar to A. In fact, this PN D is an STG because its transitions are interpreted as the changes of binary signals. We will eliminate all CSC conflicts from D (and hence from A) by adding to D two sets of binary signals:

1. $\pi_1, \ldots, \pi_k$, where k is the number of places in D.
   Signals $\pi_1, \ldots, \pi_k$ encode every reachable marking m of D. If in a marking m place p_i has a token, it is encoded by signal $\pi_i = 1$, otherwise $\pi_i = 0$. However, since $\pi_1, \ldots, \pi_k$ have to be implemented as internal signals of a circuit, it is impossible to model with them the atomic operation of a marking change, in which all tokens must be removed from input places and added to output places *simultaneously*. Instead, this is modeled in two steps: first tokens appear at output places of a transition t (signals π_i such that $p_i \in t\bullet$ are first set) and then tokens are removed from input places (signals π_j such that $p_j \in \bullet t$ are reset). If the atomicity of marking transformations was preserved by our circuit implementation, then signals $\pi_1, \ldots, \pi_k$ would distinguish all SG states. However due to this "imprecision", we need a second set of signals to eliminate all CSC problems.

2. $\tau_1, \ldots, \tau_l$, where l is the number of transitions in D.
   Signals $\tau_1, \ldots, \tau_l$ model the enabling of transitions. I.e. if t_i is enabled at some marking m, the corresponding signal τ_i is set to one. After the firing of transition t_i, signal τ_i is reset to 0.

The transformation that we apply to the original STG D is illustrated by Fig. 5.21.

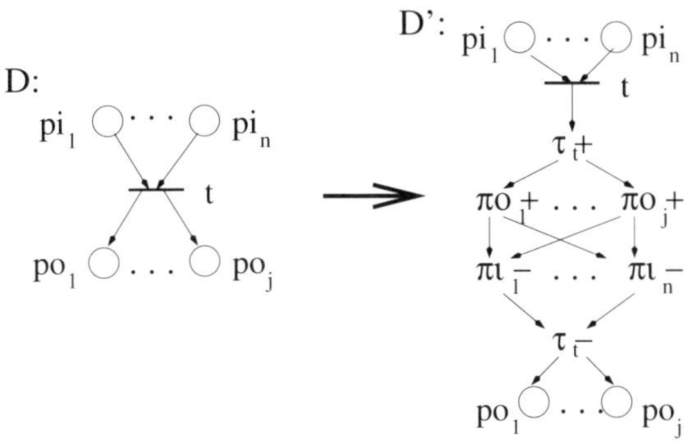

Fig. 5.21. Transformation for transition t

In Fig. 5.21 $pi_1, \ldots, pi_n$ and $po_1, \ldots, po_j$ denote the sets of input and output places of transition t.

Let us show that the STG D' obtained after the transformation is:

- trace equivalent to D,
- deterministic, commutative, consistent, output-persistent, and
- has no CSC conflicts.

1. *Trace equivalence.* This is obvious, since if all transitions of signals $\pi_1, \ldots, \pi_k$ and $\tau_1, \ldots, \tau_l$ are removed from D' (i.e. D' is projected on the original signals), then the result coincides with STG D.
2. *Consistency of state assignment in D'.* From trace equivalence follows that consistency of the original signals is preserved in D'. For signals $\tau_1, \ldots, \tau_l$ consistency directly follows from the rules of their firing.
   Let us consider the consistency of signals $\pi_1, \ldots, \pi_k$. Signal π_i changes from 1 to 0 while modeling the firing of transition for which p_i is an input place. According to the rules of transformation, at that moment signal π_i is always at 1, and so consistency cannot be violated due to falling transitions of π_i. It also cannot be violated due to rising transitions of π_i, because π_i is set to 1 only in markings where place p_i receives a token, and since STG D is safe, no new rising transition of π_i can happen before π_i- will fire.

3. *Determinism, Signal Persistency and Commutativity.* Determinism, commutativity and signal persistency of the original signals of D is guaranteed in D' by the trace equivalence between D and D'. Signal persistency of the added signals $\pi_1, \dots, \pi_k$ and $\tau_1, \dots, \tau_l$ follows from the fact that no input place of any transition of the added signals is shared by any other transition. Commutativity and determinism of the added signals are also clear.
4. *Complete State Coding.*
   Let us separate all markings of STG D' into two sets:
   a) *Settled* markings (states), in which all signals $\tau_1, \dots, \tau_l$ are equal to 0.
   b) *Transient* markings (states) in which some of signals $\tau_1, \dots, \tau_l$ are equal to 1.

   Assume that there exist states s'_1 and s'_2 that have the same binary code in D'. Let s'_1 and s'_2 correspond to the reachable markings m'_1 and m'_2 respectively. We can consider three cases.
   a) Both m'_1 and m'_2 are settled markings.
      Let $q1'$ and $q2'$ be two sequences that reach m'_1 and m'_2 respectively from the initial marking, i.e. $m'_0 \overset{q1'}{\to} m'_1$, $m_0 \overset{q2'}{\to} m'_2$. Let $q1$ and $q2$ be the sequences that are obtained by projecting $q1'$ and $q2'$ on the set of signals of the original STG D. Let $m1$ and $m2$ be the (unique) markings in D that are reached by firing $q1$ and $q2$ respectively.
      The value of signals $\pi_1, \dots, \pi_k$ in m'_1 and m'_2 is in one-to-one correspondence with m_1 and m_2. If $m_1 = m_2$ (i.e. m'_1 and m'_2 correspond to the same marking in the original PN), then settled states s'_1 and s'_2 correspond to the part of D' that models the firing of the same transition t' (see the fragment of PN in Fig. 5.21).
      The only settled states in this part are: state w'_1 in which t is enabled, w'_2 obtained after the firing of t at w'_1 in which τ_t+ is enabled, and w_3 obtained after the firing of τ_t-. Clearly all these states have different binary codes and therefore $m'_1 \neq m'_2$. On the other hand, if $m'_1 \neq m'_2$, then s_1 and s_2 cannot have the same code because they have different values of signals $\pi_1, \dots, \pi_k$.
   b) m'_1 is a settled marking and m'_2 is a transient marking.
      States s'_1 and s'_2 cannot have the same code because in s'_1 all signals $\tau_1, \dots, \tau_l$ are equal to 0, while in s'_2 at least one of them is 1.
   c) m'_1 and m'_2 are both transient markings.
      For every transition t_i of D, let us consider the set of events in D' that models the change of marking due to the firing of t_i. Let us call this set of events $+\pi_j$ and $-\pi_r$, that are "in between" $+\tau_i$ and $-\tau_i$ for which $p_j \in t_i\bullet$ and $p_r \in \bullet t_i$, the *Transient Set* of t_i, or $TrS(t_i)$.
      If the codes of s'_1 and s'_2 coincide, then obviously the same signals $E = \{\tau_{i1}, \tau_{i2}, \dots\}$ are equal to 1 in s'_1 and s'_2. Clearly the transitions $t_{i1}, t_{i2}, \dots$ of D that correspond to the signals in E are concurrent.

Transient sets of different transitions t_i and t_j ($\tau_i, \tau_j \in E$) cannot contain the same signals. Indeed, if $+\pi_r \in TrS(t_i) \cap TrS(t_j)$ ($-\pi_r \in TrS(t_i) \cap TrS(t_j)$), then two concurrent transitions t_i and t_j in D have the same output (input) place p_r and thus D is unsafe. If $+\pi_r \in TrS(t_i)$, $-\pi_r \in TrS(t_j)$ then place p_r is an output place of t_i and an input place for t_j. Thus from the concurrency of t_i and t_j it follows that p_r is unsafe. Due to the absence of self-loops each transient set cannot contain the rising and falling transitions of the same signal π_r. From this and from the non-intersection of different transient sets of concurrent transitions it follows that s_1 and s_2 cannot have the same binary code.

The consideration of these cases proves that all binary states of D' have different codes, thus satisfying the Complete State Coding property.

Note that each inserted signal in STG D defines a bipartition at the level of the original SG A. Since the final result of the insertion according to Theorem 5.5.2 is an STG satisfying the CSC requirement, we can conclude that all CSC conflicts of the original SG are solvable by signal insertions. Hence the same result (elimination of CSC conflicts) can be achieved for an excitation closed SG through an appropriate insertion of new signals based on exit and input borders. The latter means that the presented method to satisfy CSC is complete.

Of course, this result is only of theoretical interest, since the number of added signals is quite large. In the next section we discuss effective heuristics to achieve the same result by more economical means.

5.6 An Heuristic Strategy to Solve CSC

In this section we discuss the heuristic techniques to solve CSC that are used in petrify, a tool for synthesis and optimization of asynchronous control circuits [112]. It fully exploits the theoretical results presented in this book, even though it implements fewer options than the whole spectrum delivered by the STG synthesis theory.

5.6.1 Generation of I-Partitions

As described in Sect. 5.4, event insertion aims at pre-conditioning or post-conditioning some of the existing events in a TS. Fig. 5.22 illustrates different types of insertions of the event x with regard to event a. Let us assume that t_1, t_2 and t_3 are concurrent (similarly for t_4, t_5 and t_6). The insertions of the figure can be obtained as follows:

$$\text{case (b)} : \mathsf{ER}(x) = \mathsf{MWFEB}(p_1) = \mathsf{MWFEB}(p_2) = \mathsf{MWFEB}(p_3)$$
$$\text{case (c)} : \mathsf{ER}(x) = \mathsf{MWFIB}(p_1 \cap p_2)$$
$$\text{case (d)} : \mathsf{ER}(x) = \mathsf{MWFEB}(p_4 \cap p_5 \cap p_6)$$
$$\text{case (e)} : \mathsf{ER}(x) = \mathsf{MWFEB}(p_4 \cap p_5)$$

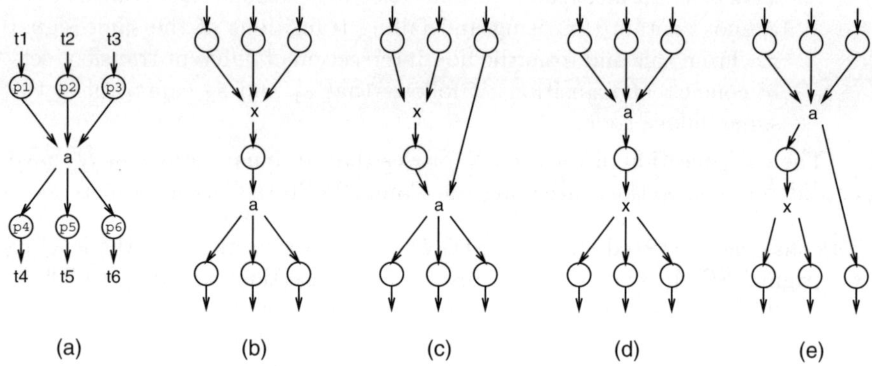

Fig. 5.22a–e. Different types of event insertions

To reduce the search space petrify uses the insertion scheme based on exit borders only. This directly covers cases b, d, and e of Fig. 5.22. Case c cannot be obtained by using only MWFEBs, since the usage of the input border is essential here. Instead of using two different schemes for insertion (by exit and input borders) petrify solves the task of capturing cases like c from Fig. 5.22 through allowing to enlarge the concurrency of the inserted event (the latter is done as a post-optimization step of the implemented algorithm). It should be obvious that case c can be obtained from case b by allowing t_3 to be concurrent with x.

5.6.2 Exploring the Space of I-Partitions

Each block of states defines an I-partition. According to the type of insertions described in Fig. 5.22, the Exit Border of a block must be either a region or the intersection of some pre-/post-regions of the same event. We consider these objects to be the "bricks" of the blocks and we explore the space of blocks by calculating unions of bricks.

Fig. 5.23 presents an algorithm similar to the $\alpha - \beta$ pruning strategy commonly used in game-playing applications [20]. Initially, all bricks of the TS are calculated by (1) obtaining all minimal regions of the TS and (2) calculating all possible intersections of pre-/post-regions of the same event. Since the number of pre- and post-regions of an event is usually small, an exhaustive generation is feasible.

The best block for event insertion is obtained as the union of adjacent bricks. At each iteration of the search, a frontier of FW (frontier width) "good" blocks is kept. Each block is enlarged by adjacent bricks and the new obtained blocks are considered candidates for the next iteration only if they are "better", according to the cost function, than their ancestors. FW is a parameter than can be tuned by the designer to define the degree of exploration of the configuration space, similar to defining the level of expertise of a chess-playing program, trading-off the quality of the solution and the computational cost to find it. Finally, the best block generated during the search is chosen.

```
bricks = calculate_all_bricks ()
frontier = good_blocks = {the best FW bricks}
repeat
    new_frontier = ∅
    for each bl ∈ frontier do
        for each br ∈ bricks adjacent to bl do
            new_bl = bl ∪ br
            if cost(new_bl) < cost(bl) then
                good_blocks = good_blocks ∪ new_bl
                new_frontier = new_frontier ∪ new_bl
    frontier = select the best FW blocks from new_frontier
until new_frontier = ∅
return the best block in good_blocks
```

Fig. 5.23. Heuristic search to find a block for event insertion

The execution of the previous algorithm would give a connected block of states as depicted in Fig. 5.24b (block b_1). In the most general case, a disconnected set of states may be appropriate to solve CSC. For this reason, the algorithm is iteratively executed with the rest of bricks of the TS (not intersecting with previously calculated blocks) until all states with CSC conflicts have been covered by some block (e.g. blocks $b_1 - b_5$ in Fig. 5.24c).

The final block for insertion is calculated as the union of disconnected blocks. A greedy block merging approach guided by the cost function is used. In Fig. 5.24d a block $b = b_1 \cup b_5$ has been obtained.

5.6.3 Increasing Concurrency

Given a block b, S^+ and S^- are initially calculated as the MWFEB of b and $\overline{b}$ respectively. This leads to a solution with minimum concurrency of the inserted event. Concurrency can be increased by enlarging S^+ and/or S^-. This is illustrated in the example of Fig. 5.25. Let us assume that $b = p_1 \cup p_2$. In this case $\mathsf{MWFEB}(b) = p_1 \cap p_2$, which produces the event insertion of Fig. 5.25c. By enlarging $\mathsf{ER}(x)$, e.g. $\mathsf{ER}(x) = p_2$, event x is made concurrent with event a (Fig. 5.25d).

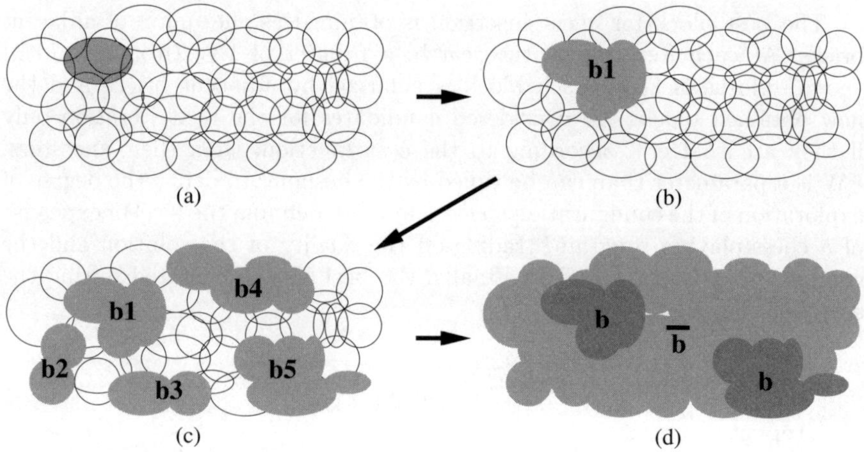

Fig. 5.24. (**a**) Brick, (**b**) block as the union of adjacent bricks, (**c**) connected blocks, (**d**) final block after the union of disconnected blocks

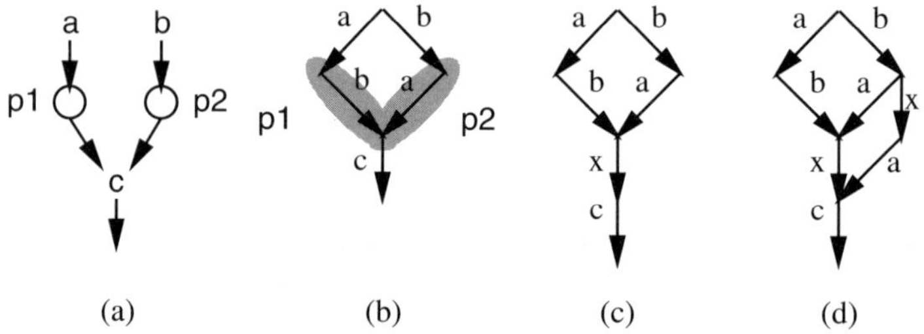

Fig. 5.25. (**a**) Petri Net, (**b**) transition system, (**c**) insertion with $\mathsf{ER}(x) = p_1 \cap p_2$, (**d**) insertion with $\mathsf{ER}(x) = p_2$

Increasing concurrency is performed at the very last stage as a post-optimization step. After the best configuration for event insertion has been calculated, S^+ and S^- are greedily enlarged by adding bricks that are adjacent to them. The enlargement is only accepted if the new configuration improves the cost of the solution.

5.7 Cost Function

During the exploration of the space of configurations to solve CSC, a cost function is used to determine the candidates that must survive. This cost function is used for building connected blocks, merging disconnected blocks and increasing concurrency.

The main objective of the cost function is to guide the search toward a correct and cheap solution of the CSC problem. Given that a great amount of configurations are explored, the cost function must not be computationally expensive.

Rather than being a real-valued function, the cost function is an algorithm that considers several implementation factors when comparing two different configurations. The following factors are considered for the insertion of signal x (in order of priority):

- ER(x+) and ER(x−) must be SIP blocks.
- The insertion of x must not modify the specification of the environment (e.g. x cannot be inserted before input events).
- The number of solved CSC conflicts must be maximized.
- The estimated complexity of the logic of the circuit must be minimized.

In the evaluation of the last two factors, some degree of freedom is allowed. For example, if the relative difference of CSC conflicts disambiguated by two configurations is similar, the one with the cheapest circuit complexity is considered to be better.

With this approach, the most computationally expensive criterion (estimation of logic) is only evaluated when configurations are guaranteed to be correct and make a tangible progress toward solving CSC.

5.7.1 Estimation of Logic

For each TS generated after the insertion of a new state signal, an approximate estimation of the complexity of the circuit is calculated.

The estimation is oriented toward a speed-independent realization based on monotonic covers (see Sect. 6.2) and is done as follows. For each ER of an output signal a sum-of-products expression is calculated, minimized with the don't-care set of the SG and the quiescent region that follows the ER. Even though the calculated cover may not necessarily be monotonic, our experiments have shown that it is monotonic in more than 80% of cases, and can be considered as a good estimation of the complexity of a monotonic cover in the rest of the cases.

The complexity of the circuit is estimated as the sum of the numbers of literals of each cover.

5.7.2 Examples of CSC Conflict Elimination

Let us return to our motivating example from "Alice's Adventures in Wonderland" and show how the presented technique works for it.

Example 5.7.1. A constrained model for the Mad Hatter's tea party Let us start from the simplified model of the behavior of Alice and March Hare during their tea party (see Fig. 5.7a). There are two pairs of CSC conflicts in this specification: {10*, 1*0*} and {0*1, 0*1*}. Construction of a bipartition

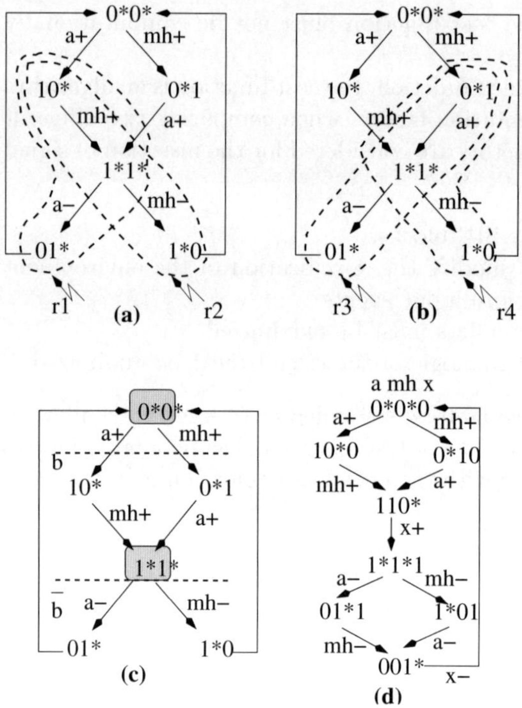

Fig. 5.26a–d. Elimination of CSC conflicts in the constrained Mad Hatter's tea party model

separating the conflicts could be done based on regions and their intersections. Indeed sets of states $r1$ and $r2$ in Fig. 5.26a are post-regions for event $a+$ and their intersection gives set of states $r1 \cap r2 = \{10^*, 1^*1^*\}$. Similarly the intersection of $r3$ and $r4$ (post-regions of $mh+$, see Fig. 5.26b) gives set of states $r3 \cap r4 = \{0^*1, 1^*1^*\}$. The union of these two produces a bipartition $\{b, \overline{b}\}$ that separates all the CSC conflicts (Fig. 5.26c). Exit borders of the bipartition (shadowed in Fig. 5.26c) are SIP-sets and well-formed. They give proper ERs for new signal x. The insertion of x transforms the original specification to an equivalent SG satisfying the CSC property (Fig. 5.26d). From the latter it is easy to get the C-element based implementation that was shown in Fig. 5.7c.

Example 5.7.2. The original model for the Mad Hatter's tea party In the original model for the Mad Hatter's tea party Alice and March Hare are more independent from each other, which results in larger number of CSC conflicts (see Fig. 5.6c). Transforming this specification to CSC satisfying SG is more elaborate and requires insertion of at least three additional signals. For simplicity we will skip the details on how the corresponding bipartitions are constructed from regions and intersections of regions (it is not much different from the previous example) and will illustrate the insertion procedure using larger steps (Fig. 5.27).

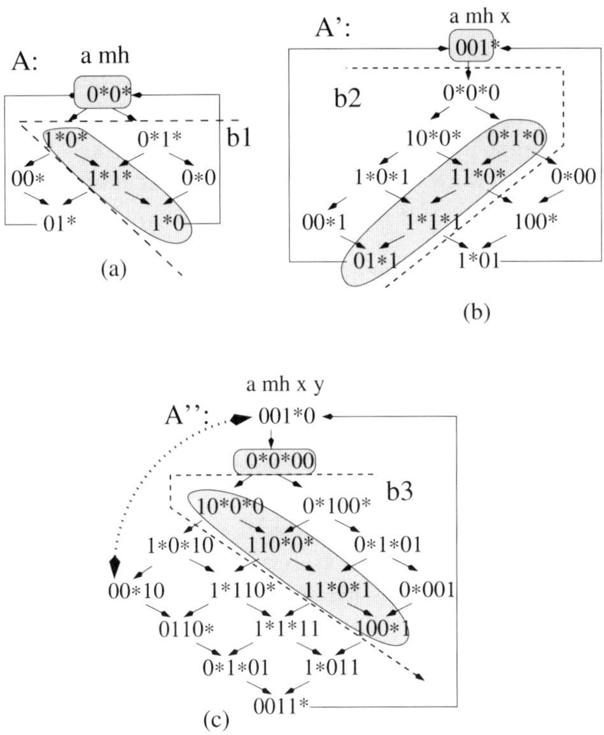

Fig. 5.27a–c. Elimination of CSC conflicts in the non-constrained Mad Hatter's tea party model

First, in SG A bipartition $\{b1, \overline{b1}\}$ is chosen (Fig. 5.27a). Exit borders of the bipartition are SIP-sets and well-formed. The insertion of the new signal x by exit borders results in a new SG A' (Fig. 5.27b) in which the CSC conflict between states 0*1* and 01* is resolved (the images of 0*1* and 01* in A' get different values of signal x). Insertion of another signal y by well-formed and SIP exit borders of bipartition $\{b2, \overline{b2}\}$ gives a SG A'' (Fig. 5.27c). Only one CSC conflict exists in A'' between states 001*0 and 00*10 (indicated by the dotted edge). This conflict remains unresolved by signal y, because it originates from states 001* and 00*1 in A', and 001* belongs to the exit border of $\overline{b2}$. From Theorem 5.5.1 we know that conflicts involving states in MWFEB are not distinguished when the insertion is done by exit borders. One more state signal is needed to disambiguate 0001* and 0*001 in A''. This signal can be inserted by exit borders of bipartition $\{b3, \overline{b3}\}$ (Fig. 5.27c), because this bipartition distinguishes the considered conflict (both states are outside the exit borders).

5.8 Related Work

A number of methods for solving the CSC problem in asynchronous specifications are known to date [113, 64, 114, 115, 116, 117, 118, 111, 119, 110]. State encoding methods for burst-mode FSMs are presented in [120].

Methods from [116, 117, 118] work at the STG level without doing state traversal. They avoid state explosion and therefore can process large specifications if some additional constraints on the structure of the STG are given. Such constraints (e.g. being free-choice, having exactly one rising and falling transition for each signal, etc.) severely limit the design space and do not produce solutions for many practical specifications.

[115] solves the CSC problem by mapping the initial SG into a flow table and then using classical flow table minimization and state assignment methods. This method is restricted to live and safe free-choice STGs and cannot process large SGs due to limitations of the classical state assignment methods.

In [118] a very general framework for state assignment is presented. The CSC problem is formulated as a search for a state variable assignment on the state graph. The correctness conditions for such an assignment are formulated as a set of Boolean constraints. The solution can be found using a Boolean satisfiability solver. Unfortunately, this approach allows handling only relatively small specifications (hundreds of states) because the computational complexity of this method is double exponential in the number of signals in the SG. Although [113] presented a method to improve effectiveness by means of preliminary decomposition of the satisfiability problem, decomposition may produce sub-optimal solutions due to the loss of information incurred during the partitioning process. Moreover, the net contraction procedure used to decompose the problem has never been formally defined for non-free-choice STGs.

In [64, 114] another method based on state signals insertion at the SG level was given. The problem is solved by constructing a CSC conflict graph between excitation regions of SG and then coloring the graph with binary encoded colors. Each bit of this code corresponds to a new state signal. After that, new state signals are inserted into the SG using the excitation regions of the original or previously inserted signals. The main drawback of this approach was its limitation to STGs without choices.

This limitation was overcome in [119] which suggested to partition the state space into blocks with no internal CSC-conflicts and then insert new signals based on excitation regions of these blocks. However restricting an insertion to excitation regions significantly limits the capabilities of the method, which is inefficient for resolving conflicts even in simple cases like that shown in Fig. 5.6c.

The insertion of new signals by regions and intersections of regions was first suggested in [121, 110]. The method is well suited for symbolic BDD representation of the main objects in the insertion procedure. This feature expands the capabilities of the state-based approach, and allows a designer to solve CSC problems for SGs with hundreds of thousands of states.

5.9 Summary

This chapter presented the theoretical framework and a practical approach for state encoding of asynchronous specifications. The theory is based on the combination of two fundamental concepts. One is the notion of regions of states in a TS (see Chap. 3). The second concept is a speed-independence preserving set (SIP-set), which is strongly related to the implementability of the specification in logic. Regions and their intersections can serve as bricks for efficient generation of SIP-sets.

Though the suggested approach is quite different from (and more complex than) the well known methods for state encoding conflict elimination for asynchronous FSMs, the final result is similar: for a large class of specifications all CSC conflicts can be successfully resolved. The refinement of a specification to an implementable form (free from CSC conflicts) is done by preserving observational equivalence, while keeping the encoding of the original signals unchanged.

6. Logic Decomposition

This chapter tackles one of the main problems of asynchronous circuit design: that of decomposing a complex Boolean function into elementary gates from a given library. In the synchronous case this is traditionally solved as two sub-problems. During the technology-independent phase [7, 8, 122, 10] one applies the theorems of Boolean algebra, and in particular Boolean and algebraic division operations, to optimally decompose the logic with a technology-independent cost function (e.g. literals for area and levels for delay). The result of this phase is a netlist of "canonical" technology-independent basic gates (e.g. inverters and 2-input *nand* gates). During the technology-dependent phase one maps the decomposed logic to the gates that are available in the library [123, 124]. The cost function at this stage may include more precise area and delay information, possibly including the effect of capacitive load and wiring estimates derived from approximate placement. Throughout this chapter we will assume a good knowledge of combinational logic synthesis techniques, as described in the above references.

In the asynchronous case this flow suffers from a basic problem: by breaking up a large function into small gates one may introduce hazards. For this reason, every application of logic division techniques must be *checked* for validity, using techniques that are closely related to those discussed in Sect. 4.2.

The key problem in this case, as will be discussed in Sect. 6.3, is to avoid the creation and analysis of a new SG each time a new candidate factor must be checked, since these operations are very expensive. Building upon the ideas presented in Burns' seminal paper [125], we illustrate how the effect of decomposition on speed-independence can be evaluated by means of *local checks* on the SG *before decomposition*. However, asynchronous circuits have also more freedom than synchronous ones when implementing a sequential specification: state-holding elements and state transitions in general are not fixed in advance. In the STG-based methodology, only the external behavior of the I/O signals is specified. The techniques of Chap. 5 insert only a minimal number of state-holding signals to ensure implementability. In Sect. 6.4 we show how both technology-independent and technology-dependent asynchronous circuit synthesis can also exploit sequential gates (such as latches, flip-flops, and so on) to minimize cost and maximize performance. Of course, this sequential optimization is more complex than the extraction of combinational factors described in Sect. 6.3. Hence it should be used only for relatively small circuits where minimization must be achieved with maximum effort, as will be illustrated by the experimental results in Sect. 6.5.

The logic decomposition of the non-input signals discussed in Sect. 6.3 and 6.4 is completed by a technology mapping step aimed at recovering area and delay based on a technology-dependent library of gates. These reductions are achieved by collapsing small fanin gates into complex gates, provided that the gates are available in the library. The collapsing process is based on the Boolean matching techniques proposed by Mailhot et al. [124], adapted to the existence of asynchronous memory elements and combinational feedback in speed-independent circuits. The interested reader is referred to [126, 127, 128] for more details on asynchronous technology mapping, in particular oriented to maximizing average case performance.

6.1 Overview

Let us start from a motivating example, that illustrates differences and problems in asynchronous logic decomposition.

Example 6.1.1. A Priority Encoder Let a 3-bit field $<x, y, z>$ be part of an instruction format, with a priority between bits from left to right (x has the highest priority while z has the lowest). An instruction decoder translates the content of the field into one-hot codes $<a, b, c, d>$ and implements a priority encoding function (see Fig. 6.1a). The formal specification of this process in VHDL is shown in Fig. 6.1b.

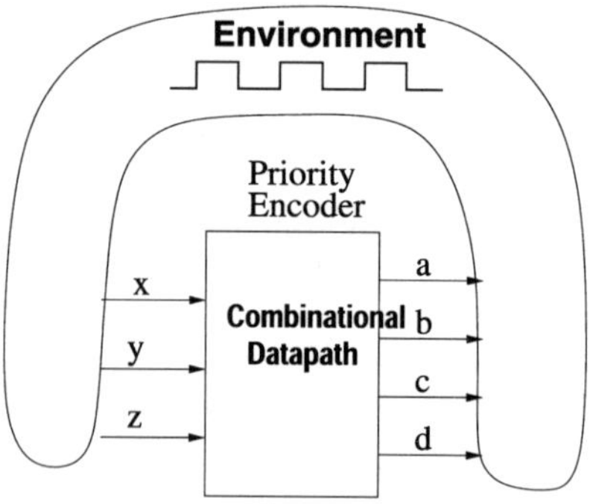

(a)

```
if    x = '1'  then  a <= '1';
elsif y = '1'  then  b <= '1';
elsif z = '1'  then  c <= '1';
else  d <='1';
endif;
```

(b)

Fig. 6.1a,b. Priority encoding

From this specification one can derive a synchronous implementation for the encoder, as shown in Fig. 6.2a. The circuit in Fig. 6.2a can be decomposed to be mapped onto a limited library, with 2-input gates only, as in Fig. 6.2b.

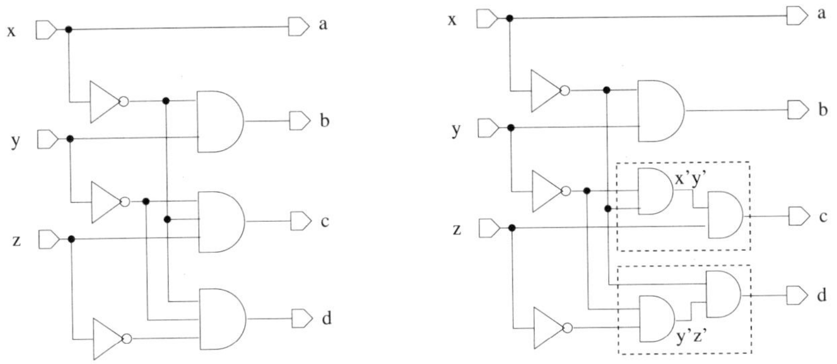

(a) (b)

Fig. 6.2a,b. Synchronous implementation of the encoder

Since timing and functionality of a circuit are separated in a synchronous implementation, the only correctness requirement for the decomposition in this case is to preserve the output logic functions. This is the main advantage of the synchronous approach, but with a small caveat[1]. One can safely replace the circuit in Fig. 6.2a with the implementation in Fig. 6.2b only if its computational delay falls within the clock margin. If due to decomposition the delay of the circuit becomes larger than the clock period, the synchronous implementation breaks down. The problem can be solved either by stretching the clock or by further restructuring the circuit to reduce the critical path delay. The first option can violate other specification requirements, while the second one may lead to timing convergence problems if the delays depend substantially on the result of the following physical design steps. Nevertheless, logic decompositions of synchronous circuits is considered to be a "solved" problem, with extensive commercial EDA tool support, which usually does not pose significant difficulties to a designer.

Let us now consider the design of the priority encoder in an asynchronous context. To ensure a self-timed behavior, an asynchronous system must have the capability to recognize when the proper data pattern has been received at its inputs, and to inform its environment about the completion of its computation. The first task can be solved by using a delay-insensitive encoding at the primary inputs [34]. For example in Fig. 6.3a inputs x, y and z are encoded using a dual-rail scheme: $x = 0 \Leftrightarrow x0 = 1, x1 = 0;\;\; x = 1 \Leftrightarrow x0 = 0, x1 = 1$). In addition, a two-phase discipline of operation is assumed:

1. computation is performed when a valid delay-insensitive code appears at the inputs (i.e. either 01 or 10 in the dual-rail case),
2. then the system simply resets to prepare for the next computation when a spacer code appears on the inputs (e.g. 00 in the dual-rail case).

[1] This caveat may not be small at all, especially with the growing scale of integration, and it has inspired asynchronous research for several decades.

Such two-phase operation is also used in synchronous systems, and the phase when no computation can be performed corresponds to the window of setup and hold times around the active clock edge.

Informing the environment about output readiness requires to introduce additional completion detection circuitry into the system. The output of completion detection is denoted by the *done* signal in Fig. 6.3a. We will not examine in detail the issue of implementing completion detection, since this task is relatively simple, as discussed in [35]. We will directly proceed to the synthesis of the circuit, using the STG-based formalism introduced earlier in this book (see [50, 129] for direct mapping methods from an HDL specification).

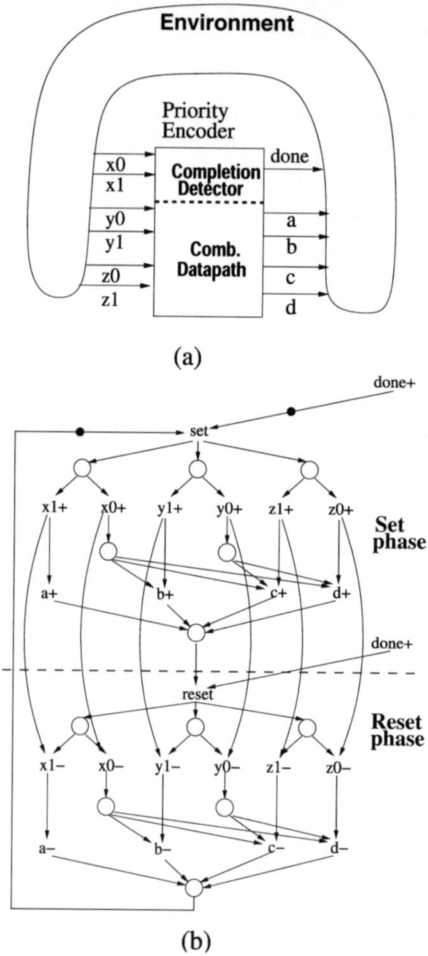

Fig. 6.3a,b. Asynchronous priority encoder

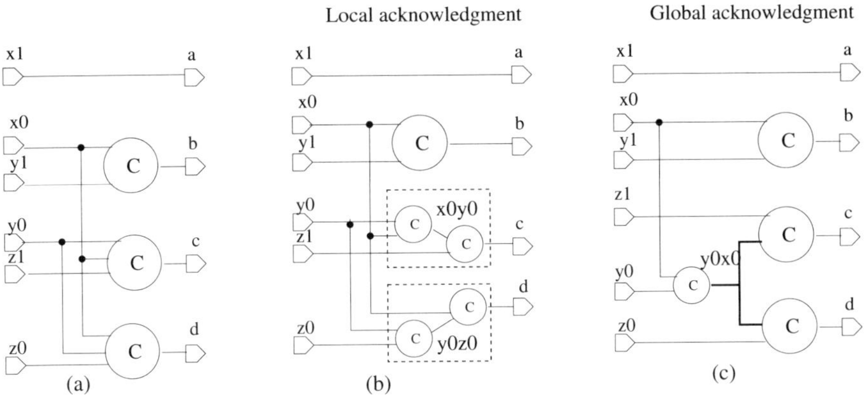

Fig. 6.4a–c. Datapath implementation for the asynchronous encoder

The behavior of an asynchronous encoder is specified by the STG in Fig. 6.3b, and a corresponding implementation is derived in Fig. 6.4a. This implementation is very similar to the synchronous one in Fig. 6.2b, with the exception that all AND gates are replaced by C-elements. Transformation into an implementation using a 2-input gate library requires to decompose the C-elements with outputs c and d. A local decomposition of 3-input C-elements into a tree of 2-input C-elements is shown in Fig. 6.4b. Unfortunately, this decomposition leads to hazardous behavior. Let us assume that the inputs of the encoder change from the spacer all-zero code to the valid dual-rail code ($x0{=}1, x1{=}0$); ($y0{=}1, y1{=}0$); ($z0{=}1, z1{=}0$). The circuit encodes it by output values $a{=}b{=}c{=}0$, $d{=}1$. After signal d switches to 1, the completion signal *done* rises, informing the environment that all transients in the circuit are finished. However this information is imprecise, because due to input $x0 = y0 = 1$, the output of C-element $x0y0$ will at some point rise, but its switching has no effect on circuit outputs. Therefore, if the delay of this C-element is large enough, it may not finish its switching before the next reset phase. Thus the circuit is hazardous, because it has a time-dependent behavior. A similar problem affects also C-element $y0z0$ in the decomposition of d. In fact, it can be shown that any local decomposition of functions c and d leads to a hazardous behavior, due to the fact that some transitions of the internal gates are not *acknowledged* by the outputs.

This problem can be solved by providing additional paths from unacknowledged gates to primary outputs. Therefore one could think of acknowledging the various switchings (e.g. rising and falling) of a gate through *multiple outputs*. An example of global acknowledgment is illustrated in Fig. 6.4c, where C-element $y0z0$ is shared by outputs c and d (as shown by the thick wires). In this case every transition of $y0z0$ indeed propagates to some output, either c or d. The obtained implementation is speed-independent.

The considerations above allow us to draw several important observations about the decomposition of asynchronous circuits:

- Decomposition is not a correctness-preserving transformation in general for asynchronous circuits, because it can lead to violations of speed-independence. For a given speed-independent asynchronous circuit there is no guarantee that a correct speed-independent decomposition exists in an arbitrary library (e.g. 2-input gates).
- Contrary to synchronous circuits, asynchronous ones need both sequential and combinational decomposition, as shown by the example of decomposing the C-elements in the priority encoder.
- The structure of the circuit around an extracted factor matters. The same gate can be hazardous when it fans out only to the function from which it has been factored (local acknowledgment), and correct when it is used also by another gate (global acknowledgment). Hence factoring some sub-function out of a particular gate can affect the cost of other gates, since also their implementation may need to be changed, sometimes for the worse, in order to acknowledge it. This is a very particular feature of asynchronous circuits, which never happens in the synchronous domain.

Clearly the theory behind logic decomposition for asynchronous circuits is much more involved than that for synchronous ones. Moreover, these difficulties translate to a much higher computational complexity. Decomposition can indeed become a bottleneck in the design flow, and a lot of efforts in this chapter are related to the identification of efficient ways to estimate the quality and feasibility of a given decomposition.

The general flow to apply different decomposition approaches is shown in Fig. 6.5.

Strictly speaking, the architecture-dependent, or syntax-directed, methods shown in the left half of the flow in Fig. 6.5 are not decomposition methods. In these methods synthesis is performed having a particular target circuit structure in mind. For example, in the standard C-architecture discussed in Sect. 6.2 each output signal is implemented by a C-element, whose inputs implement the set and reset functions of the signal. In general these functions are simpler than its complex gate implementation, and thus the cost is reduced. Moreover, under conditions formulated in Sect. 6.2, each of the set or reset functions can be safely split into orthogonal sub-functions, thus further simplifying the implementation. However the lack of an explicit decomposition step in this flow implies that there in no guarantee that each logic function within the set and reset functions is mappable to a single library gate. Of course, the likelihood of success is high for small control circuits, but drops significantly for larger ones. Though these methods are capable of locally improving the decomposition quality by a guided insertion of additional signals, as shown in the loop in Fig. 6.5, the control over final results is very loose. Due to the simplicity of this approach, it might be worthwhile to try it as

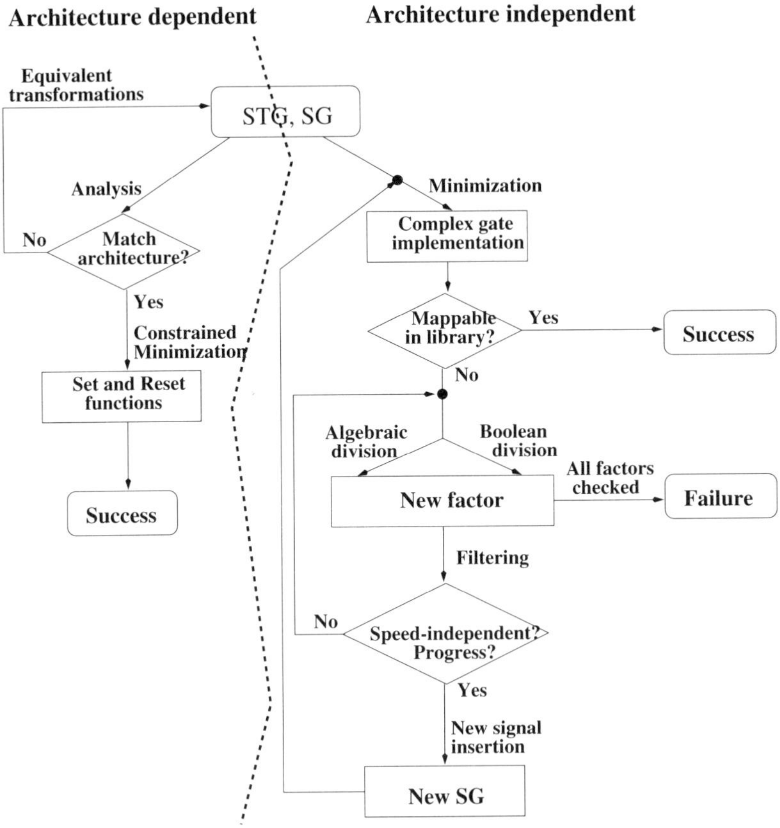

Fig. 6.5. Decomposition techniques

the first approximation, and to switch to more complicated strategies in case of failure.

The architecture-independent, or synthesis-based, methods shown in the right half of the flow in Fig. 6.5 start from a complex gate implementation derived from the original specification. If some signal of the implementation is not mappable to library gates, then its function is decomposed into two sub-functions. A decomposition might rely either on algebraic factorization, by using the well-developed kernel theory [10], or on Boolean factorization, by solving Boolean relations [130]. In both cases the number of candidates to consider is huge, and pruning of the solution space is necessary. Two filters are used:

1. the first one checks that factoring a new signal does not produce hazards in the circuit (speed-independence condition),
2. the second one tests that factorization indeed simplifies the Boolean functions in the implementation (progress condition).

If these two conditions are met, a new signal corresponding to the factor is inserted in the SG, and signal functions are re-synthesized from scratch. Note that the method does still not guarantee convergence, because no factor may satisfy both conditions. However, as experimental results in Sect. 6.5 show, the effectiveness of the method is rather high in practice.

The rest of the chapter discusses the details of both decomposition approaches, with an emphasis on architecture-independent methods.

6.2 Architecture-Based Decomposition

In this section we develop a decomposition method starting from *standard architectures*. In particular, we concentrate on the *standard-C* architecture, which is described in Fig. 6.6a (multiple AND-OR gates can exist for both setting and resetting the output). A synthesis method based on this architecture was first suggested in [131, 132]. This method defines an implementation condition that is equivalent to the Monotonic Cover conditions [133] that we use, but only for the case of decomposition into simple gates. In our work we use the more general Monotonic Cover conditions because they allow one to: (1) consider a wider class of specifications (allowing both AND and OR causality), and (2) extend the basic theory to support more aggressive optimizations detailed in [134].

In the next section we will show how to use only *implementable* gates, that is gates which exist in the chosen library, instead of the unbounded fanin gates assumed by the standard architecture-based methods.

Let us illustrate the correctness conditions for the standard-C architecture implementation by means of the example in Fig. 6.7. The complex gate implementation for output signal a, derived from the Karnaugh map of Fig. 6.7b, $a = a\overline{c} + \overline{b}\overline{c}$, is shown in Fig. 6.7c. If the library does not include AND-OR gates, this implementation is invalid and we need to decompose it, e.g. by using the standard-C implementation. The set (reset) function of signal a must be ON in all the states of $ER(a+)$ ($ER(a-)$) to implement the rising (falling) transition of A, and must be OFF in all the states where a keeps value 0 (1), i.e. in $QR(a-) \cup ER(a-)$ ($QR(a+) \cup ER(a+)$), as discussed in Sect. 4.5. The rest of the reachability space gives don't care conditions for the set (reset) function. Based on this information one can derive functions S and R from the Karnaugh map in Fig. 6.7b as $S = \overline{b}\overline{c}$ and $R = c$, resulting in the standard C-architecture implementation shown in Fig. 6.7d. A closer look at this circuit reveals a potential hazard, because starting from initial state 0000, with signal a at 0 and set function at 1, the circuit switches the output a to 1, thus acknowledging value 1 of the set function, then turns off the set function in state 1100, that is not covered by cube $\overline{b}\overline{c}$, and then turns it on again in state 1001. The last rising transition of the set function is unobservable at the outputs of the circuit, and thus its behavior is hazardous, since it shortly rises in 1001 and falls in 1011, without being acknowledged

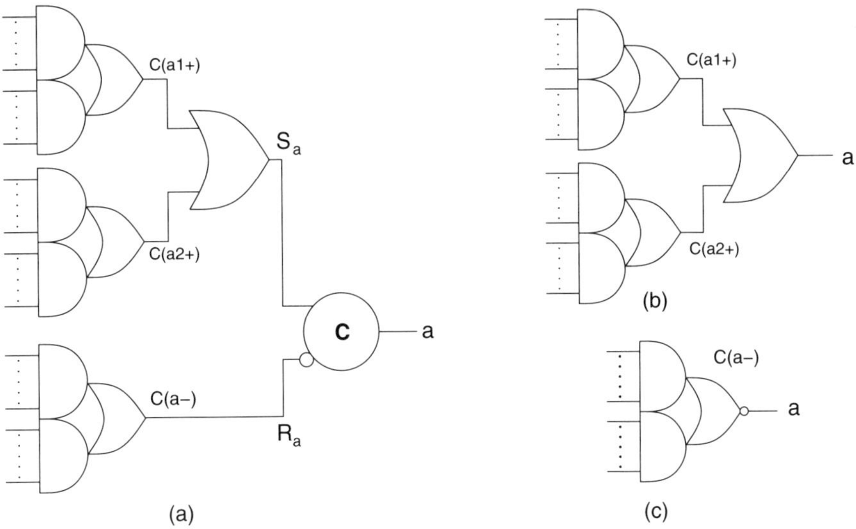

Fig. 6.6. The standard-C architecture extended for complex gates (**a**) and its possible optimizations (**b**) and (**c**)

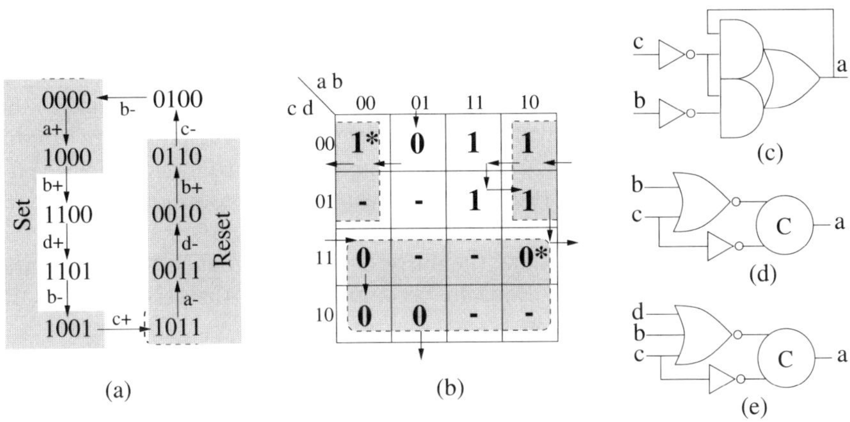

Fig. 6.7a–c. Implementation of SG (**a**) in standard-C architecture

by output a. Such a non-monotonous behavior of the set function in the quiescent region $QR(a+)$ is not speed-independent. To avoid the malfunction one needs to restrict the ON-set of the set function, by excluding state 1001, e.g. by using the redundant cover $\overline{b}\overline{c}\overline{d}$ e.g.). The resulting standard-C implementation, shown in Fig. 6.7e, is hazard-free.

In general an output signal can have several rising and falling transitions, and the principle of acknowledging the set and reset functions only through the output of the corresponding C-element requires more correctness conditions. These are called the *Monotonic Cover conditions* or the *MC-conditions* and are given below.

Let $C_j(a^*)$ denote one of the first-level AND-OR gates in the standard-C architecture. $C_j(a^*)$ is a *correct monotonic cover* for the excitation region $ER_j(a^*)$ if the following three conditions are satisfied:

1. *Cover condition:* $C_j(a^*)$ covers all states of $ER_j(a^*)$ (i.e., $C_j(a^*)$ evaluates to 1 in all states of $ER_j(a^*)$).
2. *One-hot condition:* $C_j(a*)$ does not cover any state outside $ER_j(a*) \cup QR_j(a*)$.
3. *Monotonicity condition:* $C_j(a^*)$ can fall at most once along any state sequence within $QR_j(a^*)$.

The meaning of the Monotonic Cover conditions can be explained by considering the operation of a speed-independent circuit, as discussed above when considering Fig. 6.7. Suppose, for example, that at some point during circuit operation we enter a state belonging to $ER_j(a^+)$. The cover condition ensures that the gate implementing function $C_j(a^+)$ goes from 0 to 1 in that state. The second condition guarantees that no other gates $C_k(a^+)$ in the signal network of S_a and no gates $C_k(a^-)$ in the signal network of R_a can be at 1 at that moment. Therefore $C_j(a^+)$ is the only gate in the network of signal a with the output value 1 and the propagation of this value to the output of a (through the OR gate and the C-element) gives a complete information on (acknowledges) the switchings in the network. When signal a changes its value from 0 to 1 the circuit moves from the excitation region $ER_j(a^+)$ into the quiescent region $QR_j(a^+)$. In this region according to monotonicity condition gate $C_j(a^+)$ will be reset and this switching (the only one possible in $QR_j(a^+)$) will be implicitly acknowledged when a will go low.

Since under these conditions the outputs of the first-level AND gates are *one-hot encoded*, any valid Boolean decomposition of the second-level OR gates is speed-independent.

The standard-C architecture also permits a *combinational* implementation of a signal. If the set and reset networks are the complements of each other, then a C-element with identical inputs can be simplified to a wire (see Fig. 6.6b and 6.6c. More precise conditions for such an optimization can be formulated as: the set network must cover all the states of $ER_j(a+) \cup QR_j(a+)$ for all j, and similarly the reset network must cover all the states of $ER_j(a-) \cup QR_j(a-)$.

6.3 Logic Decomposition Using Algebraic Factorization

6.3.1 Overview

As discussed in Sect. 4.5 and 6.2, any output-persistent SG satisfying the CSC condition can be implemented using the standard-C architecture. This guarantees that a correct Boolean equation can be obtained for each cover $C(a*)$. However it does not guarantee that $C(a*)$ can be implemented by one of the gates in a given library.

In order to perform technology mapping, complex gates must be decomposed until all their fragments are mappable onto library gates. The problem of decomposition of combinational circuits is well-known but, as discussed above, the methods are not directly applicable to SI circuits. The decomposition of a gate into smaller gates implicitly introduces new internal signals (with associated delays) that may cause hazards.

The approach proposed in this section splits the problem of logic decomposition of a gate into two sub-problems:

1. combinational decomposition,
2. insertion of a new hazard-free signal.

This process is iterated until all gates of the circuit can be mapped onto library gates or no more progress can be achieved, e.g. because no hazard-free decomposition can be found for any of the complex gates. Each sub-problem is briefly described in the forthcoming sections.

6.3.2 Combinational Decomposition

As is traditionally done in multi-level combinational synthesis, algebraic division has been chosen as the main operation for logic decomposition. For each cover function $C(a*)$ we look for algebraic divisors, aiming at decompositions of the following type: $C(a*) = F \cdot G + R$ where G is the quotient $C(a*)/F$, as shown in Fig. 6.8. In this figure $C(a*)$ on the left is an atomic complex gate with function $F \cdot G + R$, while on the right it is an atomic complex gate with (simpler) function $x \cdot G + R$. This decomposition scheme reduces to AND-decomposition when $R = 0$ and OR-decomposition when $G = 1$. Different examples of algebraic division are shown in Table 6.1.

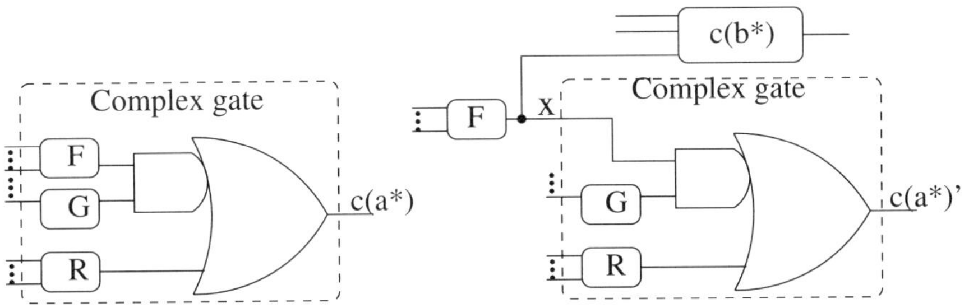

Fig. 6.8. Algebraic decomposition

Table 6.1. Examples of algebraic division

Complex gate	Divisor (x)	New gate
$abc + abd + ef$	ab (co-kernel)	$xc + xd + ef$
$abc + abd + ef$	$c + d$ (kernel)	$abx + ef$
abc	ab (AND-decomp.)	xc
$abd + ef$	abd (OR-decomp.)	$x + ef$

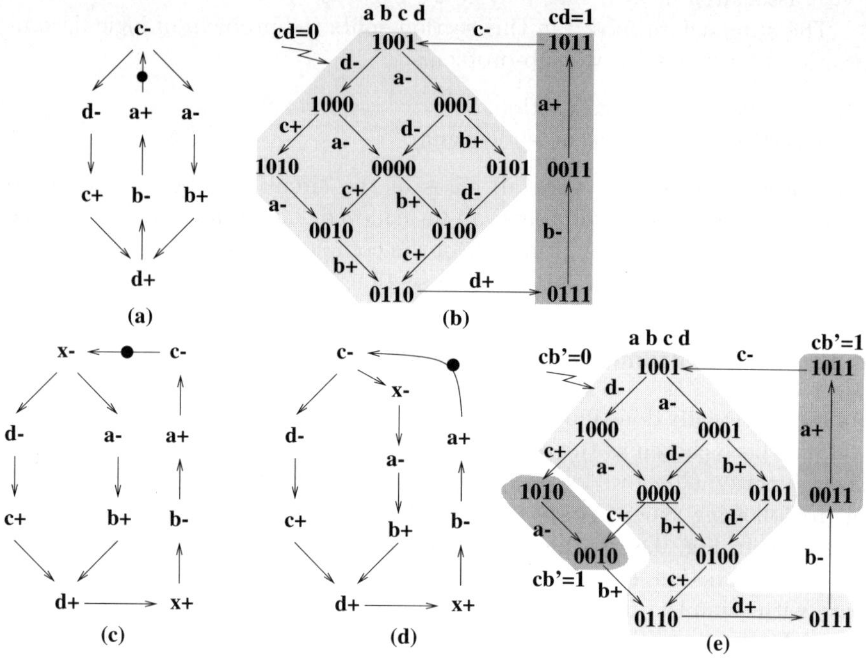

Fig. 6.9a–e. Signal insertion

Example 6.3.1. Fig. 6.9a and 6.9b depict the STG and the SG of the specification of a circuit. A complex gate implementation of the circuit is shown in Fig. 6.10a.

Let us assume that only 2-input gates are available in the library. Thus, signals a and b are not directly mappable and must be decomposed. Contrary to synchronous circuits, not every algebraic decomposition is valid. Some of them may introduce unavoidable hazards and hence violate the speed-independence requirements. To illustrate this, let us decompose the gate a in Fig. 6.10a by extracting the algebraic divisor $y = cb'$. The ON- and OFF-sets of the function for y are shown in Fig. 6.9e by shadowed areas. When the circuit enters state 0000 (underlined in Fig. 6.9e) two transitions may occur

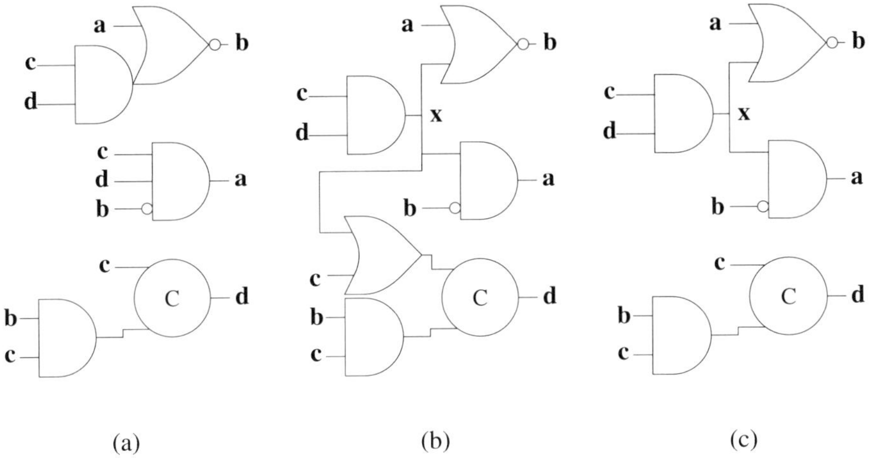

Fig. 6.10a–c. Implementations of the STGs of Fig. 6.9

concurrently: $c+$ and $b+$. Firing $c+$ first will enable gate $y = cb'$ to make a transition from low to high, while $b+$ pulls the output of the gate again to low. In a speed-independent circuit no assumptions can be made about the relative speed of concurrent transitions and therefore the considered situation is a classical illustration of hazardous behavior on the output of gate y. Hence, the decomposition $y = cb'$ is invalid and must be rejected. Early recognition of "valid factors" is a key issue in improving the efficiency of the decomposition.

6.3.3 Hazard-Free Signal Insertion

Each divisor of $C(a*)$ is a candidate function to be implemented as a new signal x of the circuit. The new signal will be hazard-free if all its transitions are *acknowledged* by other signals of the circuit. In the technique presented in this section, transitions of x may be acknowledged by *several* signals. This is more general and powerful than [96, 125] where transitions of x must be acknowledged locally, only by the same signal a from whose cover x was extracted.

Multiple acknowledgment offers two advantages:

- the same signal x can be shared by several cover functions (this corresponds to the extraction of common divisors in classical multi-level decomposition)
- a correct SI decomposition can be found even if it does not exist for solutions with single acknowledgments (as shown by the experimental results, in particular the priority encoder example).

Hazard-freedom is guaranteed for the new signal x as follows. Two new events, namely $x+$ and $x-$, are inserted in the SG so that the properties for SI implementability are preserved. The new events are defined in such a way that the implementation of signal x corresponds to the selected divisor for decomposition. If $x+$ and $x-$ can be inserted under such conditions, x is hazard-free. Now x can be used as a new signal in the support of any function cover and contribute to derive simpler equations. Care must be taken not to increase the complexity of other cover functions (Sect. 6.3.6).

Example 6.3.2. (Example 6.3.1 continued) Let us consider again the example of Fig. 6.9 and look for a hazard-free decomposition. Among the different algebraic divisors for a and b, there is one that looks especially interesting for a possible sharing of logic: $x = cd$.

The insertion of the events $x+$ and $x-$ must be done according to the implementation of the signal as $x = cd$. The shadowed areas in Fig. 6.9b indicate the sets of states in which the Boolean function cd is equal to 0 and 1 respectively. $x+$ must implement the transition from the states in which cd is equal to 0 to the states in which cd is equal to 1, i.e. $x+$ must be a successor of $d+$, whereas $x-$ must implement the opposite transition and therefore is inserted after $c-$.

Fig. 6.9c and 6.9d depict two possible insertions of signal x at the STG level. Both insertions result in specifications that are implementable as different SI circuits (shown in Fig. 6.10b and 6.10c respectively). Interestingly, both can be implemented with only 2-input gates. However, the insertion of $x-$ as a predecessor of $a-$ and $d-$ (Fig. 6.9c) changes the implementation of signal d, because the fact that $x-$ triggers $d-$ forces x to be in the support of any realization of d. A simpler circuit can be obtained if $x-$ is made concurrent with $d-$ and thus only triggers $a-$ (Fig. 6.9d). In the resulting circuit, signal x is only in the support of a and b, i.e. of those signals that acknowledge the transitions of x.

Therefore, the insertion of new signals for logic decomposition can be done by exploring different degrees of concurrency with regard to the behavior of the rest of the signals. Finding the best trade-off between concurrency and logic optimization is one of the crucial problems in the decomposition of SI circuits.

6.3.4 Pruning the Solution Space

The generation of divisors for decomposition should be pruned to avoid an explosion of candidates for complex functions.

Two conditions help in constructing an efficient filter of solutions in the huge decomposition space. Only those decompositions are considered valid which:

1. do not introduce hazards (i.e. preserve speed-independence), and
2. heuristically guarantee progress in mapping the circuit to the given library.

The above conditions could be verified in a straightforward (and inefficient) way for every function F used for decomposition, as was proposed in [135]. One could explicitly insert a new signal x, with logic function F, into the original SG and then check whether the modified SG satisfies these conditions. However, there are several reasons that make such a naive approach hardly acceptable. It was already mentioned that for complex functions the number of divisors can be huge. The situation is even worse because another dimension of complexity arises from the fact that for the same function F a new signal $x = F$ can be inserted in many different ways (see, e.g. two different insertions for $x = cd$ in Fig. 6.9c and 6.9d). Taking into account that the construction of a new SG is computationally expensive, there is no way to get an efficient implementation by the above straightforward approach.

Better results can be obtained if one checks both speed-independence (as proposed in [125] and discussed in Sect. 5.4) and progress conditions (as discussed in Sect. 6.3.6) directly on the original SG.

6.3.5 Finding a Valid Excitation Region

Given a function F that must be extracted and implemented as a separate signal x, one can use the same technique as described in Sect. 5.4 to find valid sets $ER(x+)$ and $ER(x-)$. First of all, F induces a natural bi-partition on the SG states, those in which F has value 1 and 0 respectively, as shown in Fig. 6.11a. Then $ER(x+)$ is initialized to be the input border $IB(F)$ of F. If the initial $ER(x+)$ is consistent, preserves I/O interface and is a SIP-set, then a valid insertion has been found. If one of the validity conditions does not hold, the $ER(x+)$ is expanded toward the states with $F = 1$. Similarly, $ER(x-)$ is initialized to be the input border $IB(\overline{F})$ of the complement of F.

As shown in Fig. 6.11b, by expanding $ER(x+)$ one can:

1. either obtain a well-formed set, which gives the required valid insertion,
2. or reach the boundaries of the set of states with $F = 1$ with some remaining violations.

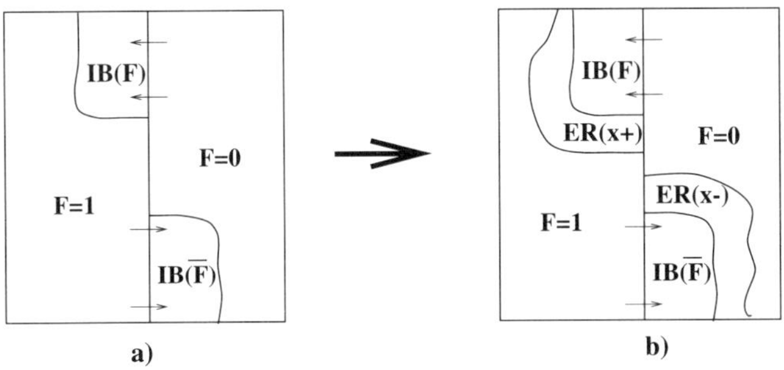

Fig. 6.11a,b. Partition of states induced by a function F

The latter implies that there exists no valid insertion of signal x with the partition implied by logic function F, and the decomposition based on F must be rejected.

Note that the solution found in the expansion of $ER(x+)$ (if any) produces a unique valid $ER(x+)$ which has the minimum size, as discussed in Sect. 5.4. This solution, however, may not be optimal, and further expansion can be applied, e.g. to increase the amount of concurrency for $x+$, whilst preserving the above conditions.

Example 6.3.3. (Example 6.3.1 continued) Figure 6.9 shows the decomposition based on the insertion of a new signal x with function $F = cd$. Let us explore different ways to select the excitation regions of x. The set of states with $cd = 0$ is entered through state 1001, while the set with $cd = 1$ is entered through 0111. Hence $IB(\overline{cd}) = 1001$ while $IB(cd) = 0111$. Both input borders are SIP-sets, satisfy consistency and their exit events correspond to output signals. Therefore $IB(\overline{cd})$ and $IB(cd)$ give valid excitation regions for the insertion of signal x and these regions have the minimum size among all valid insertions. The corresponding STG and implementation of signal x were shown in Fig. 6.9c and 6.10b respectively. The implementation of x requires the acknowledgment of transitions of x by gates a, b and d. This makes the function of d more complex than in the original SG.

To simplify the implementation let us consider the expansion of $ER(x-)$ within the set of states $cd = 0$.

The first case of expansion is shown in Fig. 6.12b where by including the state 1000 $ER(x-)$ becomes $\{1001, 1000\}$. $c+$ is an exit transition from $ER(x-)$, and c is an input signal. Hence $ER(x-)$ is an invalid selection because it does not preserve I/O interface for the original circuit (input signals cannot be delayed). The violation can be fixed by adding state 1010 to $ER(x-)$. This gives a valid selection of $ER(x-)$ with the corresponding STG and implementation of signal x shown in Fig. 6.9d and 6.10c. This circuit is simpler than that of Fig. 6.10b.

$ER(x-)$ can be expanded further, e.g. by including state 0001 as shown in Fig. 6.12c. This, however, leads to an illegal intersection with the state diamond $\{1001, 1000, 0001, 0000\}$, which violates speed-independence. To solve this problem state 0000 must also be included into $ER(x-)$. After that, $ER(x-)$ illegally intersects the state diamond $\{1000, 1010, 0000, 0010\}$, which in its own turn can be fixed by adding 0010 to $ER(x-)$. The latter gives a valid selection of $ER(x-)$ with the corresponding STG and circuit shown in Fig. 6.12d and 6.12e. This example shows how, starting from $IB(x-)$, the excitation region for transition $x-$ is expanded to satisfy the conditions of Sect. 5.4. It results in a successful decomposition because in the procedure of expansion no states in which $F = 1$ were required to be included in $ER(x-)$.

We have thus identified a way of finding a correct position in the state graph to insert a new signal for a given Boolean decomposition. In the following section we shall look at the conditions that heuristically guarantee

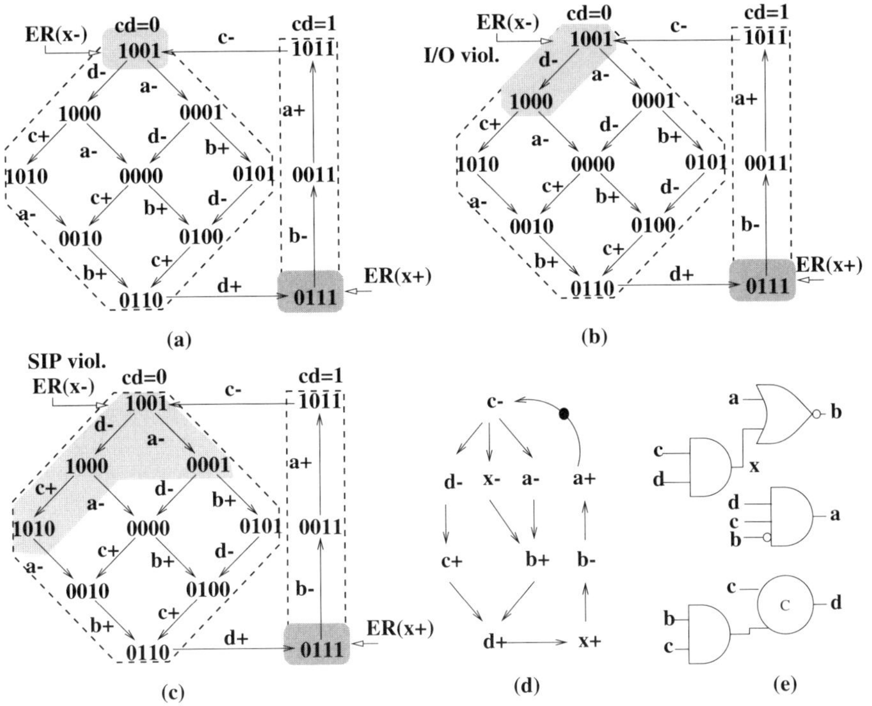

Fig. 6.12a–e. Selection of valid excitation regions

progress toward the overall goal of decomposing all gates that do not belong to the target library.

6.3.6 Progress Analysis

Even if for a new signal x, with a function F, a speed-independence preserving insertion does exist, a decomposition using x might be undesirable due to two possible complications:

1. The check for speed-independence guarantees that there exists an insertion of signal $x = F$ which does not produce hazards in the circuit, but it does not tell whether in this particular insertion the cover $C(a*) = F \cdot G + R$ can be simplified to $C(a*)' = x \cdot G + R$. It is quite possible that keeping speed-independence would require a more complicated cover implementation than $C(a*)'$.
2. The need to acknowledge the transitions of x might influence the complexity of other signals different from the factorized signal a. The cost of acknowledgment should be estimated. Allowing some increase in cost is sometimes acceptable in order to escape from local minima.

The heuristic procedure that we propose to explore the huge optimization space operates by quasi-greedy optimization of a cost function that takes into account both aspects above. Both cost estimations must be performed on the *original* SG, without explicitly adding the new signal, in order to keep the execution time of the decomposition algorithm within reasonable limits. We will call a reduction of the former *local progress* and a reduction of the latter *global progress*, and examine each one in turn after a motivating example.

Let the target cover function be $C(a*) = F \cdot G + R$, in which F is the candidate for extraction. At first, one should find valid excitation regions for the new signal $x = F$. If such $ER(x+)$ and $ER(x-)$, can be derived, as discussed in Sect. 5.4, then there is a speed-independent implementation of the SG with a new signal x. The purpose of the insertion of signal x is to simplify the cover function $C(a*) = F \cdot G + R$ by substituting F with x.

This *purely algebraic* simplification is, however, not always possible in the asynchronous case, since in order to preserve speed-independence, $C(a*)$ may now require more fan-in signals. Let us illustrate this by considering our example.

Example 6.3.4. (Example 6.3.1 continued) Fig. 6.13 shows one of the speed-independent insertions of signal x based on extracting function cd out of the functions for signals a and b for the initial specification given in Fig. 6.9. This insertion corresponds to selecting excitation regions as follows (see Fig. 6.13a): $ER(x-) = \{1001, 0001\}$ and $ER(x+) = \{0111\}$. The STG with the new signal inserted, the new state graph and the new circuit are shown in Fig. 6.13b, 6.13d and 6.13c respectively.

Although function cd has been extracted, it does not help to reduce the literal count for the target cover of signal a. It still has three literals: $a = \overline{b}cx$, because c must be an input to both gates for a and x. To avoid confusion, note that the circuit with a two input gate implementation for signal a shown in Fig. 6.10b does not correspond to the selected $ER(x-)$, as can be seen by comparing STGs from Fig. 6.9c and Fig. 6.13d. In the former, $x-$ precedes $a-$, while in the latter $a-$ is concurrent with $x-$.

The most natural implementation (based on algebraic factoring) $a = \overline{b}x$ is unfortunately incorrect, as shown in Fig. 6.13d, because function $\overline{b}x$ covers four states of the new state graph, while there are only two states in which signal a has an implied value equal to 1. Hence function $\overline{b}x$ does not provide a correct cover for signal a.

6.3.7 Local Progress Conditions

The local progress condition has the purpose of verifying that the algebraic decomposition is indeed a valid speed-independent decomposition.

Insertion of the new signal $x = F$ produces SG A'. Covers $C_{A'}(a*) = x \cdot G + R$ in SG A' and $C_A(a*) = F \cdot G + R$ in SG A might have different

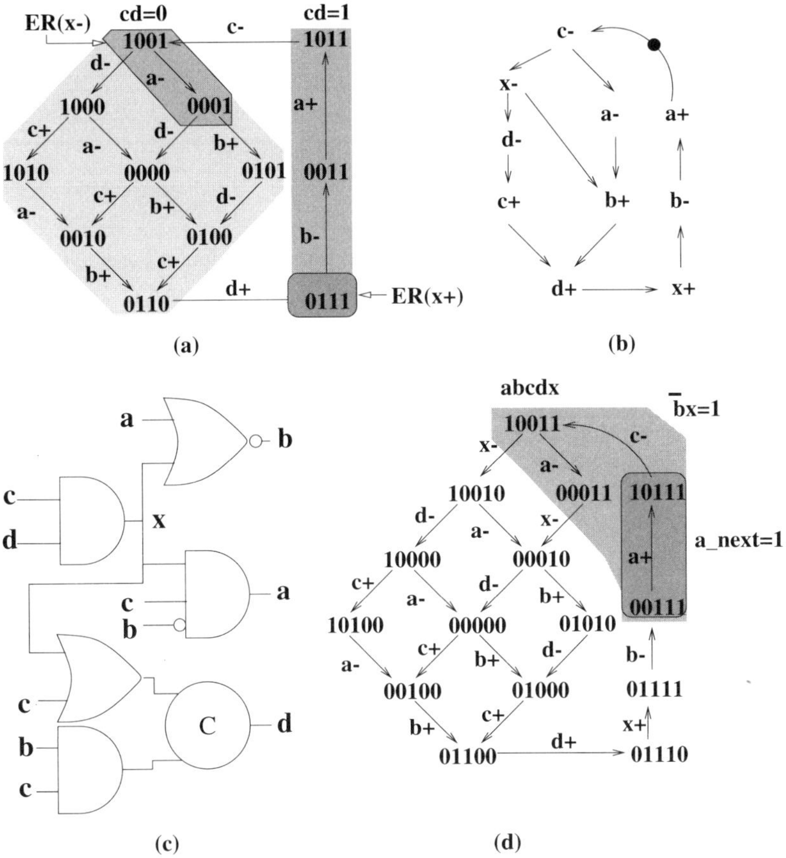

Fig. 6.13a–d. Solution with no progress

behavior only in states that correspond to $F \cdot G \cdot \overline{R}$, because $C_{A'}(a*)$ and $C_A(a*)$ obviously have the same behavior in states that correspond to R (both include that term) and outside $F \cdot G + R$. This is illustrated graphically in Fig. 6.14 through the darker shading.

Thus one must check that $C_{A'}(a*) = x \cdot G + R$ satisfies in the new SG A' all three Monotonic Cover conditions defined in Sect. 6.2. This is true, informally, if and only if:

1. Cover $x \cdot G$ (effectively, set $ER(x+)$) completely covers the intersection of $ER(a*)_{A'}$ and $F \cdot G \cdot \overline{R}$. This guarantees that no transition $a*$ is inside $ER(x+)_{A'}$ and thus ensures the Cover Condition for $C_{A'}(a*)$.
2. Cover $x \cdot G$ does not evaluate to 1 outside the union of $ER(a*)_{A'}$ and $QR(a*)_{A'}$. This ensures the One-hot Condition for $C_{A'}(a*)$.

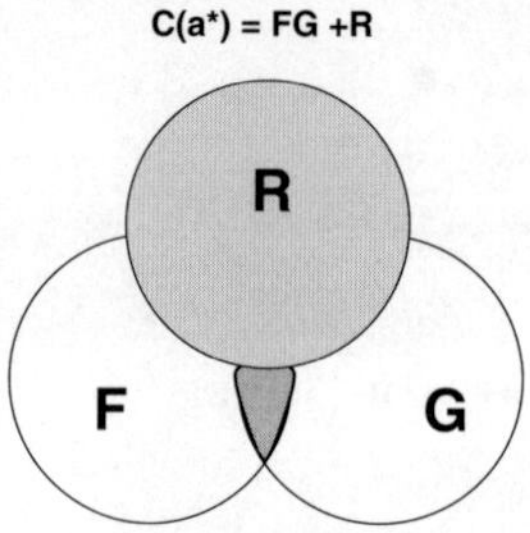

Fig. 6.14. Graphical representation of cover $C(a*)$

3. a) Cover $x \cdot G$ does not change its value from 0 to 1 in the intersection of $QR(a*)_{A'}$ and $F \cdot G \cdot \overline{R}$. This guarantees the absence of non-monotonic "1-0-1" transitions along any path in $ER(a*) \cup QR(a*)$ in A'.
   b) No transition of cover $x \cdot G$ from 1 to 0 can occur inside $QR(a*)_{A'}$, unless R already evaluates to 1. This ensures the absence of non-monotonic "0-1-0" transitions along any path in $ER(a*) \cup QR(a*)$ in A'.

In the above example function $\overline{b}x$ is not a correct cover for a because the chosen $ER(x-)$ contains transition $a-$. Due to this $x-$ cannot trigger $a-$ (similarly, transitions of b cannot trigger $a-$). Hence function $\overline{b}x$ cannot implement a trigger signal for $a-$ and signal c should be added for a proper implementation of gate a. Hence no local progress is achieved for the target of the decomposition, gate a.

Note that, even though Conditions 1-3 are formulated in terms of objects of the new SG A', it is possible to check them directly on the original SG A. The precise formulation of this check requires the notion of *state image* introduced in Sect. 5.2.

The validity of substituting a new signal x in a cover function $C(a*)$ is checked by considering the inverse images of $ER(a*)_{A'}$ and $QR(a*)_{A'}$. By construction, $ER(a*)_A$ is the inverse image of $ER(a*)_{A'}$. The inverse image for quiescent regions can include additional states because some original signal transitions are delayed by x. For example, $QR(d+)_A = \{0111, 0011, 1011\}$ (see Fig. 6.13a) while in the expanded SG $QR(d+)_{A'} = \{01110, 01111, 00111, 10111, 10011, 00011\}$ (see Fig. 6.13d) with an inverse image $QR(d+)_{A'}^{-1} = \{0111, 0011, 1011, 1001, 0001\}$. In general, computing the inverse image $QR(a*)$ is easy by starting with $QR(a*)_A$ and expanding it forward inside the next excitation region of signal a, if it has non-empty intersection with $ER(x-)$ or $ER(x+)$.

We will now formulate the local progress condition by presenting requirements that preserve the monotonic cover conditions after substituting function F with one literal x in the cover function $C(a*)$. If these requirements, formulated in terms of the original state graph are satisfied, then simplification for the gate implementing signal a is guaranteed to be possible.

Let $C_A(a*) = F \cdot G + R$ be a monotonic cover of $ER(a*)$ in SG A. Let $ER(x+)$ and $ER(x-)$ be selected for inserting a signal x. The function $C_{A'}(a*) = x \cdot G + R$ satisfies the three Monotonic Cover conditions in the new SG A', if and only if, as informally explained above[2]:

1. *Cover condition:* if $s1 \in (ER(a*) \cap F \cdot G \cdot \overline{R}) \cap ER(x+)$ and $s1 \stackrel{a*}{\to} s2$ then state $s2 \notin ER(x+)$.
2. *One-hot condition:* for all s: if $s \notin ER(a*) \cup (QR(a*)_{A'})^{-1}$, then $s \notin ER(x-) \cap G$
3. *Monotonicity conditions:*
   a) $\forall s : s \in (QR(a*) \cap F \cdot G \cdot \overline{R}) \Rightarrow s \notin ER(x+)$, and
   b) $\forall s, s1 : s \in (QR(a*)_{A'})^{-1} \cap ER(x-) \cap G;\ s1 \to s \Rightarrow s1 \in G + R$.

6.3.8 Global Progress Conditions

The local progress conditions (if satisfied) guarantee that the implementation of cover function $C(a*)$ will be simplified as a result of decomposition. However, to accept a decomposition we need to ensure that it does not significantly increase the complexity of logic for *other* signals. The solutions in Fig. 6.10b and Fig. 6.13c increase the size of the cover for d. This may or may not be considered acceptable. We use a conservative estimate of the increase of logic complexity, based on the calculation of the number of trigger signals before and after the insertion of a new signal. This is based on the following facts:

- trigger signals of a signal x (signals whose transitions are immediate predecessors of transitions of x) are always in the support of the function implementing x, and
- the function implementing a signal with no new trigger signals after the insertion will never be more complex than the original one.

Algebraic techniques for generating factors are fast and efficient. However, sometimes they are too coarse and do not deliver the full flexibility in decomposition. Boolean division suggests an alternative way for factors generation. Generally it results in better decomposition quality, but at the expense of significantly higher complexity. Boolean division *in the asynchronous context discussed below* makes no distinction between extracting combinational and sequential factors. It thus naturally handles both *combinational and sequential decomposition in a uniform way.*

[2] In the formulae we use F, G and H to denote the sets of states in which these functions evaluate to 1, and hence liberally apply Boolean operations, such as $\cdot, +$, and set operations, such as $\cap, \cup$, interchangeably.

6.4 Logic Decomposition Using Boolean Relations

We now describe a very general framework which can exploit a library of *arbitrary combinational and sequential gates* to decompose a complex gate function, as shown in Fig. 6.15.

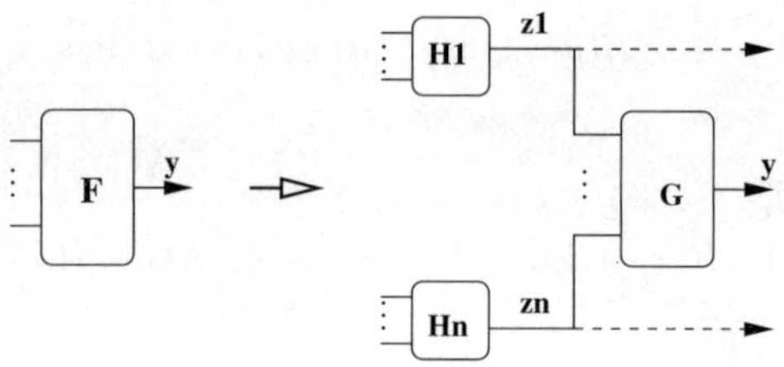

Fig. 6.15. General framework for speed-independent decomposition

Consider a possibly sequential complex gate implementing function F. The result of decomposition is a library gate designated by G and a set of (possibly complex) gates labeled $H_1, H_2, \ldots H_n$. The latter are decomposed recursively until all elements are found in the library and optimized to achieve the lowest possible cost. We thus by and large put no restrictions on the implementation architecture in this section. However, as will be seen further, for the sake of practical efficiency, our implemented procedure deals only with 2-input gates and various kinds of registers and latches to act as a gate G in the decomposition. More complex combinational gates can then be used by means of a standard technology mapping step.

An important advantage of this approach with respect to that described in the previous section is that it uses full scale Boolean decomposition, rather than just algebraic factorization. This allows us to widen the scope of implementable solutions and improve on area cost.

Our second goal in generalizing the C-element based decomposition is to allow the designer to use more *conventional types of latches*, e.g. D-latches and SR-latches, instead of C-elements that may not exist in conventional standard-cell libraries. Furthermore, as our experimental results show (see Sect. 6.5), in many cases the use of standard latches instead of C-elements helps improving the circuit implementation considerably.

The power of the method can be appreciated by looking at the example shown in Fig. 6.16a and 6.16b. The initial implementation using the "standard C-architecture" and its decomposition using two input gates by the method described in Sect. 6.3 are shown in Fig. 6.16c and 6.16d. The Boolean relation-based method produces a much cheaper solution with just two D-latches, shown in Fig. 6.16e. Despite the apparent triviality (for an experienced human designer!) of this solution, none of the previously existing automated tools had been able to obtain it. Also note that the D-latches are used in a *speed-independent* fashion, and are thus free from meta-stability and hazards. For

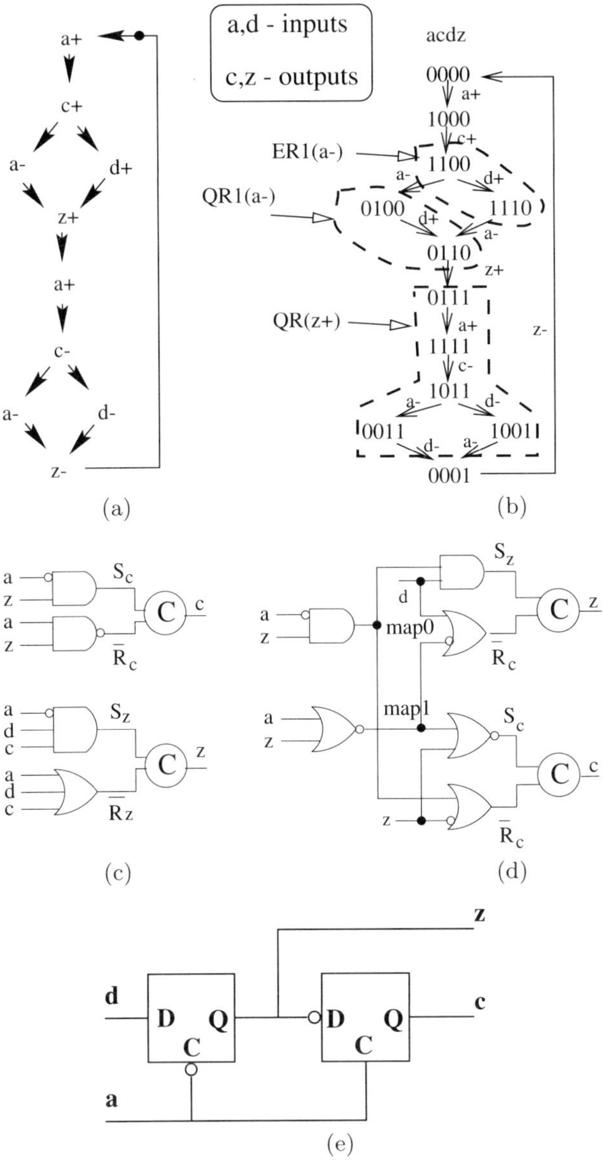

Fig. 6.16. An example of Signal Transition Graph (**a**), State Graph (**b**) and their implementation (**c**), (**d**), and (**e**) (benchmark `hazard.g`)

example, all transitions on the input must be acknowledged by the output before the clock can fall and close the latch. E.g. there is no problem with setup and hold times as long as the D to Q delay is larger than both setup and hold times, which is generally the case.

6.4.1 Overview

The method for library-driven sequential decomposition of speed-independent circuits consists of three main steps:

1. Synthesis via decomposition based on Boolean relations;
2. Signal insertion and generation of a new SG;
3. Library matching.

The pseudo-code for the decomposition algorithm is given in Fig. 6.17. At each iteration of the main loop, a new signal is inserted. To select this new signal, a set of valid decompositions is calculated for each non-input signal (line 2 in Fig. 6.17) and the best one is kept (line 3). Finally the most complex function from all the decompositions is implemented as a new signal (lines 5 and 6). Choosing the most complex function at each step allows this function to become a candidate for decomposition in the next iteration and, thus, *decompose the largest gates first*.

The algorithm terminates when all non-input signals are implementable with gates from the library or when no more signals can be further decomposed (line 4).

```
   do
1:   foreach non-input signal x do
       solutions(x):=∅;
2:     foreach gate G ∈ {latches, and2, or2} do
         solutions(x):=solutions(x) ∪ decompositions (x,G);
       endfor
3:     best_H(x) := Best SIP candidate from solutions(x);
     endfor
4:   if foreach x, best_H(x) is implementable
     or foreach x, best_H(x) is empty then exit loop;
5:   Let H be the most complex best_H(x);
6:   Insert new signal z implementing H and derive new SG;
   forever
7:Library matching;
```

Fig. 6.17. Algorithm for logic decomposition and technology mapping

The proposed method breaks an initial complex gate implementation of an SG, *starting from the gate outputs*, by using *sequential* (if the original gate function is self-dependent, i.e. it has internal feedback) or *combinational* gates. Note that after decomposition terminates, technology mapping can be performed indifferently starting from the inputs or from the outputs.

This method is complementary to the one described in Sect. 6.3, in which the decomposition was performed by using algebraic divisors of the current implementations of the output signals, and thus decomposition was performed *from the inputs* of complex gates.

Given a vector X of SG signals and a non-input signal $y \in X$ (in general the function $F(X)$ for y may be self-dependent), function $F(X)$ is decomposed into (line 2 of the algorithm in Fig. 6.17):

- a combinational or sequential gate with function $G(z_1, z_2, y)$, where z_1 and z_2 are newly introduced signals,
- two combinational functions $H_1(X)$ and $H_2(X)$ for signals z_1, z_2. The restriction that H_1, H_2 be combinational functions will be partially lifted in Sect. 6.4.4.

so that $G(H_1(X), H_2(X), y)$ implements $F(X)$. Moreover, the newly introduced signals must be speed-independent (line 3).

The problem of representing the flexibility in the choice of the functions H_1 and H_2 as Boolean relations has been explored, in the context of combinational logic minimization, by [136] among others. Here we extend its formulation to cover also *sequential* gates (in Sect. 6.4.2 and 6.4.4). This is essential in order to overcome the limitations of previous methods for speed-independent circuit synthesis that were based on a specific architecture. Now we are able to use a broad range of sequential elements, like set and reset dominant SR latches, transparent D latches, and so on. Apart from significantly improving some experimental results, this allows one to use a "generic" standard-cell library (that generally includes SR and D latches, but not C elements) without the need to design and characterize any new asynchronous-specific gates.

The algorithm proceeds as follows. It starts from an SG and derives a logic function for all its non-input signals (line 1). It then performs an implementability check for each such function as a library gate. The largest non-implementable function is selected for decomposition. In order to limit the search space, the candidates for G are limited to (line 2):

- all the sequential elements in the library (assumed to have two inputs at most, again in order to limit the search space),
- two-input *AND, OR* gates with all possible input inversions.

The flexibility in the choice of functions $\mathcal{H} = (H_1, H_2)$ is defined by a Boolean relation that represents the solution space of $F(X) = G(\mathcal{H}(X), y)$, as described in Sect. 6.4.2.

The set of function pairs (H_1, H_2) compatible with the Boolean relation is then checked for speed-independence (line 3), as described in Sect. 5.2. If neither is speed-independent, the pair is immediately rejected.

Then, both H_1 and H_2 are checked for implementability in the reduced library, in increasing order of estimated cost. Two cases are possible:

1. both are speed-independent and implementable: in this case the decomposition is accepted,
2. otherwise, the most complex implementable H_i is selected, and the other one is merged with G.

Choosing the most complex function, as mentioned above, experimentally helps with keeping the decomposition balanced. At this stage H_1 or H_2 can also be implemented as a *sequential gate* if the *sufficient* conditions described in Sect. 6.4.4 are met.

The procedure is iterated as long as there is progress or until everything has been decomposed (line 4). Each time a new function H_i is selected to be implemented as a new signal, it is inserted into the SG (line 6) and resynthesis is performed in the next iteration. Note that the insertion of a new signal occurs only at the very last step. The latter means that the quality of the decomposition (speed-independence and progress checks) should be estimated using the *original* SG, as in the algebraic approach described in the previous section. As a rule of thumb, one can generally avoid excessive execution times due to a reconstruction of the SG by doing it only in desperate cases, after all other options have been tried.

At the end, a Boolean matching step [124] can be used to recover area and delay (line 7). This step can merge together the simple 2-input combinational gates that have been (conservatively) used in the decomposition into a larger library gate. It is guaranteed not to introduce any hazards if the matched gates are atomic.

The method in Fig. 6.17 is incomplete. This is essentially due to the greedy heuristic search that accepts the smallest implementable or non-implementable but speed-independent solution. Therefore, a speed-independent solution could be missed, e.g. if it corresponds to a redundant cover as shown in Example 6.4.1. In order to enable the generation of redundant decompositions in our implementation, the method based on Boolean relations might be combined with the method based on monotonic covers (Sect. 6.2), which is better suited for a guided generation of redundant covers.

6.4.2 Specifying Permissible Decompositions with BRs

As discussed above, in this section we apply BRs to the following problem.

Given an incompletely specified Boolean function $F(X)$ for signal $y \in X$, decompose it into two-levels $y = G(z_1, z_2, y); z_1 = H_1(X); z_2 = H_2(X)$ such that $F(X) = G(H_1(X), H_2(X), y)$ implements $F(X)$ and functions H_1, H_2 and G have a simpler implementation than F. Any such pair $\{H_1, H_2\}$ will be called permissible.

Note that the first-level function $\mathcal{H}(X) = \{H_1(X), H_2(X)\}$ is a multi-output logic function, specifying the behavior of internal nodes of the decomposition, $Z = \{z_1, z_2\}$. For the sake of simplicity we consider the decomposition problem for a single-output binary function F, although generalization to multi-output and multi-valued functions is straightforward.

The final goal is decomposition to a form that is easily mappable to a given library. Hence only functions available in the library are selected as candidates for G. Then at each step of decomposition a small mappable piece (function G) is cut from the potentially complex and unmappable function F. For a selected G all permissible implementations of functions H_1 and H_2 are specified with a BR and then via minimization of BRs a few best compatible functions are obtained. All of them are verified for speed-independence by checking SIP-sets. The one which is speed-independent and has the best estimated cost is selected.

Since the support of function F can include the output variable y, it can specify sequential behavior. In the most general case two-level *sequential* decomposition is performed such that both function G and functions H_1, H_2 can be sequential, i.e. they contain their own output variables in their supports. The second level of the decomposition is made sequential by selecting a latch from the library as a candidate gate, G. The technique for deriving a sequential solution for the first level H_1, H_2 is described in Sect. 6.4.4.

The next example shows how all permissible implementations of a chosen decomposition can be expressed with BRs.

Example 6.4.1. Consider the STG in Fig. 6.18a, whose SG appears in Fig. 6.19a. Signals a, c and d are inputs and y is an output. A possible implementation of the logic function for y is $F(a, c, d, y) = ac\overline{d} + y(\overline{c} + \overline{d})$. Let us decompose this function using as G a reset-dominant Rs-latch represented by the equation $y = G(R, S, y) = \overline{R}(S + y)$ (see Fig. 6.18b). The first step specifies the permissible implementations for the first level functions $R = H_1$ and $S = H_2$ by using the BR given in the table of Fig. 6.19b. Consider, for example, vector $a, c, d, y = 0000$. It is easy to check that $F(0, 0, 0, 0) = 0$. Hence, for vector 0000 the table specifies that $(R, S) = \{11, 10, 00\} = \{1-, -0\}$. In other words, any implementation of R and S must keep for this input vector *either* R at 1 *or* S at 0, since these are the necessary and sufficient conditions for the Rs-latch to keep the value 0 at the output y, as required by the specification. On the other hand, only one solution $R = 0$, $S = 1$ is possible for the input vector 1100 which corresponds to setting the output of the Rs-latch to 1. The Boolean relation solver will find, among others, the two solutions illustrated in Fig. 6.18c and 6.18d: (1) $R = cd + \overline{y}\,\overline{c}$; $S = a$ and (2) $R = cd$; $S = ac$. Solution (2) is not speed-independent and, therefore, only solution (1) will be included as a SIP candidate. Another speed-independent solution, (3) $R = cd$; $S = ac\overline{d}$ (shown in Fig. 6.19f) corresponding to a non-prime cover will be included into a list of SIP candidates by the monotonic cover method. It differs from solution (2) by adding a redundant literal to function S. The monotonic cover method discussed in Sect. 6.2 performs direct search of speed-independent cubes and covers for each excitation region of a signal, assuming a restricted decomposition structure based on a C-element or an Rs-latch. The best solution is selected among (1) and (3) depending on the cost function.

Table 6.2 specifies compatible values of BRs for different types of gates: a C-element, a D-latch, a reset-dominant Rs-latch, a set-dominant Sr-latch,

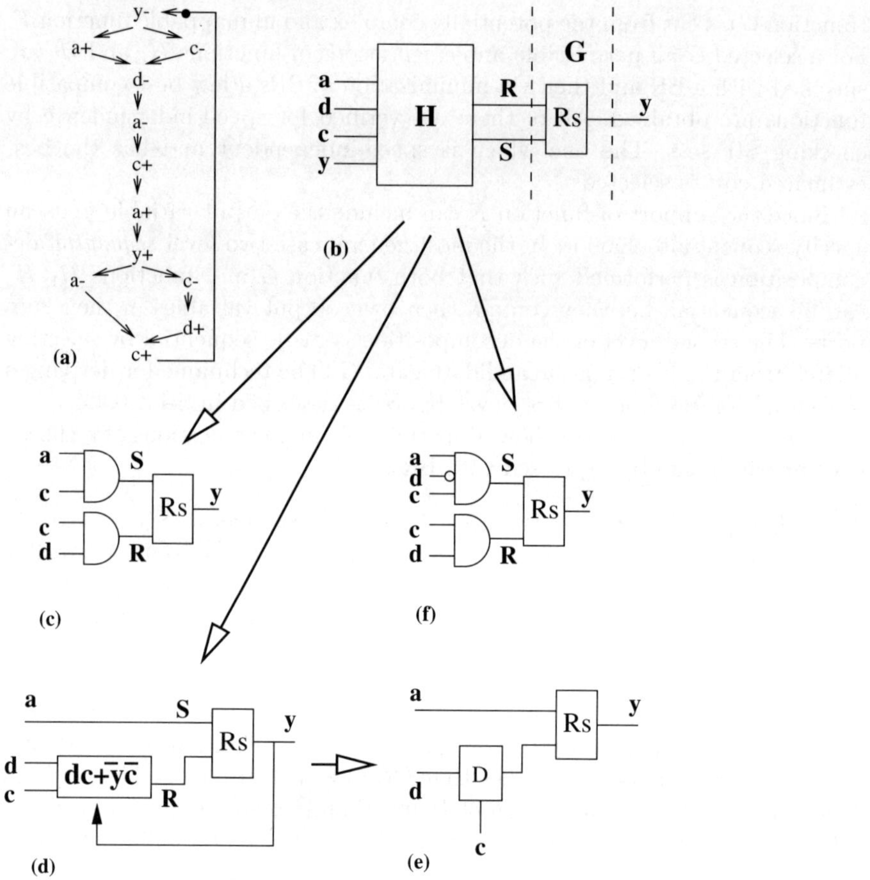

Fig. 6.18a–f. Sequential decomposition for function $y = ac\overline{d} + y(\overline{c} + \overline{d})$

a two input AND gate and a two input OR gate. All states of an SG are partitioned into four subsets, $ER(y+), QR(y+), ER(y-)$, and $QR(y-)$, with respect to signal $y = F(X)$ which is being decomposed. All states that are not reachable in the SG form a DC-set for the BR.

The second column of the first line of Table 6.2 comes from the fact that for each state, $s \in ER(y+)$ only one compatible solution, 11, is allowed for input functions H_1, H_2 of a C-element. This is because the output of a C-element in all states, $s \in ER(y+)$ is at 0 and $F(s) = 1$. Under these conditions the combination 11 is the only possible input combination that implies 1 at the output of a C-element. On the other hand, for each state $s \in QR(y+)$, the output $y = 1$ and $F(s) = 1$, hence it is enough to keep at least one input of a C-element at 1. This is expressed by values $\{1-, -1\}$ in the second column of the second line of the table. Similarly all other compatible values are derived.

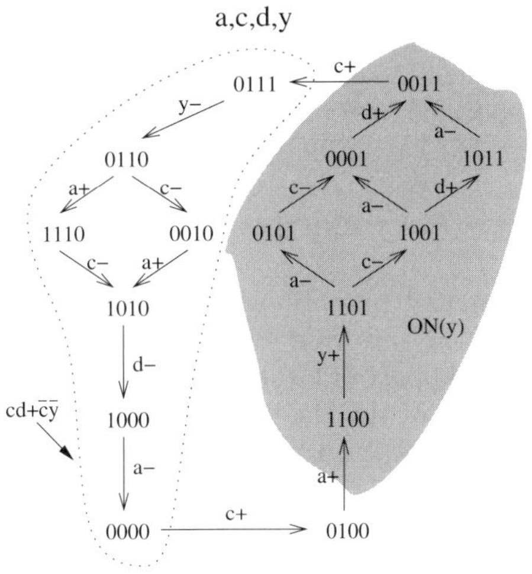

(a)

$acdy$	F	$R\ S$
0000	0	1–,–0
0001	1	0–
0010	0	1–,–0
0011	1	0–
0100	0	1–,–0
0101	1	0–
0110	0	1–,–0
0111	0	1–
1000	0	1–,–0
1001	1	0–
1010	0	1–,–0
1011	1	0–
1100	1	01
1101	1	0–
1110	0	1–,–0
1111	–	–

(b)

Fig. 6.19. (**a**) State graph, (**b**) Decomposition of signal y by an RS latch

Table 6.2. Boolean relations for different gates

Region	C-element H_1, H_2	D-latch C, D	Rs R, S
$ER(y+)$	11	11	01
$QR(y+)$	$\{1-, -1\}$	$\{0-, -1\}$	0–
$ER(y-)$	00	10	1–
$QR(y-)$	$\{0-, -0\}$	$\{0-, -0\}$	$\{1-, -0\}$
unreachable	– –	– –	– –

Region	Sr S, R	AND H_1, H_2	OR H_1, H_2
$ER(y+)$	1–	11	$\{1-, -1\}$
$QR(y+)$	$\{1-, -0\}$	11	$\{1-, -1\}$
$ER(y-)$	01	$\{0-, -0\}$	00
$QR(y-)$	0–	$\{0-, -0\}$	00
unreachable	– –	– –	– –

6.4.3 Functional Representation of Boolean Relations

Given an SG satisfying the CSC requirement, each output signal $y \in X$ is associated with a unique *incompletely* specified function $F(X)$, whose DC-set represents the set of unreachable states. $F(X)$ can be represented by three *completely* specified functions, denoted $ON^y(X)$, $OFF^y(X)$ and $DC^y(X)$ representing the ON-, OFF-, and DC-set of $F(X)$, such that they are pairwise disjoint and their union is a tautology.

Let a generic 2-input gate be represented by a Boolean equation $q = G(z_1, z_2, q)$, where z_1, z_2 are the inputs of the gate, and q is its output[3]. The gate is sequential if q belongs to the true support of $G(z_1, z_2, q)$.

We now give the characteristic function of the Boolean relation for the implementation of $F(X)$ with gate G. This characteristic function represents *all* permissible implementations of $z_1 = H_1(X), z_2 = H_2(X)$ that allow F to be decomposed by G.

$$\begin{aligned} BR^y(X, z_1, z_2) = {} & ON^y(X) \cdot G(z_1, z_2, y) + \\ & OFF^y(X) \cdot \overline{G(z_1, z_2, y)} + \\ & DC^y(X) \end{aligned} \tag{6.1}$$

Intuitively, this equation specifies the relation that associates:

- every X-minterm in the ON-set of signal y with every z_1, z_2, y-minterm that makes $G(z_1, z_2, y) = 1$, and
- every X-minterm in the OFF-set of signal y and every z_1, z_2, y-minterm that makes $G(z_1, z_2, y) = 0$,

In the example of Fig. 6.19b, the minterms $\bar{a}\bar{c}\bar{d}\bar{y}$, $\bar{a}\bar{c}\bar{d}y$ and $acdy$ belong to the OFF-, ON- and DC-set respectively. Thus, the Boolean relation for the function $G(R, S, y) = \overline{R}(S + y)$ will be:

$$\begin{aligned} & BR^y(a, c, d, y, R, S) = \\ & \bar{a}\bar{c}\bar{d}\bar{y} \cdot \overline{\overline{R}(S + y)} + \bar{a}\bar{c}\bar{d}y \cdot \overline{R}(S + y) + \cdots + acdy \; = \\ & \bar{a}\bar{c}\bar{d}\bar{y} \cdot (R + \overline{S}) + \bar{a}\bar{c}\bar{d}y \cdot \overline{R} + \cdots + acdy \end{aligned}$$

Given the characteristic function (6.1), the corresponding table describing the Boolean relation can be derived using cofactors. For each minterm m with support in X, the cofactor BR^y_m gives the characteristic function of all compatible values for $z_1, \ldots, z_n$.

Finding a decomposition of F with gate G is reduced to finding a set of 2 functions $H_1(X), H_2(X)$ such that

$$BR^y(X, H_1(X), H_2(X)) = 1 \tag{6.2}$$

---

[3] As we already mentioned, in the context of Boolean equations representing gates we use the "=" sign to denote "assignment", rather than mathematical equality. Hence q in the left-hand side of this equation stands for the *next* value of signal q while the one in the right-hand side corresponds to its *current* value.

Example 6.4.2. (Example 6.4.1 continued) The SG shown in Fig. 6.19a corresponds to the STG in Fig. 6.18. Let us consider how the implementation of signal y with a reset-dominant Rs-latch can be expressed using the characteristic function of the BR. Recall that the table shown in Fig. 6.19b represents the function $F(a,c,d,y) = ac\overline{d} + y(\overline{c}+\overline{d})$ and the permissible values for the inputs R and S of the Rs-latch. The ON-set, OFF-set, and DC-set of function $F(a,c,d,y)$ are defined by the following completely specified functions:

$$\begin{aligned} ON^y &= y(\overline{c}+\overline{d}) + ac\overline{d} \\ OFF^y &= \overline{y}(\overline{a}+\overline{c}+d) + \overline{a}cd \\ DC^y &= acdy \end{aligned}$$

The set of permissible implementations for R and S is characterized by the following characteristic function of the BR specified in the table. It can be obtained using Equation 6.1 by substituting expressions for ON^y, OFF^y, DC^y, and the function of an Rs-latch, $\overline{R}(S+y)$:

$$\begin{aligned} &BR^y(a,c,d,y,R,S) = \overline{R}Sac\overline{d} + \overline{R}y(\overline{c}+\overline{d}) + \\ &(R+\overline{S})\overline{y}(\overline{a}+\overline{c}+d) + \overline{a}cd(R+\overline{S}\overline{y}) + acdy \end{aligned} \tag{6.3}$$

This function has value 1 for *all* combinations represented in the table of Fig. 6.19b and value 0 for all combinations that are not in the table (e.g. for $(a,c,d,y,R,S) = 000001$). For example, the set of compatible values for $acdy = 0110$ is given by the cofactor

$$BR^y_{\overline{a}cd\overline{y}} = R + \overline{S}$$

which correspond to the terms $1-$ and -0 given for the Boolean relation for that minterm.

Two possible solutions for the equation $BR^y(a,c,d,y,R,S) = 1$ corresponding to Fig. 6.18c and 6.18d are:

$$\begin{aligned} R &= cd \; ; \; S = ac \\ R &= cd + \overline{y}\,\overline{c} \; ; \; S = a \end{aligned}$$

6.4.4 Two-Level Sequential Decomposition

Accurate estimation of the cost of each solution produced by the Boolean relation minimizer is essential in order to ensure the quality of the final result. The minimizer itself can only handle combinational logic, but often (as shown below) the best solution can be obtained by replacing a combinational gate with a sequential one. This section discusses some heuristic techniques that can be used to identify when such a replacement is possible without altering the asynchronous circuit behavior, and without undergoing the cost of a full-blown sequential optimization step. In particular, it discusses how a decomposed combinational function $z = H(X,y)$ can be replaced by a sequential function $z = H'(X,z)$ without changing the behavior of the circuit.

Example 6.4.3. (Example 6.4.1 continued) Let us assume that the considered library contains three-input AND, OR gates and Rs-, Sr- and D-latches. Implementation (1) of signal y by an Rs-latch with inputs $R{=}cd$ and $S{=}ac\overline{d}$ matches the library and requires two AND gates (one with two and one with three inputs) and one Rs-latch. The implementation (2) of y by an Rs-latch with inputs $R{=}cd + \overline{y}\,\overline{c}$ and $S{=}a$ would be rejected, as it requires a complex AND-OR gate which is not in the library. However, when input $\overline{y}$ in the function $cd + \overline{y}\,\overline{c}$ is replaced by signal R, the output behavior of R will not change, i.e. function $R{=}cd + \overline{y}\,\overline{c}$ can be safely replaced by $R{=}cd + R\overline{c}$. The latter equation corresponds to the function of a D-latch and gives the valid implementation shown in Fig. 6.18e.

The technique to improve the precision of the cost estimation step, by partially considering sequential gates, works as follows:

1. Produce permissible functions $z_1 = H_1(X)$ and $z_2 = H_2(X)$ via the minimization of Boolean relations (z_1 and z_2 are always combinational, since $z_1, z_2 \notin X$).
2. Estimate the complexity of H_1 and H_2:
   if H_i matches the library **then** *Complexity* = cost of the gate **else** *Complexity* = literal count
3. Estimate the possible simplification of H_1 and H_2 due to adding signals z_1 and z_2 to their supports, i.e. estimate the complexity of the new pair $\{H_1', H_2'\}$ of permissible functions $z_1 = H_1'(X, z_1, z_2)$, $z_2 = H_2'(X, z_1, z_2)$.
4. Choose the best complexity between H_i and H_i'.

Let us consider the task of determining H_1' and H_2' as in step 3. Let A be an SG encoded by variables from set V and let $z = H(X, y)$, such that $X \subseteq V, y \in V$, be an equation for the new variable $z \notin V$ which is to be inserted in A. The resulting SG is denoted $A'{=}Ins(A, z{=}H(X, y))$ (sometimes we will write simply $A'{=}Ins(A, z)$ or $A'{=}Ins(A, z_1, \ldots, z_k)$ when more than one signal is inserted).

A solution for Step 3 of the above procedure can be obtained by minimizing functions for signals z_1 and z_2 in SG $A' = Ins(A, z_1, z_2)$. However, according to the rule of thumb discussed in Sect. 6.4.1, this is rather inefficient, because the creation of SG A' is computationally expensive. Hence, instead of looking for an exact estimation of complexity for signals z_1 and z_2, a heuristic solution could be used, following the ideas on input resubstitution presented in Example 6.4.3. For the sake of computational efficiency, the formal conditions on input resubstitution should be formulated in terms of the original SG A, rather than in terms of the SG A' obtained after the insertion of new signals. Note that this heuristic estimation covers only the cases when one of the input signals for a combinational permissible function H_i is replaced by the feedback z_i from the output of H_i itself. Other cases can also be investigated, but checking them would be too complex.

Lemma 6.4.1. *Let Boolean function $H(X, y)$ implement the inserted signal z and be positive (negative) unate in y. Let $H'(X, z)$ be the function obtained from $H(X, y)$ by replacing each literal y (or $\overline{y}$) by literal z. The SGs $A'=Ins(A, z=H(X, y))$ and $A''=Ins(A, z=H'(X, z))$ are isomorphic, i.e. have the same states and arcs, iff the following condition is satisfied: $\delta H(X, y)/\delta y \cdot S^* \equiv 0$, where S^* is the characteristic function describing the set of states $ER(z+) \cup ER(z-)$ in A.*

Informally Lemma 6.4.1 states that resubstitution of input y by z is permissible if in all states where the value of function $H(X, y)$ depends on y, the inserted signal z has a stable value.

The intuition behind this lies in the way of constructing SGs A' or A'' when signal z is inserted in SG A. Any state $s \in A$ which belongs to excitation region $ER(z*)$ of z generate two states s' and s'' in A' or A'' ($s' \overset{z*}{\rightarrow} s''$). For A' and A'' to be isomorphic, in both these states the value of functions $H(X, y)$ and $H'(X, y)$ must coincide. Otherwise the enablings of signal z would be different in these states, meaning that either s' or s'' would have different output arcs in A' and A''. However, if the condition of Lemma 6.4.1 is violated, then $H(X, y)$ will keep the same value in both s' and s'', while $H'(X, z)$ will change its value due to the change of z. Hence A', A'' cannot be isomorphic if the conditions of the Lemma are violated.

Example 6.4.4. (Example 6.4.1 continued) Let input R of the Rs-latch be implemented as $cd + \overline{y}\,\overline{c}$ (see Fig. 6.18d). The ON-set of function $H=cd+\overline{y}\,\overline{c}$ is shown by the dashed line in Fig. 6.19a. The input border of H is the set of states by which its ON-set is entered in the original SG A, i.e. $IB(H) = \{0111\}$. By similar consideration we have that $IB(\overline{H}) = \{0100\}$. These input borders satisfy the SIP conditions and hence $IB(H)$ can be taken as $ER(R+)$, while $ER(R-)$ must be expanded beyond $IB(\overline{H})$ by state 1100 so that it does not delay the input transition $a+$ ($ER(R-) = \{0100, 1100\}$).

The set of states where the value of function H essentially depends on signal y is given by the function $\delta H(X, y)/\delta y = a\overline{c}$. H is negative unate in y and cube $a\overline{c}$ has no intersection with $ER(R+) \cup ER(R-)$. Therefore by the condition of Lemma 6.4.1 literal $\overline{y}$ can be replaced by literal R, thus producing a new permissible function $R=cd + R\overline{c}$.

This result can be generalized for binate functions, as follows.

Lemma 6.4.2. *Let Boolean function $H(X, y)$ implement the inserted signal z and be binate in y. Function H can be represented as $H(X, y) = Fy + G\overline{y} + R$, where F, G and R are Boolean functions not depending on y. Let $H'(X, z) = z(F + G) + R$, then SGs $A'=Ins(A, z=H(X, y))$ and $A''=Ins(A, z=H'(X, z))$ are isomorphic iff the following conditions are satisfied:*

(1) $\delta H(X, y)/\delta y \cdot S^* \equiv 0$ (2) $F * G * \overline{R} \cap S^+ \equiv 0$,

where S^* and S^+ are characteristic Boolean functions describing sets of states $ER(z+) \cup ER(z-)$ and $ER(z+)$ in A, respectively.

Proof.
⇒ To prove that SGs $A'=Ins(A, z=H(X,y))$ and $A''=Ins(A, z=H'(X,z))$ are isomorphic under Conditions (1) and (2) we will show that starting from the same initial state all the corresponding states of A' and A'' have the same set of enabled signals. As the construction of SG proceeds by switching enabled signals, the latter clearly means that A' and A'' have the same sets of states and arcs, i.e. are isomorphic in the graph sense.

The only signal function that is different in A' and A'' is the function for signal z. Therefore to prove the isomorphism it is sufficient to show that enablings of signal z in all reachable states of A' and A'' are the same.

Let us consider all possible combinations of values for R, F, G, y and the implied values for $H(X,y), z$ and $H'(X,z)$ in SG A'. They are presented in Table 6.3. Let us analyze the various possible cases.

1. $R = 1, F = G = y = -$ (State $s = 1{-}{-}{-}$), where – stands for any value. Clearly functions $H(X,y)$ and $H'(X,z)$ will exhibit the same behavior in these states of SG A'.
2. $s = 000-$. Then both $H(X,y)$ and $H'(X,z)$ will have value 0 independent of z.
3. $s = 0010$. The implied value of $H(X,y)$ is 1 due to $G\overline{y}$. The implied value of z in SG A' is either 1 or 0*. Let us consider state $s1 = 0011$, adjacent to s by signal y. In $s1$ $H(X,y) = 0$ and therefore $\delta H(X,y)/\delta y = 1$ in both s and $s1$. Then according to Condition 1 z must be stable in s. If $z = 1$ then $H(X,y) = H'(X,z) = 1$.
4. $s = 0011$. In this state z is stable (see the consideration above) and therefore $H(X,y) = H'(X,z) = 0$.
5. For states 0100 and 0101 we can apply considerations similar to items 3 and 4.

Table 6.3. Substitution of input signal in binate function

$RFGy$	$H(X,y)$	Value of z in SG A'	$H'(X,z)$
1—	1	1 or 0*	1
000–	0	0 or 1*	0
0010	1	1 or 0*	1
0011	0	0 or 1*	0
0100	0	0 or 1*	0
0101	1	1 or 0*	1
0110	1	1 or 0*	1
0111	1	1 or 0*	1

6. $s = 0110$. The implied value for $H(X,y)$ is 1. The implied value for z in SG A' is either 1 or $0*$. $s \in F * G * \overline{R}$, then according to Condition 2 z must be stable 1. In such case $H(X,y) = H'(X,z) = 1$.
7. $s = 0111$. Similar to item 6.

Hence by exhaustive consideration of all possible cases, we can conclude that when Conditions (1) and (2) are satisfied the values of $H(X,y)$ and $H'(X,z)$ coincide. This ensures the same enablings of signal z in the states of A' and A''. Therefore A' and A'' are isomorphic.

$\Leftarrow$. Let us consider the consequences of violations of Condition (1) or (2).

1. Assume that Condition (1) is violated. Then in the original SG A there exist two states $s1$ and $s2$ that are different only in the value of signal y, such that at least for one of them $\delta H(X,y)/\delta y \cdot S^* = 1$. I.e. for $s1$ we have $\delta H(X,y)/\delta y = 1$ and $s1 \in ER_A(z+)$ (the case $s1 \in ER_A(z-)$ is similar). From $\delta H(X,y)/\delta y = 1$ it follows that in $s1$ the value of function R ($H(X,y) = Fy + G\overline{y} + R$) should be 0, while one of the functions F or G is at 1.
   From $s1 \in ER_A(z+)$ it follows that in SG A'=$Ins(A, z$=$H(X,y))$ there exist two states $s1'$ and $s1''$ which correspond to $s1$ and are separated by the firing of signal z, $s1' \xrightarrow{z+} s1''$. In both these states the value of functions F, G and R is the same because z is the only signal that changes and F, G and R do not depend on z. This means that $H(X,y)$ also has the same value in both states $s1'$ and $s1''$.
   Let us consider states $s1'$ and $s1''$ in SG A''=$Ins(A, z$=$H'(X,z))$. From $H'(X,z) = z(F+G) + R$ it follows that $H'(X,z)$ has different values in $s1'$ and $s1''$ because of the change of signal z. Hence in the corresponding states of A' and A'' the functions $H(X,y)$ and $H'(X,z)$ have different values that lead to different enablings of signal z. We can conclude that A' and A'' are not isomorphic.
2. Assume that Condition (2) is violated. Then in the original SG A there exists state $s \in ER(z+)$ such that $F * G * \overline{R} = 1$ in s. Consider states s' and s'' that correspond to s in SG A'=$Ins(A, z$=$H(X,y))$, such that $s' \xrightarrow{z+} s''$. In both states $H(X,y)$ has the same value, while $H'(X,z)$ has different values. Arguments similar to those used in the previous case prove that A' and A'' are not isomorphic.

The conditions of Lemma 6.4.2 can be efficiently checked within our BDD-based framework. They require to check two tautologies involving functions defined over the states of the original SG A. This heuristic solution is a trade-off between computational efficiency and optimality. Even though the estimation is still not completely exact (the exact solution requires the creation of A'=$Ins(A,z)$), it allows one to discover and use the implementation of Fig. 6.18e.

6.4.5 Heuristic Selection of the Best Decomposition

Each generated decomposition for an output signal consists of a (possibly sequential) output gate with behavior $y = G(z_1, z_2, y)$ and a set of decomposed functions $z_1 = H_1(X)$ and $z_2 = H_2(X)$. From the selected decomposition, only one of the $H_i(X)$ functions will be chosen for its implementation as a new signal.

Thus, once a set of compatible solutions has been generated for each output signal, the best candidate is selected according to the following criteria (in priority order):

1. At least one of the decomposed functions $H_i(X)$ must be speed-independent. This means that at least one new signal that contributes to decompose an output signal can be inserted.
2. The acknowledgment of the decomposed functions must not increase the complexity of the implementation of other signals (see Sect. 6.4.6).
3. Solutions in which all decomposed functions $H_i(X)$ are implementable in the library are preferred.
4. Solutions in which the complexity of the largest non-implementable function $H_i(X)$ is minimized are preferred. This criterion helps to balance the complexity of the decomposed functions and derive balanced tree-like structures rather than linear ones[4].
5. The estimated savings obtained by sharing a function for the implementation of several output signals is also considered as a second order priority criterion.

Among the best candidate solutions for all output signals, the function $H_i(X)$ with the largest complexity, i.e. the farthest from implementability, is selected to be implemented as a new output signal of the SG.

The complexity of a function is calculated as the number of literals in factored form. In case it is a sequential function and it matches some of the latches of the gate library, the implementation cost is directly obtained from the information provided by the library.

6.4.6 Signal Acknowledgment and Insertion

For each function delivered by the BR solver, an efficient SIP insertion must be found, as described in Sect. 6.3.5. We use a heuristic technique to explore different $ER(x+)$ and $ER(x-)$ sets for each function, then makes the final selection according to the following criteria:

- Sets that are only acknowledged by the signal that is being decomposed (i.e. local acknowledgment) are preferred.

[4] Different criteria, of course, may be used when we also consider the delay of the resulting implementation, since in that case keeping late arriving signals close to the output is generally useful and can require unbalanced trees.

- If no set with local acknowledgment is found, the one with least acknowledgment cost is selected, i.e. with the least number of signals delayed by the new inserted signal.

The selection of the $ER(x+)$ and $ER(x-)$ sets is done independently. The cost of acknowledgment is estimated by considering the influence of the inserted signal x on the implementability of the other signals. The cost can be either increased or decreased depending on how $ER(x+)$ and $ER(x-)$ are selected, and is calculated by incrementally deriving the new SG after signal insertion.

6.5 Experimental Results

The methods for logic decomposition presented in the previous sections have been implemented in the tool petrify and applied to a set of benchmarks called *async_stg*, containing 32 specifications of asynchronous controllers. Only 5 out of the 32 examples were not implementable with 2-input gates, and both Boolean and algebraic decomposition failed to decompose exactly the same examples. Only 2 out of these 5 examples could not be decomposed when attempting to implement them with 3-input gates. These results are significantly better than those reported earlier in [96]. In [96] 18 controllers from the same *async_stg* set were tried, and only 8 of them were successfully decomposed into 2-input gates. In general the obtained improvement is due to allowing global acknowledgment of signals, as discussed about the priority encoder example in Sect. 6.1.

Global-acknowledgment allows the method to effectively decompose complex gates with high fan-in (6 or 7 literals). Fig. 6.20 illustrates this by showing implementations of the circuit *mr1* before and after logic decomposition into 2-input gates.

The results of a comparison between the algebraic and Boolean decomposition techniques are shown in Table 6.4.

The columns "literals/latches" report the complexity of the circuits derived after logic decomposition into 2-input gates. The complexity of each gate is measured as the number of literals required to implement it as a sum-of-product gate, ignoring the cost of input and output inversions. Thus both a 2-input *exor* gate ($a\bar{b} + \bar{a}b$) and a gate implementing function $ab + ac + db + dc = \overline{\bar{a}\bar{d} + \bar{b}\bar{c}}$ are considered to be 4-literal gates. The results obtained by the method presented in Sect. 6.4 ("Br") are significantly better than those obtained by the method presented in Sect. 6.3 ("Alg"). This improvement is mainly achieved because of two reasons:

- The superiority of Boolean methods versus algebraic methods for logic decomposition.

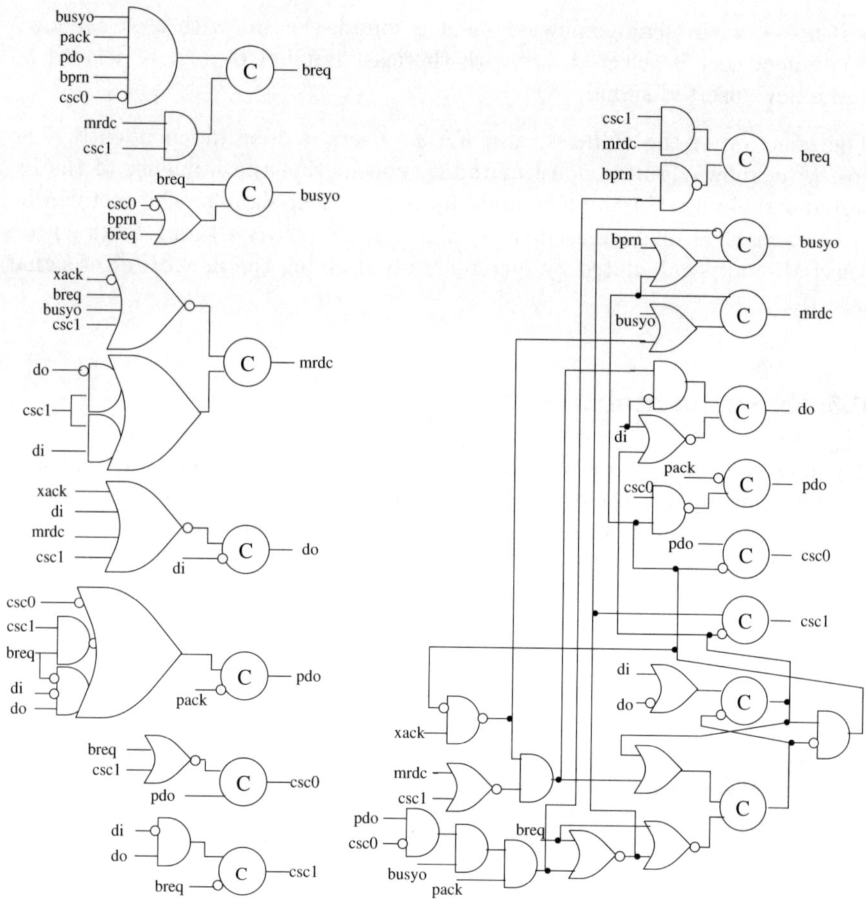

Fig. 6.20. *mr1* before and after logic decomposition into 2-literal gates

- The intensive use of different types of latches to implement sequential functions, with respect to the C-element-based implementation in column "Alg".

Note that the library used for the "Br" experiments was deliberately restricted to D, Sr and Rs latches (i.e. without C-elements, since they are generally not part of standard cell libraries).

Boolean decomposition does not contribute significantly to the decomposability of the circuit, since all the reported examples are also decomposable by using algebraic methods. Thus, we can conclude that Boolean methods mainly affect the quality of the circuits (area and delay) whereas the method for signal insertion with multiple acknowledgment mainly contributes to their decomposability.

Table 6.4. Comparison between Boolean and algebraic decomposition

Circuit	signals I/O	literals/latches Alg	Br	CPU (secs) Alg/Br
chu133	3/4	12/1	10/2	2/4
chu150	3/3	14/2	10/1	2/18
converta	2/3	12/3	8/3	2/14
dff	2/2	12/2	0/2	4/1
drs	2/2	n.a.	0/2	n.a./7
ebergen	2/3	20/3	6/2	2/4
hazard	2/2	12/2	0/2	1/1
half	2/2	2/2	6/2	1/4
mp-forward-pkt	3/5	14/3	14/2	3/31
mr1	4/7	36/9	28/2	126/456
nak-pa	4/6	20/4	18/2	4/441
nowick	3/3	16/1	16/1	3/170
rcv-setup	3/2	10/1	8/1	2/10
sbuf-ram-write	5/7	22/6	20/2	23/696
trimos-send	3/6	36/8	14/10	129/2071
vbe5b	3/3	10/2	10/2	1/10
vbe5c	3/3	4/3	4/3	1/12
Total		252/52	172/36	306/3960

In fact, the improved results obtained by using the Boolean relation method are paid in terms of a significant increase in CPU time (about one order of magnitude).

6.5.1 The Cost of Speed Independence

Table 6.5 is an attempt to evaluate the cost of preserving speed independence during the decomposition of an asynchronous circuit.

The experiments have been done as follows. For each benchmark, the following script has been run in SIS [137], using the library `asynch.genlib`: `astg_to_f; source script.rugged; map`. The resulting netlists could be considered a lower bound on the area of the circuit regardless of its hazardous behavior (i.e. the circuit only implements the correct function for each output signal, without regard to hazards). `script.rugged` is the best known general-purpose optimization script for combinational logic.

The columns labeled `SI` report the results obtained by the method proposed in Sect. 6.4. Two decomposition strategies have been experimented before mapping the circuit onto the library:

- Decompose all gates into 2-input gates (`2 inp`).
- Decompose only those gates that are not directly mappable into gates of the library (`map`).

Table 6.5. The cost of speed-independence

Circuit	Area non-SI			Area SI		
	lib 1	lib 2	best	2 inp	map	best
chu133	224	216	216	216	208	208
chu150	192	160	160	160	168	160
converta	352	312	312	216	224	216
dff	144	96	96	88	88	88
drs	112	112	112	80	80	80
ebergen	184	160	160	160	144	144
hazard	144	120	120	104	104	104
half	184	184	184	154	154	154
mp-forward-pkt	232	232	232	256	256	256
mr1	656	624	624	480	982	480
nak-pa	256	248	248	250	344	250
nowick	248	248	248	232	256	232
rcv-setup	120	120	120	136	128	128
sbuf-ram-write	296	296	296	360	338	338
trimos-send	576	480	480	786	684	684
vbe5b	208	216	208	202	224	202
vbe5c	160	160	160	178	208	178
Total	4288	3984	**3976**	4058	4590	**3902**

There is no clear evidence that performing an aggressive decomposition into 2-input gates is always the best approach for technology mapping. The insertion of multiple-fanout signals offers opportunities to share logic in the circuit, but also precludes the mapper from taking advantage of the flexibility of mapping tree-like structures.

Looking at the best results for non-SI/SI implementations, one can conclude that preserving speed independence does not involve a significant overhead. The experiments show that the reported area is similar. Some benchmarks are even more efficiently implemented by using the SI-preserving decomposition, due to the efficient mapping of functions into latches by using Boolean relations.

These results show that eliminating hazards in an asynchronous circuit by ensuring speed-independence is not costly in terms of area overhead. However the requirement of speed-independence has a significant impact on the effectiveness of decomposition, making some specification non-decomposable in a hazard-free manner.

6.6 Summary

In this chapter we described two main techniques for decomposition of asynchronous logic circuits. Both are based on a three-step approach: (1) candidate selection, (2) decomposition, and (3) library matching.

1. The first step chooses a *candidate* for decomposition.
   a) In the first method, this is done by using the set of algebraic divisors of the target function F.
   b) In the second method, this is done by using Boolean relations. Thus each complex gate F is iteratively split into a two-input combinational or sequential gate G available in the library and two gates H_1 and H_2 that are simpler than F, while preserving the original behavior and speed-independence of the circuit. The best candidates for H_1 and H_2 are heuristically selected for the next step, attempting to minimize the cost in terms of implementability and new signal insertion overhead.
2. The second step performs the actual decomposition.
   a) In the first method, this is done by implementing the candidate function as a new signal x that can be used to re-synthesize the whole circuit. Multiple acknowledgments for x appear automatically at this function generation step and help to guarantee the hazard-freedom of the decomposed function.
   b) In the second method, this is done by implementing H_1 and/or H_2 as new signals into the state graph specification, and re-synthesizing logic from the latter.

   Steps 1 and 2 are iteratively applied to each complex gate that cannot be mapped into the library.
3. The third step, library matching, is used to recover area and delay. It can collapse into a larger library gate the simple 2-input combinational gates that have been (conservatively) used in decomposing complex gates. No violations of speed-independence can arise if the matched gates are atomic.

Especially the second method yields a significant improvement over previously known techniques. This is due to the significantly larger optimization space exploited by using Boolean relations for decomposition and a broader class of latches. In principle, any sequential gate could be used, including, e.g. asymmetric C elements, the only limit being the size of the space to be explored. Furthermore, the ability to implement sequential functions with SR and D latches significantly improves the practicality of the method. Indeed one should not completely rely, as earlier methods did, on the availability of C-elements in a conventional library.

Additionally, the methods proposed in this chapter have both been aimed at area minimization. The fact that both methods generate several candidates for the decomposition of each output signal suggests the possibility of defining a tunable cost function (trading-off area and delay) that could improve the quality of the circuit according to the designer's preferences.

7. Synthesis with Relative Timing

This chapter presents a synthesis approach for asynchronous circuits that takes timing into account. A fundamental problem appears when synthesis with timing is performed: the delays of the components are not known because the system has not been synthesized yet, and the system cannot be synthesized by using timing information because the components of the system are not known yet. This poses a chicken-and-egg problem that can be solved with the notion of *relative timing* and an iterative design flow.

A formal model, *lazy transition systems*, is introduced to represent the timing information required for synthesis. With that information, optimized circuits can be obtained. This chapter also proposes a method for back-annotation that derives a set of sufficient timing constraints that guarantee the correctness of the synthesized circuits.

7.1 Motivation

The theory and methods presented in the previous chapters of this book aim at the synthesis of speed-independent circuits. This means that the timing of the events of the system is determined by causality relations and the correctness of the circuit does not depend on the actual delays of the gates. Even though the delay model for speed-independent circuits is optimistic with regard to the wires, it is highly pessimistic with regard to the gates, since their delays are usually bounded and short.

When time comes into play, as a new dimension in the state space, more aggressive optimizations can be performed. An example was already presented in Sect. 2.6, when a slow environment was assumed for the design of a VME bus controller.

The timed behavior of the circuit can be used to enforce some of the ordered firing of events instead of using explicit causality. This aspect is illustrated in the example of Fig. 7.1. After the occurrence of $y+$, the behavior represented by the STG of Fig. 7.1b is enabled. The delays of the gates are represented in brackets. A delay interval $[d, D]$ indicates that the output of the gate with output signal x will change $\delta(x)$ time units after the gate has been enabled, with $d \leq \delta(x) \leq D$.

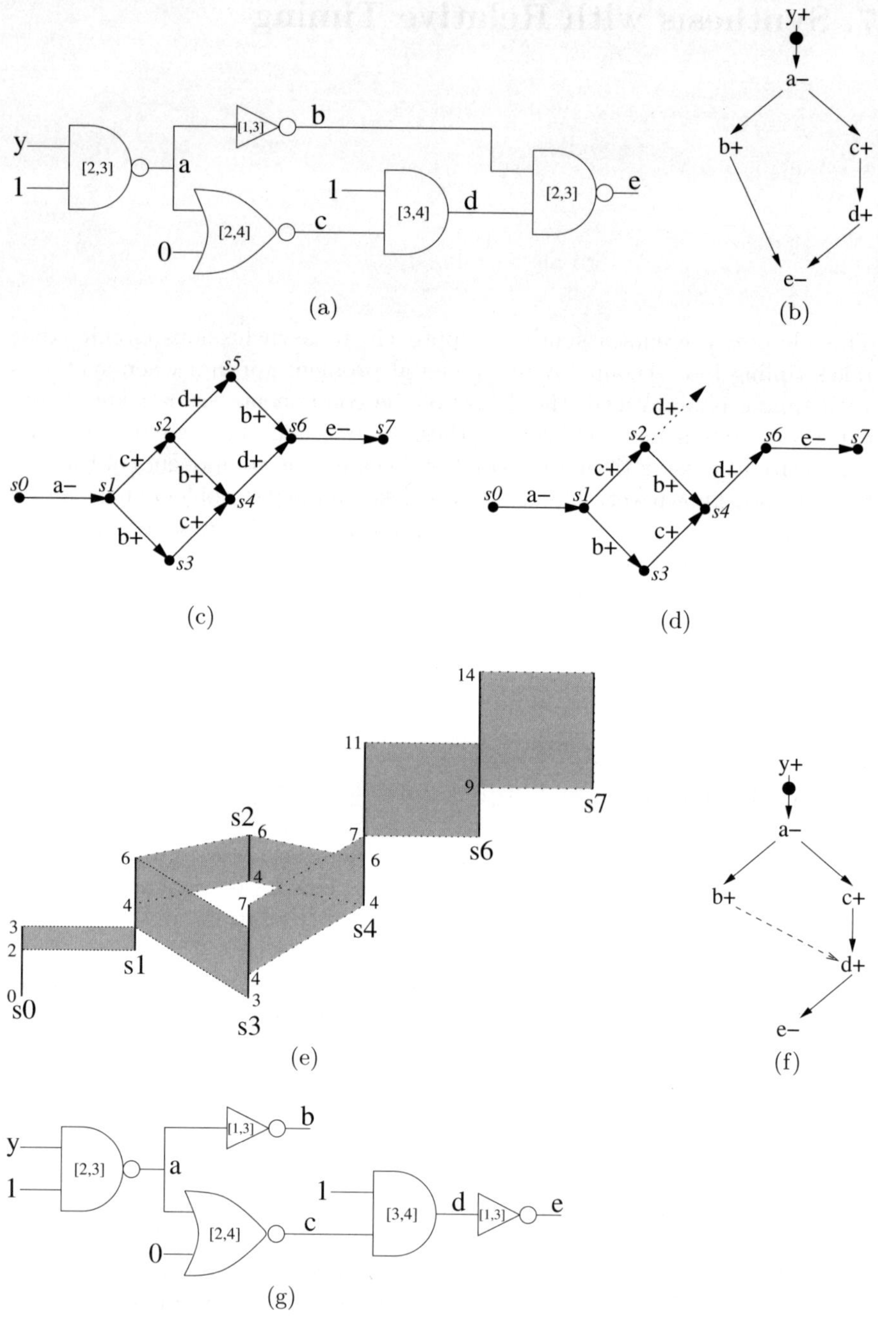

Fig. 7.1. (**a**) Timed circuit, (**b**) STG for the untimed behavior, (**c**) SG for the untimed behavior, (**d**) SG for the timed behavior, (**e**) SG with absolute timing, (**f**) STG for the timed behavior, (**g**) optimized circuit

By analyzing the actual delays of the circuit, one can realize that $b+$ will always occur before $d+$. To prove that, one only needs to show that the earliest time $d+$ can occur is 5 times units after the firing of $a-$, while the latest time $b+$ can occur is 3 time units after the firing of $a-$. Therefore, no time assignment can be found in such a way that $d+$ occurs before $b+$. That makes state $s5$ unreachable (Fig. 7.1d).

The diagram in Fig. 7.1e depicts the state space of the system with absolute timing. The vertical axis represents time. Each discrete state is represented by a vertical bar that indicates the time interval in which that state can occur, assuming that the event $y+$ occurs at time 0. For example, the circuit can be in state $s4$ only in the interval $[4, 11]$, whereas the event $s4 \xrightarrow{d+} s6$ can only fire in the interval $[7, 11]$.

Given that $b+$ will always fire before $d+$, the causality arc $b+ \rightarrow e-$ will be always guaranteed by the actual gate delays and the causality arc $d+ \rightarrow e-$. Hence, a potential optimization of the circuit may consider this phenomenon and ignore the explicit causality $b+ \rightarrow e-$, thus leading to the circuit depicted in Fig. 7.1g.

7.1.1 Synthesis with Timing

Using timing information during synthesis is a common practice. In fact, any non-delay-insensitive circuit assumes some restricted timed behavior. For example, the correctness of speed-independent circuits is based on the fact that some of the wire forks of the circuit are isochronic [43].

In Sect. 1.3.1, several delay models and modes of operation were presented. Their combination determines a classification of different types of circuits, each one considering different timing assumptions to ensure correct behaviors. The circuits we are referring to in this chapter combine the *input/output mode*, as in speed-independent circuits, with the *bounded path delay model* (as discussed in Sect. 1.3.1).

A similar class of circuits has been used in the synthesis approach presented for *timed circuits* [138]. The aim of the approach is similar to the one presented in this chapter: derive optimized circuits by taking timing information into account. The way timing information is handled is different: in [138], absolute timing information about the components of the circuit is used, whereas the approach presented in this chapter uses relative timing information.

7.1.2 Why Relative Timing?

At the initial steps of the synthesis flow, it is difficult to know the timed behavior of a circuit for a variety of reasons:

- Asynchronous specifications are often incomplete and require the addition of state signals, for which no absolute timing information is available.

- Even after state encoding, no absolute timing information about non-input signals of the circuit is known before both technology independent (logic synthesis) and technology dependent (technology mapping) optimizations have been performed. This leads to a chicken-and-egg problem in any method based on absolute timing information: for efficiency, synthesis needs delay bounds, but delay bounds are unknown before synthesis is completed. In timed synthesis this is solved by iterating delay guessing and synthesis without guarantee of convergence.
  Relative timing also may require iteration, but constraints of the kind "*signal a must be faster than signal b*" are generally easier to satisfy than the specific time intervals used by absolute timing methods.
- All modern synthesis flows both for custom and ASIC design include transistor or gate sizing, buffer insertion and selection of parameters (e.g. threshold voltage Vt) with the goal of meeting timing constraints and optimizing different design aspects (power, area, delay, etc.) A netlist can be sized differently depending on a given set of constraints, and the resulting gate delays may differ by an order of magnitude depending on the sizes of devices and other selected parameters.
- Placement and routing may further change absolute delay information associated with circuit elements.

For this reason the use of relative delay information between circuit events follows the established engineering practice of many high-speed circuit design groups (see e.g. design of pulse-domino logic in [139]).

7.1.3 Abstraction of Time

Rather than calculating the exact time intervals in which each state can be visited by any valid run, it is sufficient for synthesis to know whether each state is visited by some time-consistent run and what the enabling conditions for every visited state are. In other words, only *the set of reachable states in the timed domain* and the values of the next-state function for every signal in every reachable state are needed. This information can be represented by abstracting absolute timing out of the model, leading to the definition of a new computational model called a *lazy transition system* [140], in which timing information is only represented by making a distinction between the enabling and the firing of an event.

While absolute timing requires complex techniques to represent the space of reachable timed regions or states [141] (e.g. difference bound matrices, polyhedra, etc.) the generation of the reachable state space for relative timing is of the same complexity as for untimed systems.

Fig. 7.1d represents the lazy transition system associated to Fig. 7.1e. The dashed arc with event $d+$ from state $s2$ indicates that $d+$ is enabled in that state, but it cannot actually fire due to its delay.

This chapter proposes a synthesis flow in which timing information is specified as a set of assumptions that relate the firing order of concurrently enabled events, such as *"event b+ will always fire before event d+"*. Lazy transition systems are used as the computational model for synthesis.

7.1.4 Design Flow

The synthesis flow with relative timing follows the paradigm *"assume and, if useful, guarantee"*. Similar principles have already been used in some asynchronous designs [142, 143, 144, 27]. Given an untimed computational model, e.g. a transition system, the synthesis of an asynchronous circuit is performed as follows (see Fig. 7.2):

1. Provide a specification and a set T of relative timing assumptions, either derived by the user or generated automatically, on the behavior of the system.
2. Synthesize the circuit by using (possibly a subset of) the don't care conditions induced by T.
3. Derive a set C of sufficient timing constraints that guarantee the correctness of the circuit's behavior.
4. Perform transistor sizing and parameter selection to satisfy the set of constraints C (and possibly some other design constraints).
5. If C cannot be satisfied, calculate a less stringent set T and go to step 2.

In step 1, timing assumptions can be either provided by the designer or generated automatically [145]. In the first case, the assumptions typically come from the knowledge of the temporal behavior of the environment, e.g. some of the input events are slow. In the second case, realistic assumptions on the implementation of a circuit can be considered, e.g. the delay of one gate is typically shorter than the delay of two gates.

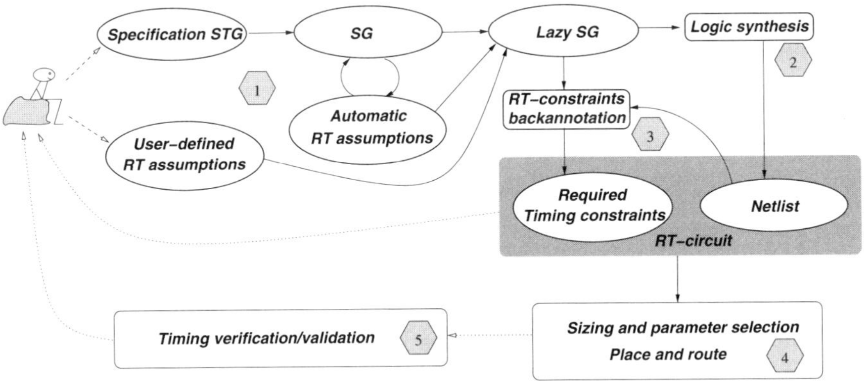

Fig. 7.2. Design flow for synthesis with relative timing

Not all the timing assumptions in T may be needed to improve the quality of the circuits. During synthesis, only a subset of the don't cares induced by them may be used for optimization.

The goal of step 3 is to find a less restrictive set of constraints that guarantee the circuit's correctness. These constraints may not necessarily match the timing assumptions in T.

Once the circuit and the set C have been derived, the designer must guarantee that the required timing constraints are met. This can be achieved, if necessary, by modifying the actual delays of the components, for example by delay padding or transistor sizing.

Finally, step 5 is required to converge in the chicken-and-egg problem when the initial set T of assumptions results in a circuit that cannot meet the set C of constraints.

7.2 Lazy Transition Systems and Lazy State Graphs

In this section the formal model used for synthesis with relative timing is introduced. It is an extension of the transition system model by introducing the notion of laziness.

A *lazy transition system* (LzTS) is a transition system in which there is an explicit distinction between the enabling and the firing of events. Fig. 7.1d is an example of LzTS in which event $d+$ is lazy, since it is enabled in states $s2$ and $s4$ but it can only fire in state $s4$. This distinction enables us to model the delay of events qualitatively, but not quantitatively.

Formally, a lazy transition system is a pair $A = (A', \mathsf{EnR})$, where $A' = (S, E, T, s_{in})$ is a transition system and $\mathsf{EnR} : E \to 2^S$ is a function that defines the *enabling region* of each event. For each event $e \in E$, the *firing region of* e is defined as

$$\mathsf{FR}(e) = \{s \in S \mid s \xrightarrow{e}\}.$$

For any event e of a LzTS, the condition $\mathsf{FR}(e) \subseteq \mathsf{EnR}(e)$ must hold. An event e is said to be *lazy* if $\mathsf{EnR}(e) \neq \mathsf{FR}(e)$. Any state of a LzTS in $\mathsf{EnR}(e) \setminus \mathsf{FR}(e)$ is a state in which the event e is enabled but cannot fire due to the delays associated to the system's events. In the example of Fig. 7.1d we have

$$\mathsf{EnR}(d+) = \{s2, s4\}$$
$$\mathsf{FR}(d+) = \{s4\}$$

The binary interpretation of a LzTS is a *lazy state graph* (LzSG) $G = (A, X, \lambda_S, \lambda_E)$, where A is a LzTS and X, λ_S and λ_E have the same interpretation as in the definition of SG (see Definition 4.1.2).

The concept of *lazy quiescent region* (LzQR) is useful for the synthesis of circuits. It is derived from the concept of quiescent region in TSs. It is defined as follows:

$$\mathsf{LzQR}(a+) = \mathsf{QR}(a+) \setminus \mathsf{EnR}(a-)$$
$$\mathsf{LzQR}(a-) = \mathsf{QR}(a-) \setminus \mathsf{EnR}(a+)$$

where QR refers to the underlying TS.

7.3 Overview and Example

This section gives an intuitive description of the optimizations based on timing assumptions. It is illustrated by an implementation of the xyz specification shown in Fig. 7.3a, that describes an autonomous circuit with three output signals. The starting point is given by the speed-independent implementation shown in Fig. 7.3d.

From here, more precise timing relationships, considering the actual delays required by a signal to propagate through different stages of logic, can be expressed. For example, one can assume that a signal propagates through a single gate faster than through k gates ($k > 1$), where k is an implementation and/or technology dependent parameter. This can be formalized in terms of delay range for gates: if the delay range for all gates is $[\delta_{min}, \delta_{max}]$ then the assumption can be posed as $k * \delta_{min} > \delta_{max}$. Similar assumptions were successfully exploited in [146] for area and performance optimization.

7.3.1 First Timing Assumption

Let us assume that the delay of two gates is always longer than the delay of one gate in the circuit for the xyz example, using a given technology. Under this assumption, even though the transitions $y+$ and $x-$ are potentially concurrent in the STG, $y+$ would always occur *before* $x-$ in a circuit. In the STG, this timing assumption can be expressed by a special *timing arc* going from $y+$ to $x-$ [147] (denoted by a dashed line in Fig. 7.4a). Timing restricts possible behaviors of the implementation. In particular, state 001 becomes *unreachable* because it can only be entered when $x-$ fires before $y+$. In unreachable states, the next state logic functions for all signals can be defined arbitrarily. Therefore, the use of timing assumptions increases the DC-set for output functions, thus giving extra room for optimization.

For the xyz example, moving state 001 into the DC-set of z simplifies its function from $z = x + \overline{y}z$ to a buffer ($z = x$), as shown in Fig. 7.4c and 7.4d. State 101 can be included into the enabling region of $x-$. The selected implementation for signal x, $x = \overline{z}(x + \overline{y})$, is the same as for the untimed specification and corresponds to the $\mathsf{EnR}(x-) = \mathsf{FR}(x-) = \{111\}$. Signal x in this implementation is not lazy and no timing constraints are required. An alternative implementation could have been taken with $x = \overline{y} + x\overline{z}$ corresponding to $\mathsf{EnR}(x-) = \{101, 111\}$ and $\mathsf{FR}(x-) = \{111\}$. It might have shorter latency for $x-$, but requires timing constraint $y+$ before $x-$ to ensure the correct operation of signal x.

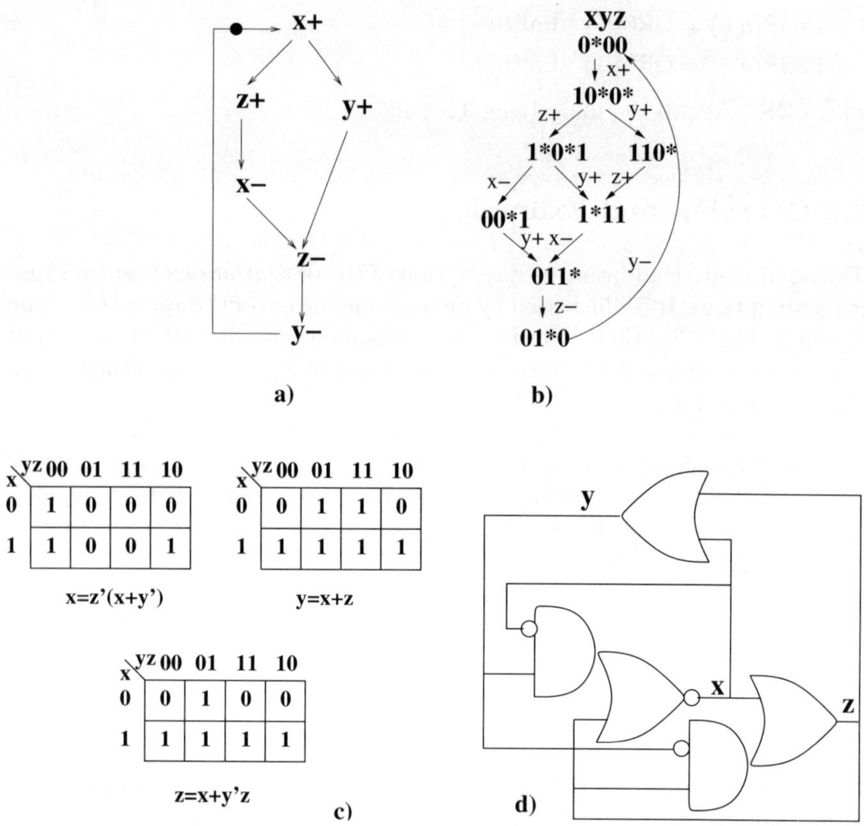

Fig. 7.3. (**a**) STG, (**b**) SG, (**c**) next-state functions, (**d**) complex-gate implementation

7.3.2 Second Timing Assumption

Let us consider the concurrent transitions $z+$ and $y+$. They are triggered by the same event $x+$ and, because of the timing assumption $2 * \delta_{min} > \delta_{max}$, no other gate can fire until both outputs y and z are high. Therefore, for all other signals of the circuit, the difference in firing times of $y+$ and $z+$ is negligible. This means that, from the point of view of the rest of the circuit, the firings of $y+$ and $z+$ are *simultaneous* and *indistinguishable*, and they can replace each other in the causal relations with other events.

In the xyz example, $x-$ is the only transition that is affected by $z+$ or $y+$. The dashed hyper-arc from $\langle z+, y+\rangle$ to $x-$ (see Fig. 7.5a) represents the simultaneity of $y+$ and $z+$ with respect to $x-$. Formally it means that, for the triggering of $x-$, any non-empty subset of the set of events $\{y+, z+\}$ can be chosen. This gives a set of states in which $x-$ can be enabled, $\mathsf{EnR}(x-)$, which is shadowed in Fig. 7.5b.

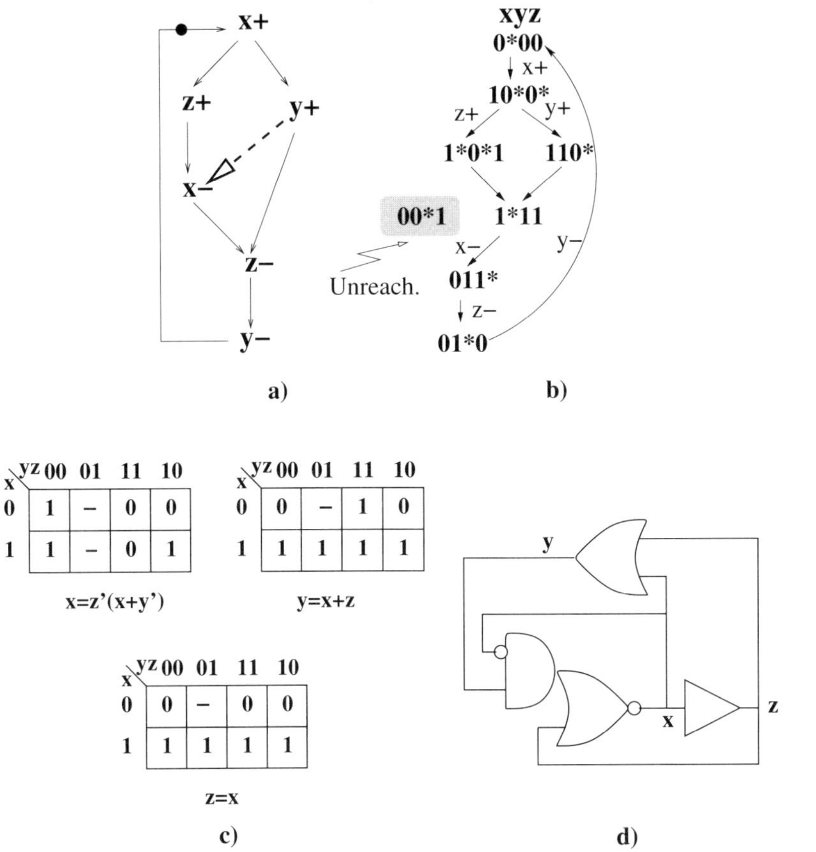

Fig. 7.4a–d. Optimization by using timed unreachable states

It is important to note that:

- Even though $x-$ might be enabled in any state of $\mathsf{EnR}(x-)$, its firing (due to timing assumptions) can occur only after $y+$ and $z+$ have fired. This defines $\mathsf{FR}(x-) = \{111\}$. This behavior is called *lazy* because a signal is not eager to fire immediately after its enabling, but waits until some other events have fired.
- Performance can be slightly affected, either positively or negatively, by the fact that the arrival time of the new trigger signals may be different from the ones in the specification.
- The specified EnR gives an *upper bound* for the set of states in which a signal can be enabled. In a particular implementation, the actual enabling region can be a subset of the specified enabling region. By exploring different subsets, several implementations can be obtained and evaluated according to some given cost criteria (e.g. area and performance).

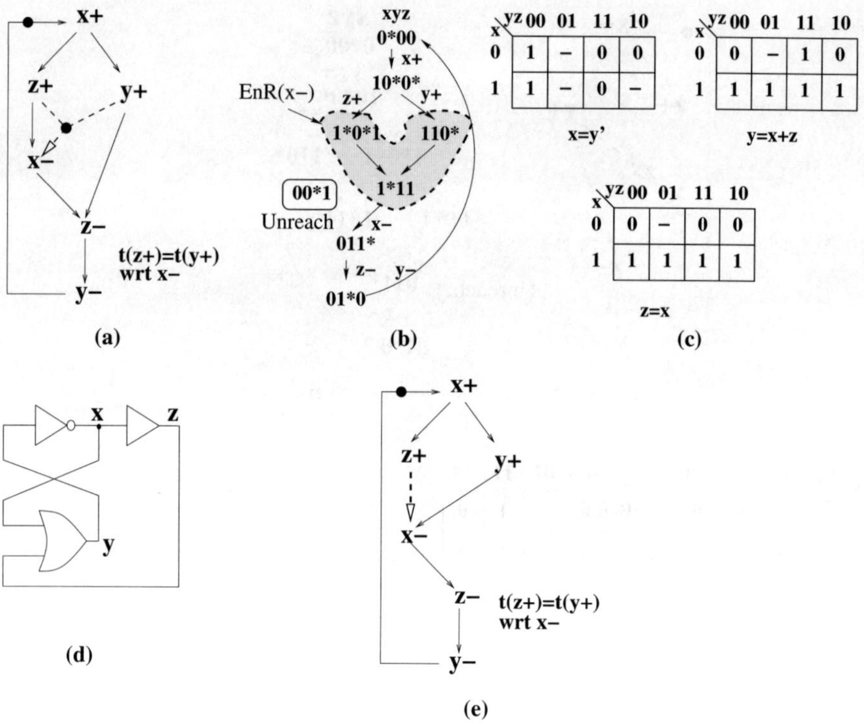

Fig. 7.5a–e. Timing optimizations for the xyz example

7.3.3 Logic Minimization

The enabling region of a signal implicitly results in a set of vertices in the DC-set of the corresponding logic function. For the enabling of $x-$ in the xyz example, different subsets of $\{101, 110, 111\}$ can be chosen. Transition $x-$ fires at state 111, i.e. $\mathsf{FR}(x-) = \{111\}$ and, therefore, any definition of $\mathsf{EnR}(x-)$ should cover state 111, since $\mathsf{FR}(x-) \subseteq \mathsf{EnR}(x-)$. Enabling $x-$ in the other two states 101 and 110 can be chosen arbitrarily, i.e. these states can be moved into the DC-set of the function for x (see Fig. 7.5c). After logic minimization, the function for x, which simply becomes an inverter, is defined to be 0 in state 110 and 1 in 101, i.e. the enabling region corresponding to the implementation is $\mathsf{EnR}(x-) = \{110, 111\}$. The back-annotation of this implementation is shown in the STG of Fig. 7.5e in which $x-$ is triggered by $y+$ instead of $z+$. This change of causal dependencies is valid under the assumption that $y+$ and $z+$ are simultaneous with respect to $x-$, and results in $z+$ firing before $x-$. This is indicated by a timing (dashed) arc.

The timed circuit in Fig. 7.5d is much simpler than the speed-independent one in Fig. 7.3d. Moreover, if just a single timing constraint *"the delay of $z+$ is less than sum of the delays of $y+$ and $x-$"* is satisfied, then the optimized circuit is a correct implementation of the original specification.

Section 7.7 discusses how to derive, from the untimed specification and logic implementation, a *reduced* set of constraints that are sufficient to guarantee its correctness.

7.3.4 Summary

Two potential sources of optimizations based on timing assumptions can now be applied:

1. Unreachability of some states due to timing (*timed unreachable states*).
2. Freedom in choosing enabling regions for signals due to early enabling or simultaneity of transitions (*lazy behavior*).

In both cases, the DC-set for the logic functions increases, thus leading to simpler implementations. Unreachable states provide global don't cares (DC for all next state functions), while lazy enabling provides additional local don't cares (DC for the corresponding lazy signal only).

7.4 Timing Assumptions

Timing assumptions are generally defined in the form of a partial order in the firing of sets of events, e.g. event a fires before event b. However, this form is ambiguous for cyclic specifications because their transitions can be instantiated many times and different instances may have different ordering relations. More rigor can be achieved by unfolding the original specification into an equivalent acyclic description [148]. The theory of timed unfoldings is however restricted to simple structural classes of STGs and the timing analysis algorithms are computationally expensive [149, 150]. This work relies on a more conservative approximation of timing assumptions, as modeled by LzTSs.

We also do not consider explicitly the case of specifications that have multiple instances of the same event, e.g. `lds+/1` and `lds+/2` in Fig. 2.3c, with different causality and concurrency relations. For simplicity in the nomenclature, this work considers that the same timing assumptions are applied to all instances of the same event. Extending the approach to different assumptions for different instances at the STG level is quite straightforward.

Some ordering relations between events of a LzTS $A = ((S, E, T, s_{in}), \mathsf{EnR})$ are first introduced. They are straightforward extensions to LzTSs of similar definitions for TSs given in Sect. 4.2.

Definition 7.4.1 (Conflict). *An event* $e_1 \in E$ disables *another event* $e_2 \in E$ *if* $\exists s_1 \xrightarrow{e_1} s_2$ *such that* $s_1 \in \mathsf{EnR}(e_2)$ *and* $s_2 \notin \mathsf{EnR}(e_2)$. *Two events* $e_1, e_2 \in E$ *are in* conflict *if* e_1 *disables* e_2 *or* e_2 *disables* e_1.

Definition 7.4.2 (Concurrency). *Two events* $e_1, e_2 \in E$ *are concurrent (denoted by* $e_1 \parallel e_2$*) if*

1. $\mathsf{EnR}(e_1) \cap \mathsf{EnR}(e_2) \neq \emptyset$ *and they are not in conflict, and*
2. $\forall s \in \mathsf{FR}(e_1) \cap \mathsf{FR}(e_2)$:
   $(s \xrightarrow{e_1} s_1) \in T \ \wedge \ (s \xrightarrow{e_2} s_2) \in T \implies$
   $\exists s_3 \in S : (s_1 \xrightarrow{e_2} s_3) \in T \ \wedge \ (s_2 \xrightarrow{e_1} s_3) \in T.$

The second condition is the analogue of the non-conflict requirement, but is applied to the FR rather than the EnR. It also requires a "diamond" shaped organization of the FR (sometimes called local confluence).

Definition 7.4.3 (Trigger). *An event* $e_1 \in E$ *triggers another event* $e_2 \in E$ *(denoted by* $e_1 \rightsquigarrow e_2$*) if* $\exists s_1 \xrightarrow{e_1} s_2$ *such that* $s_1 \notin \mathsf{EnR}(e_2)$ *and* $s_2 \in \mathsf{EnR}(e_2)$*.*

This section proposes three types of timing assumptions. Each assumption enables transformation of a LzTS $A = ((S, E, T, s_{in}), \mathsf{EnR})$ into another LzTS $A' = ((S', E, T', s_{in}), \mathsf{EnR}')$ in which the set of events and the initial state remain the same, but there is typically more freedom for logic optimization. In A' enabling regions are defined by EnR' as *upper bounds* of enabling regions in all possible implementations according to the considered timing assumption.

For the formalization of the timing assumptions, the following definitions are also used:

Definition 7.4.4 (Reachable states). *Given a transition system* $A = (S, E, T, s_{in})$*, the set of* reachable states *from state* s *is recursively defined as:*

$$\mathsf{Reach}(s, T) = \{s\} \cup \bigcup_{s \to s' \in T} \mathsf{Reach}(s', T)$$

Henceforth, it is assumed that $S = \mathsf{Reach}(s_{in}, T)$ for any TS.

Definition 7.4.5 (Backward reachable states). *Given a transition system* $A = (S, E, T, s_{in})$*, and two subsets of states* $Y \subseteq X \subseteq S$*, the set of states backward reachable within* X *from* Y *is recursively defined as*

$$\mathsf{BackReach}(X, Y) = Y \cup \bigcup_{s \to s' \in T, s \in X, s' \in Y} \mathsf{BackReach}(X, \{s\})$$

In other words, $\mathsf{BackReach}(X, Y)$ are the states in X that have a path within X to some state in Y.

The previous definitions can be easily extended to LzTSs in a straightforward manner.

7.4.1 Difference Assumptions

Given two concurrent events a and b, a *difference assumption* $b < a$, assumes that b fires earlier than a. Formally, it can be defined through the *maximum separation* $Sep_{max}(b, a)$ between both events [150, 151]. The maximum separation gives an upper bound on the difference between the firing times of b and a. If $Sep_{max}(b, a) < 0$ then b always fires earlier than a.

As discussed in [150, 151], firing times pertain to event *instances* in a given run of a cyclic event-based specifications, rather than to *events*. However, for the sake of simplicity in this section we will always consider cyclic specifications in which for all the instances of pairs of events it is possible to identify pairs of instances that are "directly related" together (e.g. one always causes the other). Thus, following the notation of [150, 151] we will always discuss about the maximum separation of events, while implicitly referring to the maximum separation of such related pairs.

In a LzTS, assumption $b < a$ can be represented by the *concurrency reduction* of a with respect to b. The new LzTS A' is obtained from A as follows:

- Let $C = \mathsf{EnR}(a) \cap \mathsf{EnR}(b)$.
- $T' = T \setminus \{s \xrightarrow{a} s' | s \in \mathsf{BackReach}(\mathsf{FR}(a), C)\}$.
- $S' = \mathsf{Reach}(s_{in}, T')$.
- For any $e \in E$: $\mathsf{EnR}'(e) = \mathsf{EnR}(e) \cap S'$.

C is the set of states in which a and b are both enabled (concurrent). The transformation removes the arcs labeled with event a that start in states from C or states from $\mathsf{EnR}(a)$ preceding C.

All timing assumptions can be formalized by using the notion of event separation. However, intuition on local timing behavior is enough to reason about the assumptions presented in this work.

Let us illustrate the application of a difference assumption $b+ < d+$ in the example of Fig. 7.6a and 7.6b. $C = \{1010\}$ and $\mathsf{BackReach}(\mathsf{FR}(d+), C) = \{1010\}$. Thus, the arc $1010 \xrightarrow{d+} 1011$ is removed from T. After that, the set of states $\{1011, 1001\}$ becomes unreachable. The resulting LzSG is depicted in Fig. 7.6c with lazy event $d+$, for which $\mathsf{FR}(d+) = \{1110, 0110\}$ and $\mathsf{EnR}(d+) = \mathsf{FR}(d+) \cup \{1010\}$.

Difference assumptions are the main source for the elimination of timed unreachable states [138, 152], but they cannot fully express the lazy behavior of signals.

7.4.2 Simultaneity Assumptions

Exploiting simultaneity in transition firings is a key factor in the burst-mode design methodology [153, 154]. Here, the environment is considered to be slow and therefore the skew of delays for output signals is negligible, i.e. output

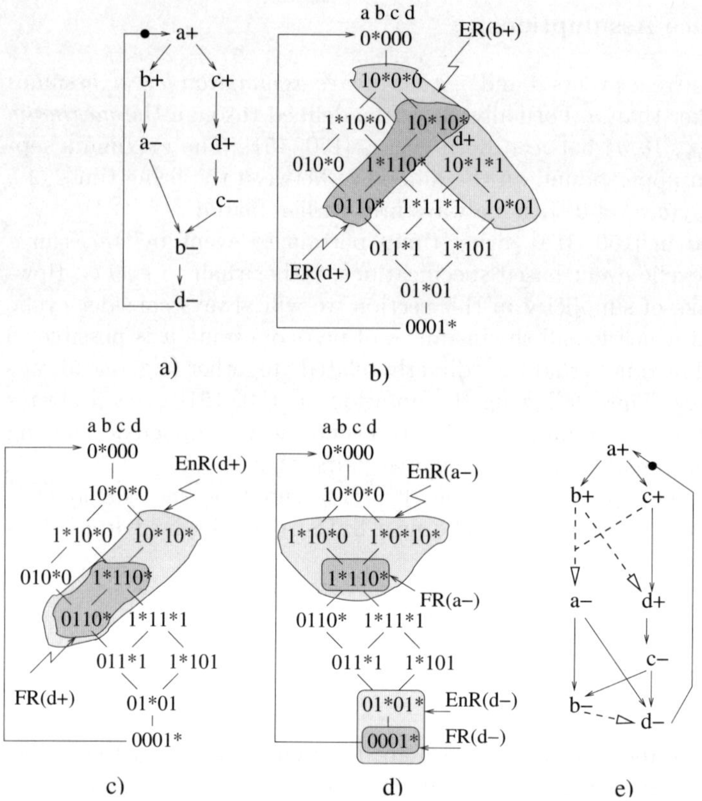

Fig. 7.6. (**a**) STG, (**b**) transition system, (**c**) LzTS after difference assumption, (**d**) LzTS after simultaneity and early enabling assumptions and (**e**) STG with timing assumptions

transitions are *simultaneous* from the point of view of the environment (this is one aspect of the so-called fundamental mode assumption). The weak point of fundamental mode is that it is applied to all circuit inputs and outputs, which essentially relies on an even distribution of propagation delays within the circuit. In order to lift this restriction, simultaneity assumptions can be considered more locally, thus introducing a *local fundamental mode* with respect to particular groups of transitions. The simultaneity assumption is a *relative notion*, which is defined on a set of events $E' = \{e_1, \dots, e_k\}$ with respect to a reference event a, triggered by some of the events in E'. From the point of view of a, the skew in firing times of events in E' is negligible. Formally this can be defined by the following separation inequalities: $\forall e_i, e_j \in E'$, $|Sep_{max}(e_i, e_j)| < \delta_{min}(a)$, where $\delta_{min}(a)$ is a lower bound for the delay of event a from its enabling (i.e. the minimum separation from any of its triggers).

The assumptions are only applicable under the following conditions:

- $\forall e_i, e_j \in E'$, $e_i \parallel e_j$,
- $\exists e \in E' : e \rightsquigarrow a$

Informally, the simultaneity conditions are only applicable when the events in E' are concurrent and at least one of them triggers a.

The new LzTS A' is obtained from A as follows:

- Let $C = \bigcup_{e_i \in E'} \mathsf{EnR}(e_i) \cap \{s | \exists\ s' \xrightarrow{e_j} s\ :\ e_j \in E'\}$.
- $T' = T \setminus \{s \xrightarrow{a} s' | s \in \mathsf{BackReach}(\mathsf{FR}(a), C)\}$.
- $S' = \mathsf{Reach}(s_{in}, T')$.
- $\mathsf{EnR}'(a) = (\mathsf{EnR}(a) \cup C) \cap S'$.
- For any $e \in E, e \neq a$: $\mathsf{EnR}'(e) = \mathsf{EnR}(e) \cap S'$

C is the set of states in which some event in E' has already fired but some other events in E' are still enabled. Let us consider the simultaneity assumption between transitions $b+$ and $c+$ with respect to $a-$, a being an output signal, in the LzSG from Fig. 7.6c. In this case, $C = \{1100, 1010\}$. This assumption influences the LzSG in two ways:

1. State 0100, which is entered when $a-$ fires before $c+$, becomes unreachable. From $|Sep_{max}(c+, b+)| < \delta_{min}(a-)$ (coming from the simultaneity assumption) and $Sep_{max}(b+, a-) < 0$ (coming from the causality between $b+$ and $a-$), the difference assumption $Sep_{max}(c+, a-) < 0$ can be inferred as well.
2. $\mathsf{EnR}(a-)$ is extended to the state 1010 (see Fig. 7.6d).

The second point implies that simultaneity assumptions, and hence the possibility of optimization based on them, are inherently more powerful than difference assumptions only (that capture only the first point).

7.4.3 Early Enabling Assumptions

The simultaneity assumptions exploit "laziness" between concurrent transitions. This idea can be generalized for ordered transitions as well. Assume that event a triggers event b and that the implementation of a is "faster" than that of b (or more formally: $\delta_{max}(a) < \delta_{min}(b)$). Then, the enabling of b could be started simultaneously with the enabling of a, and the proper ordering of a before b would be ensured by the timing properties of the implementation. In the LzTS this would result in the expansion of $\mathsf{EnR}(b)$ into $\mathsf{EnR}(a)$.

Formally, the *early enabling* of event b with respect to a can be applied when $a \rightsquigarrow b$. The new LzTS A' is obtained from A as follows:

- Let $C = \{s | \exists s \xrightarrow{a} s' : s \notin \mathsf{EnR}(b)\ \wedge\ s' \in \mathsf{EnR}(b)\}$.
- $T' = T$.
- $S' = S$.

- $\mathsf{EnR}'(b) = \mathsf{EnR}(b) \cup C$.
- For any $e \in E, e \neq b$: $\mathsf{EnR}'(e) = \mathsf{EnR}(e)$.

The early enabling of $d-$ with respect to $b-$ is illustrated in Fig. 7.6d. All the timing assumptions introduced so far are shown in the STG of Fig. 7.6e, where the dashed arc $(b+, d+)$ corresponds to the difference assumption $b+ < d+$, the hyper-arc $(b + c+, a-)$ corresponds to the simultaneity of $b+, c+$ with respect to $a-$, and the triggering of $d-$ by $a-$ and $c-$ (instead of $b-$) shows the early enabling of $d-$ (the timing arc $(b-, d-)$ is needed to keep the information about the original ordering between $b-$ and $d-$). The transformation for early enabling has been defined only in the case of one backward step, i.e. the implementation of some signal a that triggers b is faster than that of b, and hence b can be enabled at the same time as a and still fire after a purely due to timing. This definition can be generalized for multiple backward steps, i.e. the total delay of the implementations of two signals a and b such that a triggers b and b triggers c is faster than the implementation of c, that can thus be enabled together with a and still fire after b. Of course assumptions going beyond one step are often much less realistic and harder to satisfy.

The above three types of timing assumptions are the cornerstone for timing optimization. Note that difference assumptions are mainly used for removal of the timed unreachable states, while simultaneity and early enabling open a new way for simplifying logic by choosing a particularly useful lazy behavior of the signals.

7.5 Synthesis with Relative Timing

This section presents the theory for the synthesis of hazard-free asynchronous circuits with relative timing assumptions. Lazy transition systems are used as the specification model that incorporates timing. Synthesis is performed on lazy state graphs.

7.5.1 Implementability Properties

The *next-state function* of each output signal of a LzSG, in order to implement it as a circuit, is defined as follows:

$$f_a(z) = \begin{cases} 1 & \text{if } \exists s \in \mathsf{FR}(a+) \cup \mathsf{LzQR}(a+) \text{ s.t. } \lambda_S(s) = z \\ 0 & \text{if } \exists s \in \mathsf{FR}(a-) \cup \mathsf{LzQR}(a-) \text{ s.t. } \lambda_S(s) = z \\ - & \text{otherwise} \end{cases} \tag{7.1}$$

This definition generally gives a larger DC-set than the one presented in Chap. 2 for SGs (see Table 2.1, p. 21) due to two reasons:

- More states are unreachable, since timing assumptions can reduce concurrency and
- States in $\mathsf{EnR}\setminus\mathsf{FR}$ belong to neither FR nor LzQR. Hence, they are included into the DC-set.

For a LzSG to be implementable as a hazard-free circuit, the properties of CSC and output persistency must be extended.

The CSC property holds in a LzSG when f_a is well-defined, that is *if* (but *not only if*) there exists no pair of states (s, s') such that $\lambda_S(s) = \lambda_S(s')$ and $s \in \mathsf{EnR}(a+) \cup \mathsf{LzQR}(a+)$ and $s' \in \mathsf{EnR}(a-) \cup \mathsf{LzQR}(a-)$.

The notion of output persistency (see Definition 4.2.6, p. 73) can also be extended to LzTSs. If a LzTS is output persistent, then all signals are hazard-free using the bounded path delay model (Sect. 1.3.1, [5, 48]) when the bounds satisfy the timing assumptions implied by the LzTS. Output persistency also requires a monotonic behavior between the enabledness and the firing of an event, i.e. any event exiting a firing region must also exit the corresponding enabling region. This is captured by the notion of monotonic cover (the corresponding condition for TSs was discussed in Sect. 6.2).

Definition 7.5.1 (Monotonicity). *Given two sets of states S_1 and S_2 of a TS, S_1 is a* monotonic cover *of S_2 if $S_2 \subseteq S_1$ and for any transition $s \to s'$:*

$$(s \in S_1 \setminus S_2 \implies s' \in S_1) \;\wedge\; (s \in S_2 \implies s' \notin S_1 \setminus S_2)$$

Intuitively, once S_1 is entered, the only way to leave it is via a state in its subset ("exit border") S_2. In the SG of Fig. 7.3b, the set $\{101, 110, 111\}$ is a monotonic cover of $\mathsf{FR}(x-)$. However, the set $\{100, 101, 111\}$ is not, since the transition $100 \xrightarrow{y+} 110$ violates the conditions for monotonicity.

Definition 7.5.2 (Persistency). *Given a LzTS $A = (A', \mathsf{EnR})$ with $A' = (S, E, T, s_{in})$, an event $e \in E$ is persistent if e is persistent in A' and $\mathsf{EnR}(e)$ is a monotonic cover of $\mathsf{FR}(e)$.*

Intuitively, persistency in LzTS indicates that once $\mathsf{EnR}(e)$ has been entered, it can only be exited by firing e, as in Definition 4.2.6. Moreover, persistency in A' indicates that no transition can switch an event from fireable (in $\mathsf{FR}(e)$) to only enabled (in $\mathsf{EnR}(e) \setminus \mathsf{FR}(e)$).

Thus, a LzSG is implementable as a hazard-free circuit assuming delay bounds on some of its paths if the following properties, extended to LzSGs, hold:

- consistency,
- complete state coding,
- output persistency.

These conditions are an extension to circuits with inputs and relative timing of the semimodularity conditions used by Muller to guarantee hazard-freedom for autonomous circuits with unbounded delays [103, 94, 43], and discussed in detail in the previous chapters.

7.5.2 Synthesis Flow with Relative Timing

The flow for logic synthesis with relative timing assumptions is as follows:

1. Define a set of timing assumptions on a TS A and derive a specification LzTS $A_T = (A', \mathsf{EnR}_T)$ according to the defined assumptions. For example, this permits the transformation of the TS in Fig. 7.3b to the LzTS in Fig. 7.5b.
2. Insert state signals for resolving CSC conflicts and thus making an LzSG implementable. In this synthesis flow, state encoding can be automatically solved by using an extension of the method presented in Chap. 5. The differences are:
   - Only those encoding conflicts reachable in the timed domain are considered in the cost function (no effort is invested in solving unreachable conflicts).
   - Timing assumptions can be generated for inserted state signals using the rules introduced in Sect. 7.6, thus implying that the events of new inserted signals can also be lazy.
3. Derive another implementation LzTS $A_I = (A', \mathsf{EnR}_I)$ in which the implementability conditions hold and $\mathsf{EnR}_I(e) \subseteq \mathsf{EnR}_T(e)$ for any event e. A_T is the LzTS that defines the upper bounds on the EnRs of the events, i.e. how early each event can be enabled without firing. A_I defines a particular implementation in which the enabling of each event cannot be earlier than the one defined by A_T. The method for defining A_I from A_T is based on logic minimization and is explained in Sect. 7.5.3.
4. Derive a circuit implementation for the corresponding LzSG according to the logic functions defined by Equation 7.1.
5. Back-annotate a set of sufficient timing constraints required for the correctness of the implementation.

Steps 3 and 4 are discussed in Sect. 7.5. Steps 1 and 5 are presented in Sect. 7.6 and 7.7, respectively. Step 2 is not discussed in more detail, since the basic theory is similar to that for speed-independent circuit synthesis presented in Chap. 5.

In the example of Fig. 7.5b, the only lazy event is $x-$. For signal x, the following regions are defined:

$$\mathsf{EnR}_T(x+) = \{000\}; \quad \mathsf{LzQR}_T(x+) = \{100\}$$
$$\mathsf{EnR}_T(x-) = \{101, 110, 111\}; \quad \mathsf{LzQR}_T(x-) = \{011, 010\}$$

For the circuit in Fig. 7.5e, the corresponding A_I fulfills the properties for implementability and has the following regions for signal x:

$$\mathsf{EnR}_I(x+) = \{000\}; \quad \mathsf{LzQR}_I(x+) = \{100, 101\}$$
$$\mathsf{EnR}_I(x-) = \{110, 111\}; \quad \mathsf{LzQR}_I(x-) = \{011, 010\}$$

7.5.3 Synthesis Algorithm

The timing assumptions on the behavior of the circuit and the environment can be specified by the designer or generated automatically (see Sect. 7.6). Two types of assumptions are considered:

- $\tau(a) < \tau(b)$, indicating that event a will occur before event b. In case both events are concurrent, it corresponds to a difference assumption. In case a triggers b, it corresponds to the early enabling of b with respect to a.
- $\tau(a) \simeq \tau(b)$ *wrt* c, indicating that the firing of a and b can be considered simultaneous with regard to c (simultaneity assumption).

In the example of Fig. 7.5, the following assumptions have been specified for optimization:

$$\tau(y+) < \tau(x-) \quad \text{and} \quad \tau(y+) \simeq \tau(z+) \text{ wrt } x-$$

```
ON = FR(x+) ∪ LzQR_T(x+);
OFF = FR(x−) ∪ LzQR_T(x−);
repeat
  C(x) = Boolean_minimization(λ_S(ON), λ_S(OFF));

  /* Make the ON cover monotonic*/
  EnR_I(x+) = λ_S^{-1}(C(x)) ∩ EnR_T(x+);
  H_on = {s ∈ EnR_I(x+) | ∃s → s' : s' ∈ EnR_T(x+) \ EnR_I(x+)};
  OFF = OFF ∪ H_on;

  /* Make the OFF cover monotonic*/
  EnR_I(x−) = λ_S^{-1}(¬C(x)) ∩ EnR_T(x−);
  H_off = {s ∈ EnR_I(x−) | ∃s → s' : s' ∈ EnR_T(x−) \ EnR_I(x−)};
  ON = ON ∪ H_off;
until (H_on = ∅) ∧ (H_off = ∅);
```

Fig. 7.7. Algorithm for logic synthesis of output signal x

The algorithm for the synthesis of each output signal x is shown in Fig. 7.7, in which the definition of λ_S has been extended to sets of states and Boolean vectors as follows:

$$\lambda_S(X) = \{\lambda_S(s) \mid s \in X\}$$
$$\lambda_S^{-1}(Y) = \{s \in S \mid \lambda_S(s) \in Y\}$$

The algorithm takes an LzTS, A_T, as input and generates another LzTS, A_I, and a logic function $\mathsf{C}(x)$ for each output signal, according to the design flow described in Sect. 7.5.2. In case each function $\mathsf{C}(x)$ is implemented as a complex gate, the circuit is guaranteed to be hazard-free under the given timing assumptions.

This heuristic algorithm calculates EnR_I iteratively until a monotonic cover is found. Initially, ON and OFF are defined in such a way that the states in $\mathsf{EnR}_T(x+) \setminus \mathsf{FR}(x+)$ and $\mathsf{EnR}_T(x-) \setminus \mathsf{FR}(x-)$ are not covered, i.e. their binary codes are in the DC-set. Boolean minimization is invoked by defining the ON-set and the OFF-set, and a completely specified function is obtained. Next, monotonicity of $\mathsf{C}(x)$ is checked. H_{on} is the set of states in $\mathsf{EnR}_I(x+)$ covered by $\mathsf{C}(x)$ that lead to another state in $\mathsf{EnR}_I(x+)$ not covered by $\mathsf{C}(x)$. These states are removed from $\mathsf{EnR}_I(x+)$ for the next iteration. The loop converges monotonically to a valid solution bounded by the case $\mathsf{EnR}_I(x+) = \mathsf{FR}(x+)$. A similar procedure is performed on the complement of $\mathsf{C}(x)$ for $\mathsf{EnR}_I(x-)$. Thus, the DC-set is reduced at each iteration of the algorithm to enforce the monotonicity of the cover. This reduction is illustrated in Fig. 7.8.

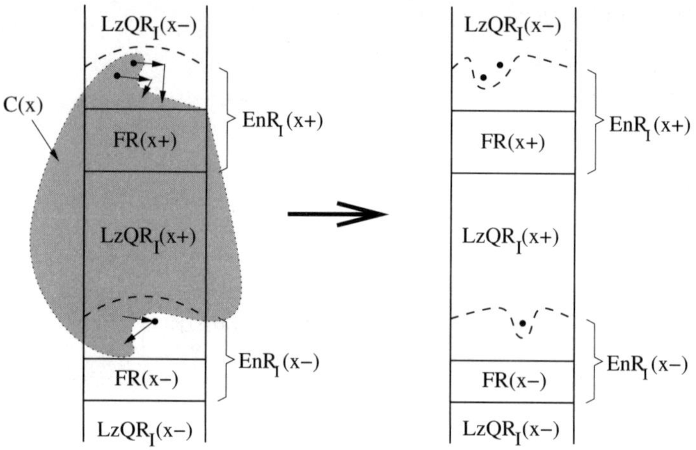

Fig. 7.8. Iteration to reduce EnR_I for non-monotonic covers

In practice, most covers $\mathsf{C}(x)$ are monotonic after the first Boolean minimization and no iteration is required. Only in some rare cases, more than two iterations are executed.

The algorithm in Fig. 7.7 generates a netlist of complex gates based on the functions $C(x)$ obtained by the minimization procedure. This algorithm can be easily extended to the synthesis of asynchronous circuits with generalized C-elements and Set/Reset functions, $S(x)$ and $R(x)$, corresponding to the enabling of $x+$ and $x-$ respectively.

7.6 Automatic Generation of Timing Assumptions

Synthesis by relative timing assumptions is an effective technique to perform circuit optimizations aiming at area and delay reductions. However, the technique may lose most of its appeal if the derivation of timing assumptions is a tedious task that requires a significant amount of effort in the analysis of the specification and the circuit.

On the other hand, the user can reasonably provide timing assumptions only for the signals that are present in the original specification of the circuit. Once synthesis starts introducing new signals to satisfy CSC or to decompose gates, the user cannot take part in the derivation of timing assumptions for them.

This section poses and attempts to positively solve the following question: can the derivation of relative timing assumptions be automated?

However, when talking about timing assumptions one may also ask: can we derive assumptions that are at the same time effective and "realistic"? The method proposed in this section relies on a simplified delay model for the circuit. This model is used as a starting point to perform a coarse timing analysis and derive assumptions. The model also assumes, as discussed in Sect. 7.1, that the designer can tune the final circuit to guarantee the timing constraints required for the correctness of the circuit. In the worst case, the designer can always reject those assumptions that cannot be met and re-synthesize the circuit. Let us illustrate this method with an example.

Fig. 7.9a and 7.9b depict a Petri net and the corresponding transition system. Let us assume that the events represent signal transitions. Let us analyze the fragment of the Petri net that specifies the behavior of events b, c and e. A possible circuit implementing that behavior could resemble the one presented in the upper picture of Fig. 7.10, in which the gate implementing c is more complex than the gate implementing e.

If we work with a design flow in which the delay of any gate can be changed by delay padding or transistor sizing, we can derive a set of assumptions that are reasonable to satisfy, and that can be used to improve the delay of the circuit. Let us consider, for example, the assumption that c will occur before e after having fired b. This assumption reduces the state space and might produce a different implementation, such as the one depicted in the lower picture of Fig. 7.10. But that implementation, in which the gates for c and e are now similar, requires c to occur before e. Therefore, the designer must ensure this behavior by analyzing the circuit and, if necessary, modifying it to enforce the correct ordering for c and e. In this example, the modification of the circuit can speed up signal c and slow down signal e. This in turn could affect either positively or negatively the overall performance of the system.

This section presents a method for automatic generation of relative timing assumptions. First, ordering relations between events are defined. Then, the intuition behind this method is explained using a simple delay model for input and non-input events and rules for deriving timing assumptions are given.

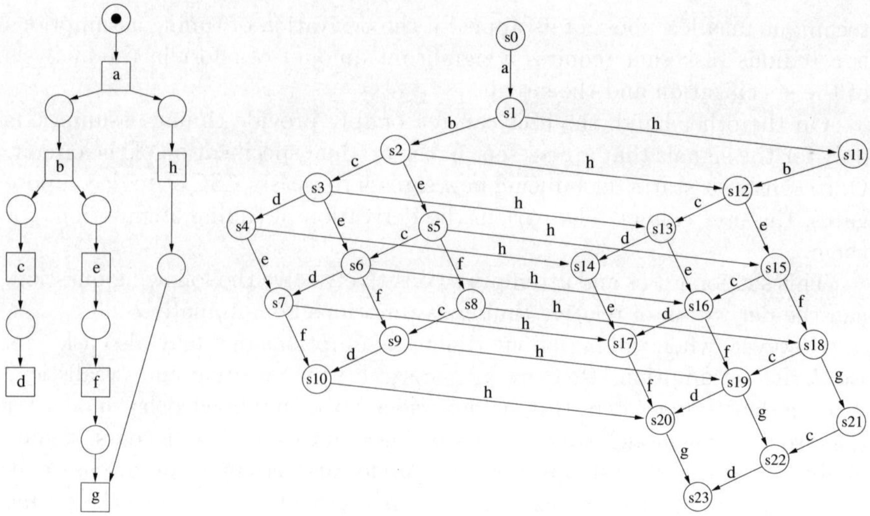

Fig. 7.9. (a) Petri net, (b) Transition System

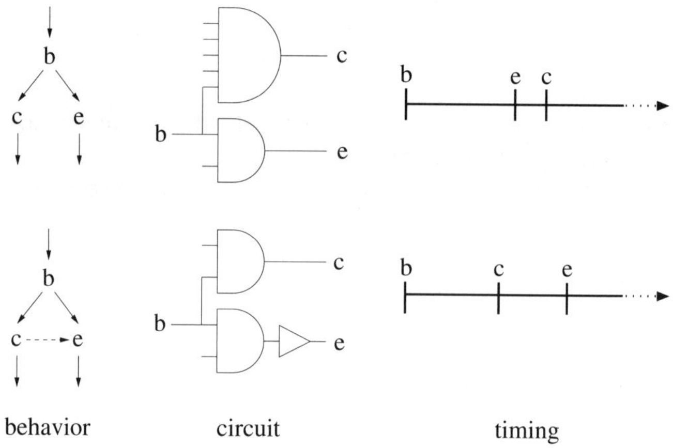

Fig. 7.10. Optimization by using difference timing assumption

7.6.1 Ordering Relations

Let $A = ((S, T, E, s_{in}), \mathsf{EnR})$ be a lazy transition system.

Definition 7.6.1 (Enabled before). *Let $e_1, e_2 \in E$ be two concurrent events. We say that e_1* can be enabled before[1] *e_2 (denoted by $e_1 \lhd e_2$) if $\exists s_1 \to s_2$ such that $s_1 \in \mathsf{EnR}(e_1) \setminus \mathsf{EnR}(e_2)$ and $s_2 \in \mathsf{EnR}(e_1) \cap \mathsf{EnR}(e_2)$.*

[1] We say "can be" because different occurrences of e_1 can be both before and after e_2. This definition is concerned only with the existence of the former.

Definition 7.6.2 (Enabled simultaneously). *Let $e_1, e_2 \in E$ be two concurrent events. We say that e_1 and e_2* can be enabled simultaneously *(denoted by $e_1 \Diamond e_2$) if $\exists s_1 \rightarrow s_2$ such that $s_1 \notin \mathsf{EnR}(e_1) \cup \mathsf{EnR}(e_2)$ and $s_2 \in \mathsf{EnR}(e_1) \cap \mathsf{EnR}(e_2)$.*

The following definition is an extension of Definition 7.6.1 to sets of events. For a proper understanding, some intuition is required. It is helpful to model the situation in which an event e is much slower than another set of events X and e is never enabled before any of the events in X. This situation occurs in systems in which the input events (environment) are much slower than the output events (see Sect. 7.6.3). The expected behavior is, thus, that the input event fires after all the output events. The definition itself, however, is concerned with the opposite case, in which an event *can be enabled* before a set of events X, and, therefore, it describes the conditions when timing optimization *cannot be applied*.

Definition 7.6.3 (Enabled before a set of events). *Let $e \in E$ be an event pairwise concurrent with all the events in the set $X = \{e_1, \ldots, e_n\} \subset E$. We say that e* can be enabled before *X (denoted by $e \lhd X$) if $\exists s_1 \xrightarrow{e'} s_2$ such that $s_1 \in \mathsf{EnR}(e) \setminus \mathsf{EnR}(X)$, $s_2 \in \mathsf{EnR}(e) \cap \mathsf{EnR}(X)$ and $e' \notin X$, where $\mathsf{EnR}(X) = \mathsf{EnR}(e_1) \cup \ldots \cup \mathsf{EnR}(e_n)$.*

Let us call $\mathsf{EnR}(X)$ the union of all enabling regions of the events in X. Since e is concurrent with all events in X, then $\mathsf{EnR}(e) \cap \mathsf{EnR}(X)$ is not empty.

Now assume that e is a slow event (e.g. from the environment). Assume that all events in X are internal/output events that are very fast, as in Fundamental Mode asynchronous circuits [5] (but *localized* to a set of events). If we know that e is never enabled before entering $\mathsf{EnR}(X)$, then we know that all events in X will fire before e. This even considers the possibility that the events in X have causality relations among them.

Fig. 7.9b depicts the transition system derived from the Petri net of Fig. 7.9a. Events a and b are not concurrent, since $\mathsf{EnR}(a) = \{s_0\}$ and $\mathsf{EnR}(b) = \{s_1, s_{11}\}$ are disjoint. Events c and f are concurrent. Moreover, c can be enabled before f since there is a transition $s_2 \rightarrow s_5$ such that $s_2 \in \mathsf{EnR}(c) \setminus \mathsf{EnR}(f)$ and $s_5 \in \mathsf{EnR}(c) \cap \mathsf{EnR}(f)$. However, f cannot be enabled before c. Events d and f are also concurrent and they can be enabled before each other (see transitions $s_3 \rightarrow s_6$ and $s_5 \rightarrow s_6$). Events c and e are also concurrent but none can be enabled before each other, i.e. they are always enabled simultaneously.

Let us now analyze the enabling relation of event d with some sets of events, to show that it is non-obvious. Event d cannot be enabled before $\{e, f\}$ but can be enabled before $\{e, f, g\}$ since there is a transition $s_9 \xrightarrow{h} s_{19}$ such that $s_9 \in \mathsf{EnR}(d) \setminus \mathsf{EnR}(\{e, f, g\})$, $s_{19} \in \mathsf{EnR}(d) \cap \mathsf{EnR}(\{e, f, g\})$ and $h \notin \{e, f, g\}$. On the other hand, d cannot be enabled before $\{e, f, g, h\}$. In other words, if we control the delay of $\{e, f\}$, then we can make sure that they fire faster than d. On the other hand, if only g is added to the "controllable delay" set, then we can no longer ensure faster firing, since g may be slowed down by h. Finally, if we add h to that set, then we can make all events in it faster than d again.

7.6.2 Delay Model

This section presents a *very simple* delay model for events of a TS that gives an intuitive motivation for the automatic generation of timing assumptions. A simple delay model is needed, similar to the literal count in combinational logic synthesis, that can be computed *before* deriving a logic implementation and that allows us to bootstrap the timing optimization process. The model, although simple, generates reasonable timing assumptions that can be satisfied by gate selection or transistor sizing. This fact will be shown by comparing with manual designs in Sect. 7.8. This delay model can be changed depending on the design requirements.

The delay $\delta(e)$ of an event e is defined, as before, as the difference between its enabling time and its firing time. Three types of events are considered [2]:

- **Non-input events**: their delay is in the interval $[1-\varepsilon, 1+\varepsilon]$
- **Fast input events**: their delay is in the interval $(1+\varepsilon, \infty)$
- **Slow input events**: their delay is in the interval $[\Delta, \infty)$

In this context, ϵ denotes the maximum allowed delay variation of each event with respect to a unit delay. The synthesis approach also assumes that:

- the delay of a gate implementing a non-input event can be increased to be larger than that of another gate by delay padding or transistor sizing.
- the delay of two gates can always be made longer than the delay of one gate. Hence, this imposes the constraint that $\varepsilon < 1/3$.
- the difference between the delays of two gates can be made shorter than the delay of one gate.
- the circuit will never take longer than Δ time units (minimum delay of a slow input event) in becoming stable from any state of the system assuming a quiescent environment (no input events firing).

The previous assumptions on the timing behavior of the circuit can be translated into assumptions on the firing order of the events.

7.6.3 Rules for Deriving Timing Assumptions

Rules for deriving timing assumptions are presented in the following format:

[2] "Very fast" input events that are not slower than some internal events can be considered as well and treated more or less like non-input events. This consideration is omitted here for simplicity.

- *Ordering relations*: ordering relations that must be satisfied in a LzTS for a rule to be applied.
- *Timing assumption*: a timing assumption that can be generated automatically.
- *Justifying delay assumptions*: Informal justification of a rule based on the above delay model.

Assumptions Between Non-input Events. Assume that $e_1, e_2, e_3 \in E$ are non-input events. The first three rules apply when events e_1 and e_2 are concurrent. The fourth one applies when e_1 triggers e_2. The following rules can be applied for deriving timing assumptions between non-input events:

I. Event enabled before another event.
- *Ordering relations*: $(e_1 \parallel e_2) \wedge (e_1 \lhd e_2) \wedge (e_2 \ntriangleleft e_1) \wedge (e_1 \not\Diamond e_2)$.
- *Difference timing assumption*: e_1 fires before e_2
- *Justifying delay assumptions*: the delay of one gate can be made shorter than the delay of two gates.

II. Events simultaneously enabled.
- **Ordering relations**: $(e_1 \parallel e_2) \wedge (e_1 \Diamond e_2) \wedge (e_2 \ntriangleleft e_1)$.
- **Difference timing assumption**: e_1 fires before e_2
- **Justifying delay assumptions**: the delay of the gate implementing e_2 can be made longer than the delay of the gate implementing e_1.

III. Event triggered by events simultaneously enabled.
- **Ordering relations**: $(e_1 \parallel e_2) \wedge (e_1 \ntriangleleft e_2) \wedge (e_2 \ntriangleleft e_1) \wedge [(e_1 \rightsquigarrow e_3) \vee (e_2 \rightsquigarrow e_3)]$.
- **Simultaneity timing assumption**: e_1 and e_2 are simultaneous with respect to e_3.
- **Justifying delay assumptions**: the difference between the delays of two gates can be made shorter than the delay of one gate.

IV. Early enabling for ordered events.
- **Ordering relations**: $(e_1 \rightsquigarrow e_2)$.
- **Early enabling timing assumption**: e_1 fires before e_2 (but e_2 can be enabled concurrently with e_1).
- **Justifying delay assumptions**: the delay of the gate implementing e_1 can be made shorter than the delay of the gate implementing e_2.

Let us illustrate the previous cases with the example of Fig. 7.9. Let us assume that all events are non-input. Timing assumptions of type I can be derived for the pairs of events (c, f), (c, g) and (e, d), where the first element of the pair is assumed to fire before the second.

Timing assumptions of type II can be applied to the pairs (b, h) and (c, e). Note that in both cases, the enabling conditions are symmetric, i.e. both events are always enabled simultaneously. However, only one firing order can be chosen by assuming that one of the events can be delayed by increasing the delay of its corresponding gate. This choice can be done heuristically by considering different implementation factors. For example, the choice of one specific firing order may make some states with encoding conflicts unreachable. Another possible heuristic would be to estimate the complexity of the logic for each event. If the gate corresponding to one event is more complex than the other, it can be assumed that the former will be slower than the latter (thus avoiding delay padding to meet the timing assumption).

Timing assumptions of type III can be applied to the events triggered by the pairs (b, h) and (c, e). Let us analyze the pair (b, h) that triggers the events c, e and g. The timing assumption informally means that the difference between the firing times of b and h is negligible from the point of view of c, e and g. This opens new possibilities for optimization by using the simultaneity constraints mentioned in Sect. 7.4.

Timing assumptions of type IV can be applied, e.g. to event d triggered by event c. For this assumption, the enabling region for d includes the states $\{s_2, s_5, s_8, s_{12}, s_{15}, s_{18}, s_{21}\}$ in addition to the states $\{s_3, s_6, s_9, s_{13}, s_{16}, s_{19}, s_{22}\}$ already in the firing region.

Assumptions Between Non-input and Input Events. Assume that $e_1, e_2 \in E$ are a non-input and an input event respectively and that they are concurrent.

V. Input not enabled before non-input event.

- **Ordering relations**: $(e_1 \parallel e_2) \ \wedge \ e_2 \not\triangleleft e_1$.
- **Difference timing assumption**: e_1 fires before e_2.
- **Justifying delay assumptions**: the delay of the environment is longer than the delay of one gate.

This assumption is similar to types I and II for the case in which e_2 is an input event. The delay assumption used in this case states that the response time of the environment (both slow and fast) will always be longer than the delay of one gate.

Assumptions Between Non-input Events and Slow Input Events. Assume that $e \in E$ is a *slow* input event, $X = \{e_1, \ldots, e_n\} \subset E$ is a set of non-input events and e is pairwise concurrent with all the events in X.

VI. Slow input not enabled before non-input events.
- **Ordering relations**: $(\forall e_i \in X : e \parallel e_i) \;\wedge\; e \not\lhd X$.
- **Difference timing assumptions**: X fires before e.
- **Justifying delay assumptions**: the delay of the slow input event is longer than Δ (the delay required by the circuit to stabilize under a quiescent environment).

To illustrate the meaning of this timing assumption, consider the example of Fig. 7.9, where h is an input event and d is a slow input event. The rest of the events are non-input. After firing the events a, b and c a state in which d, e and h are enabled is reached (s_3). At this point it can be assumed that e and f will fire before d (two gate delays vs. slow environment). However, no assumptions can be made about the firing order between d and g since g is preceded by an input event (h) for which no upper bound on the delay can be assumed. If h had been a non-input event, d would be assumed to fire after h and g also.

7.7 Back-Annotation of Timing Constraints

Logic synthesis with relative timing assumptions is able to derive a hazard-free circuit that is correct in the timed domain, i.e. in that subset of states of the untimed domain that is reachable by applying the timing assumptions. After the logic synthesis step, the assumptions that have been actually used as don't cares to obtain the synthesis results are propagated to the back-end (e.g. sizing and buffering) tools as a set of constraints to be satisfied. After the back-end step, including physical design, is completed the validity of the timing constraints must be verified to ensure the correct operation of the circuit.

Some of the timing assumptions provided by the user or automatically generated do not contribute to restricting the set of reachable states or the set of transitions and hence are redundant. Moreover, the circuit netlist derived by logic synthesis may be correct for a set of states *larger* than the one defined by the timed domain, i.e. one which can be obtained by a set of less stringent timing assumptions. In other words, some of the timing assumptions are redundant for a particular logic synthesis solution, while some other can be relaxed. This section attempts to answer the following question:

> *Can we derive a minimal set of timing assumptions sufficient for a circuit to be correct?*

This set of timing assumptions back-annotated for a given logic synthesis solution is called *timing constraints*. Timing assumptions (both manual and automatic) are part of the specification and provide additional freedom for logic synthesis, while timing constraints are a part of the implementation, since they constitute *sufficient* requirements to be met for a particular netlist solution to be valid.

Example 7.7.1. Let us analyze the example in Fig. 7.11. The shadowed states in the SG of Fig. 7.11a correspond to the timed domain determined by the timing assumptions

$$z+ < y+ \qquad \text{and} \qquad y+ < x-$$

Under these assumptions, logic synthesis can be performed by considering the states 110 and 001 unreachable.

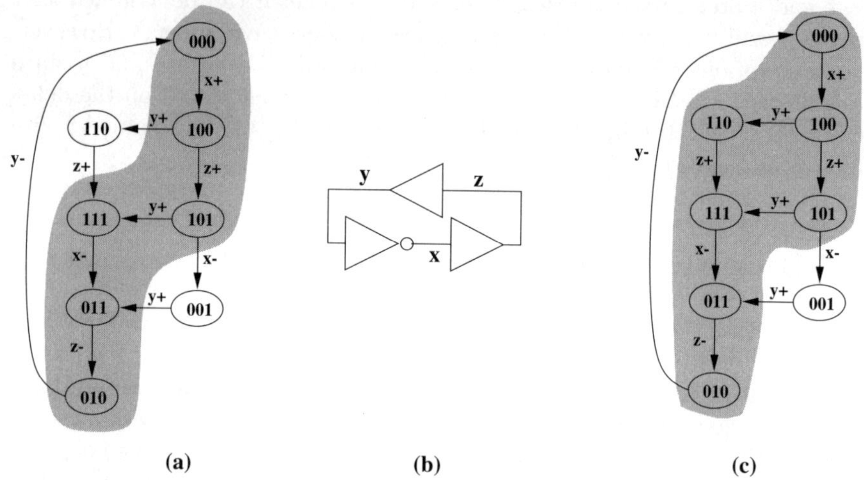

Fig. 7.11. (**a**,**c**) SGs with timed domains. (**b**) Circuit

The circuits of Fig. 7.4d and 7.11b have a correct behavior under the stated assumptions. Looking at the circuit of Fig. 7.4d it can be observed that:

- The gates $x = \overline{z + \overline{x}y}$ and $y = x + z$ are correct implementations for the whole untimed domain.
- The gate $z = x$ is a correct implementation for all the states except for 001. In this state, $z-$ is enabled according to the next state function of the implementation, but it is not enabled according to the specification.

Thus, even though the circuit has been obtained using the DC-set implied by both assumptions, only one relative timing constraint $y+ < x-$ must be ensured for the circuit to be correct, because *only part of the enlarged DC set has been used in a way that is inconsistent with the original specification*. In general, each gate of the circuit is correct for a subset of the untimed domain which is also a superset of the timed domain. The circuit is correct for those states in which all gates are correct.

Example 7.7.2. (Example 7.7.1 continued) Let us now take the implementation of Fig. 7.11b and analyze the gate $x = \overline{y}$, while ignoring the other gates for now. With regard to the untimed domain, the next-state function for x disagrees with the gate $x = \overline{y}$ in three states: 001, 110 and 101. But the consequences are different in each state. In 110, x should remain stable at 1. However, gate $x = \overline{y}$ enables $x-$ in state 110. To preserve circuit correctness two options are possible:

1. State 110 could be made unreachable by concurrency reduction. This in turn could be achieved in two ways:
   - By concurrency reduction in the untimed domain, based on changing logic (i.e. trigger) dependencies between signals as described in [155, 156].
   - By concurrency reduction in the timed domain, based on relative timing constraints that would preserve concurrency for *enabling*, but restrict concurrency for *firing* of signal transitions.
2. State 110 could remain reachable, while $x-$ would be enabled but not fireable, since another enabled transition fires before $x-$. More formally: $110 \in \mathsf{EnR}(x-) \setminus \mathsf{FR}(x-)$.

Similar considerations can be made for state 001.

State 101 illustrates a different case. According to the original specification SG, $x-$ is enabled in 101. In the implementation, however, signal x is stable in 101. This corresponds to a concurrency reduction for signal x in the untimed domain, and this is generally considered to be a valid implementation of the original specification. Concurrency is reduced because state 101 becomes a don't care for signal x when 001 is assumed to be unreachable (see Sect. 7.4). In summary, for the correctness of the gate $x = \overline{y}$, it is sufficient that states 110 and 001 are unreachable. However, gate $x = \overline{y}$ ensures that state 001 is unreachable. Hence only 110 must be made unreachable by timing constraints or by further concurrency reduction at the logic level.

A similar analysis can be done for gates $y = z$ and $z = x$. The sufficient requirements for the correctness of all three gates are summarized in Table 7.1. Interestingly, it can be concluded that the circuit is correct under any timing assumption, i.e. it is speed-independent, since all states required to be unreachable are forced to be unreachable by the concurrency reduction due to the chosen gate implementation. In particular, state 001 needs to be unreachable for gate $z = x$ to be a correct implementation of signal z and it is made unreachable by implementing signal x with gate $x = \overline{y}$.

Example 7.7.3. (Example 7.7.1 continued) Let us consider the same example under the assumption "$z+$ and $y+$ are simultaneous with respect to $x-$". Under this assumption, state 001 is unreachable. In addition, states 101 and 110 become don't cares for signal x, since both belong to $\mathsf{EnR}(x-)$ according to the semantics of the simultaneity assumption.

Table 7.1. Correctness requirements for the circuit of Fig. 7.11b

Gate	Unreachable states required	Unreachable states ensured by logic
$x = \overline{y}$	110, 001	001
$y = z$	110	110
$z = x$	001	

Only one timing constraint, $z+ < x-$, is sufficient for the circuit in Fig. 7.5d to be correct. Gate $x = \overline{y}$ is not enabled in 101, hence concurrency is reduced in this state with respect to the original specification and state 001 becomes unreachable under any gate delay. On the contrary, state 110 corresponds to the expansion of $\mathsf{EnR}(x-)$. This enabling is lazy since $110 \in \mathsf{EnR}(x-) \setminus \mathsf{FR}(x-)$.

7.7.1 Correctness Conditions

The synthesis flow presented in this chapter starts with an untimed specification $A = (S, E, T, s_{in})$. After logic synthesis with timing assumptions, a gate implementation is obtained.

Let us consider the circuit operation, ignoring timing assumptions. The untimed behavior of the gate implementation from a given initial state s_{in} can be represented by a transition system $A_G = (S_G, E_G, T_G, s_{in})$. A_G is obtained from A by substituting T with the new transition relation T_G, which coincides with T for the input events and models the behavior of the gates for the output events. Finally, T_G and S_G are calculated by only considering the reachability set from s_{in}. Obviously, circuit operation within A_G may not be correct outside the timed domain, e.g. it may be hazardous.

In the remainder of this section the following assumptions are used entirely *for the sake of simplicity*. They do not limit the applicability of the theory.

- The sets of signals of A and A_G are assumed to be the same and the states are assumed to be uniquely identified by their encodings.
- The set of states S_G reachable by circuit G in the untimed domain can be much larger than the original set S, due to the possibility of reaching incorrect behaviors. It is sufficient to calculate only a border of incorrect behaviors instead of the entire S_G.
- The original transition system A is not required to be untimed. It can include some timing assumptions (e.g. user-defined timing assumptions regarding the behavior of the environment). This helps in reducing the state space of the original specification for large circuits.

Since A_G is an untimed behavior, T_G may contain transitions not present in T, e.g. those transitions that become fireable when the timing assumptions used for synthesis are ignored. On the other hand, some transitions in T

may not belong to T_G due to the concurrency reduction imposed by the implementation.

The problem to be solved is to find a set of timing constraints for A_G, that result in a new lazy (timed) transition system $A_C = ((S_C, E, T_C, s_{in}), \mathsf{EnR}_C)$ such that $T_C \subseteq T \cap T_G$ and the gate netlist derived for A_G is still a valid implementation for A_C. The set of states is S_C implicitly induced by the initial state and the transition relation.

A *valid* implementation must satisfy three conditions:

1. The sequences of signal transitions produced by the circuit, when operated within the originally specified environment and timing constraints, are a *subset* of the sequences allowed by the STG (no new transition is allowed).
2. No new deadlocks (states in which no signal transition is enabled) are created. I.e. deadlocking specifications are considered legal, and we just do not introduce new deadlocks.
3. The implementation is hazard-free, i.e. A_C is output persistent.

Let us define three predicates characterizing the above conditions:

$$\mathit{new\_tr}(G) = \{s \xrightarrow{a_i *} s' \in T_G \mid s \xrightarrow{a_i *} s' \notin T\}$$

These are transitions that can fire in A_G (untimed circuit), but cannot fire in the original specification.

Due to concurrency reduction that might have been applied during logic synthesis, some states of S may become unreachable in S_G. Concurrency reduction eliminates some transitions in T_G, that might result in new deadlock states if all outgoing transitions from a reachable state are removed. Such deadlocks can be avoided by making them unreachable during legal circuit operation. Thus, we define

$$\mathit{to\_deadlock}(G) = \{s \xrightarrow{a_i *} s' \in T_G \mid s' \text{ is a deadlock in } A_G \text{ but not in } A\}$$

New hazardous states are captured with the following predicate

$$\mathit{to\_hazards}(G) = \{s \xrightarrow{a_i *} s' \in T_G \mid s' \text{ is output non-persistent in } A_G, \text{ but not in } A\}$$

Finally we define,

$$\mathit{valid}(G) = T_G \setminus (\mathit{new\_tr}(G) \cup \mathit{to\_deadlock}(G) \cup \mathit{to\_hazards}(G))$$

7.7.2 Problem Formulation

The problem to be solved consists of finding a set C of timing constraints, not more stringent than the ones used for synthesis, such that the set of transitions T_C obtained after applying the constraints is a subset of *valid*(G).

A trivial solution to this problem is to take the complete set of timing assumptions used for logic synthesis. Our goal, however, is to find a less stringent set of constraints sufficient to make the circuit correct. In general, we should look for such a set of constraints that "makes most sense" or that is easiest to satisfy. But the solution of this optimization problem, unfortunately characterized by a very fuzzy cost function, is left to future work.

Instead, a state-based cost function is used to guide heuristics aiming at finding the set C of timing constraints. The cost function is based on the following observation: large state spaces generally require simple constraints.

A corner case of the back-annotation problem would be the situation in which a speed-independent circuit is derived after synthesis with timing assumptions. In that case, the solution to the problem would be an empty set of timing constraints (see Example 7.7.2).

Fig. 7.12 illustrates the back-annotation problem. The arrows denote the invalid transitions of the circuit. The "timed domain" represents that state space of the circuit under all timing assumptions. $S_G \cap S$ represents the state space in which the circuit behaves correctly. Similarly for the transitions not exiting $S_G \cap S$. Constraints C_1 and C_2 are less stringent than the timed domain defined by all timing assumptions, and are enough to guarantee the correctness of the circuit. Note that the states in $S \setminus S_G$ are those eliminated by concurrency reduction. Also note that constraint $C1$ cuts one of the transitions from the timed domain to the region of incorrect behavior, which otherwise might occur due to early enabling.

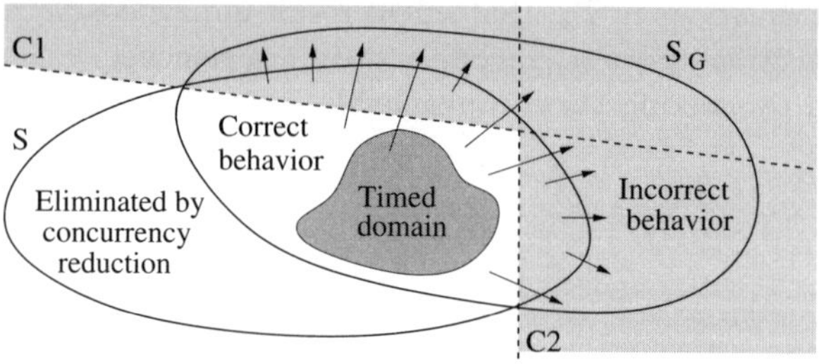

Fig. 7.12. Formulation of the back-annotation problem. $\{C_1, C_2\}$ is a set of timing constraints sufficient for the correctness of the circuit

7.7.3 Finding a Set of Timing Constraints

Relative timing constraints are defined in terms of firing order of events. Constraining the firing order between a pair of events only makes sense when they are concurrently enabled. Thus, each timing constraint C_i can be denoted by an ordered pair of concurrent events, e.g. $C_i = (e_j < e_k)$. Given a constraint $C_i = (e_j < e_k)$, the set of arcs *disabled*(C_i) are defined as

$$disabled(C_i) = \{s \xrightarrow{e_k} s' \mid \exists s \to s_1 \to \ldots \to s_n : \\ s_1, \ldots, s_{n-1} \in \mathsf{FR}(e_k) \ \wedge \ s_n \in \mathsf{FR}(e_k) \cap \mathsf{FR}(e_j)\}$$

In particular, the path $s_1 \to \ldots \to s_n$ can be empty if $s \in \mathsf{FR}(e_j) \cap \mathsf{FR}(e_k)$. $disabled(C_i)$ is the set of arcs with label e_k that must not fire in order for e_j to fire before e_k, i.e. those arcs with source states in which both events are concurrent or preceding $\mathsf{FR}(e_j) \cap \mathsf{FR}(e_k)$ inside $\mathsf{FR}(e_k)$.

Given a set of constraints $C = \{C_1, \ldots, C_p\}$, $disabled(C_i)$ can be used to compute T_C, that is the set of reachable transitions after removing the ones in

$$\bigcup_{C_i \in C} disabled(C_i).$$

Finding a set C that removes all transitions not in $valid(G)$ can be posed as a covering problem in which all possible firing order constraints of pairs of events are the covering elements.

Currently, `petrify` uses a greedy approach to solve the covering problem. It merely consists in choosing the constraint that removes the maximum number transitions not in $valid(G)$ and that have not been removed by previous constraints. This process is repeated until all reachable transitions become valid.

Example 7.7.4. Fig. 7.13 shows an example with a simplified version of the back-annotation problem, given that the removed objects are states instead of transitions. Assume that the set of states $S_G = \{s_0, \ldots, s_{10}\}$ is reachable by the untimed implementation of the circuit, and that the set of states $\{s_0, s_1, s_2, s_5, s_8, s_9, s_{10}\}$ is the one reachable after considering the delays of the circuit. However, incorrect behavior is only manifested in states s_6 and s_7. Table 7.2 contains the set of states that become unreachable by reducing concurrency between each pair of concurrent events[3]. For example, by imposing the order $d < b$, states s_2 and s_3 become unreachable. Pairs $(b, d), (b, e), (d, c), (e, c)$ can be used to preserve correctness.

The problem to be solved is the following: find a *small* set of ordering constraints between pairs of events such that the new set of reachable states does not intersect the set of incorrect states $\{s_6, s_7\}$. Moreover, we want to *maximize the set of reachable states*, i.e. to find a set of timing constraints that makes a small number of correct states unreachable and keeps the TS strongly connected. Larger sets of reachable states heuristically result in less stringent sets of constraints, thus simplifying the validation or verification of the circuit. Moreover, they often imply more concurrency, and hence heuristically result in better global performance.

[3] For simplicity, unreachable states are reported in the table for this example. In general, the analysis must be performed by calculating the removed *disabled* arcs. In this particular case, the resulting analysis is the same.

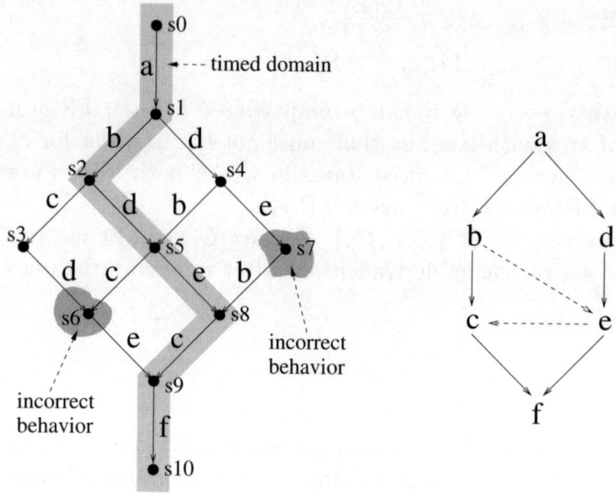

Fig. 7.13. Example for back-annotation

Table 7.2. Unreachable states for each pair of ordered events $e_1 < e_2$ in the example of Fig. 7.13

$<$	b	c	d	e
b			$\{s_4, s_7\}$	$\{s_7\}$
c			$\{s_4, s_5, s_7, s_8\}$	$\{s_7, s_8\}$
d	$\{s_2, s_3\}$	$\{s_3\}$		
e	$\{s_2, s_3, s_5, s_6\}$	$\{s_3, s_6\}$		

The problem can be posed as a *covering problem*. The bold cells of Table 7.2 correspond to those constraints that do not remove any state from the timed domain. The covering problem can be formulated as follows:

$$(e < c) \quad \wedge \quad (b < d \quad \vee \quad b < e)$$

The constraint $e < c$ is the simplest one that can remove state s_6. Any other, e.g. $e < b$, is more stringent. The constraints $b < d$ and $b < e$ are the ones that can remove the state s_7. The minimum-cost solution is

$$C = \{e < c, b < e\}$$

and

$$S_C = \{s_0, s_1, s_2, s_4, s_5, s_8, s_9, s_{10}\}$$

7.8 Experimental Results

This sections presents some results illustrating the impact of relative timing on the synthesis of asynchronous control circuits. The techniques for synthesis, automatic derivation of relative timing assumptions and back-annotation have been implemented in the tool petrify.

7.8.1 Academic Examples

The results for a well-known set of academic benchmarks are presented in Tables 7.3 and 7.4, for specifications without and with *state coding conflicts*, respectively.

Table 7.3. Results for specifications without CSC

	Area			Response time			State signals		
circuit	**SI**$_a$	**SI**$_t$	**TI**	**SI**$_a$	**SI**$_t$	**TI**	**SI**$_a$	**SI**$_t$	**TI**
adfast	18	31	13	2.17	1.00	1.00	2	2	0
alloc-outbound	20	23	22	1.50	1.11	1.00	2	2	2
master-read	65	79	45	2.29	1.33	1.29	7	7	3
mmu0	33	47	20	2.31	1.38	1.38	3	3	0
mmu1	25	32	15	1.60	1.12	1.12	2	2	1
mr0	50	51	30	1.60	1.45	1.15	3	3	2
mr1	36	39	20	2.25	1.19	1.19	4	3	0
nak-pa	24	35	24	1.25	1.00	1.00	1	1	1
nowick	18	19	16	1.50	1.17	1.00	1	1	1
ram-read-sbuf	30	26	21	1.10	1.00	1.00	1	1	0
sbuf-ram-write	24	44	24	1.63	1.00	1.00	2	2	1
sbuf-read-ctl	18	21	16	2.00	1.50	1.50	1	1	1
seq3	18	22	18	1.50	1.00	1.00	2	2	2
seq-mix	23	28	24	1.40	1.20	1.00	2	2	2
vmebus	22	33	17	2.29	1.57	1.57	1	1	0
Total	**424**	**530**	**325**	**1.76**	**1.20**	**1.15**	**34**	**33**	**16**

The experiments have been performed as follows:

- Columns labeled with **SI**$_a$ report results for speed-independent circuits derived by inserting state signals with the aim of minimizing area.
- Columns labeled with **SI**$_t$ are derived similarly, but with the aim of minimizing delay. `Petrify` tries to increase the concurrency of the newly inserted signals until they are outside the critical path of the specification. In case the original specification has no encoding conflicts (Table 7.4), there is no difference between **SI**$_a$ and **SI**$_t$.
- Columns labeled with **TI** report results for relative timing circuits. Relative timing assumptions are derived automatically by considering the environment to be *slow*. State signals are inserted aiming at delay minimization.

Table 7.4. Results for specifications with CSC

circuit	Area SI	TI
chu133	15	14
chu150	16	14
converta	19	14
ebergen	16	16
half	8	7
hazard	8	8
mslatch	24	20
trimos-send	30	21
var1	18	8
vbe5b	13	12
vbe5c	10	10
vbe6a	28	24
vbe10b	32	26
wrdatab	35	33
Total	**272**	**227**

For each experiment, area is estimated as the number of literals of the *set* and *reset* networks of generalized C elements. Delay (response time) is estimated as the average number of non-input events in the critical path between the firing of two input events. Given that the estimated response time of the specification does not change when no new signals are inserted, it is not reported in Table 7.4.

Relative timing assumptions have a crucial impact on solving state encoding, since `petrify` inserts new signals only to disambiguate conflicts in the timed domain. Reducing the number of signals also contributes to improving the area and the performance of the circuit.

Comparing the columns $\mathbf{SI}_t$ and **TI**, a reduction of about 40% in area can be observed. The reduction in response time is less than 5% if all events have a delay of one time unit. However, the performance improvement is much more significant if it is evaluated with actual delays, given that the logic of the timed implementation is much simpler. This analysis is reported in Sect. 7.8.2. The improvement obtained for specifications with complete state coding is about 17% in area. This reduction also contributes to improving the performance of the circuits. All the obtained circuits and the corresponding timing constraints were validated by simulation. Only in some cases, transistor sizing or delay padding was required to meet some stringent constraints.

7.8.2 A FIFO Controller

This section describes the development of a FIFO cell (specified in Fig. 7.14a and 7.14b), a simplified abstraction of a part of the RAPPID design [142]. The goal of the specification is to keep the left and right handshakes as decoupled as possible. The modules on the left and right hand sides of the controller have a similar speed to the controller itself. In fact, these events are generated by twin modules connected at each side. For this reason, in this case it is not wise to assume that the input events are slow.

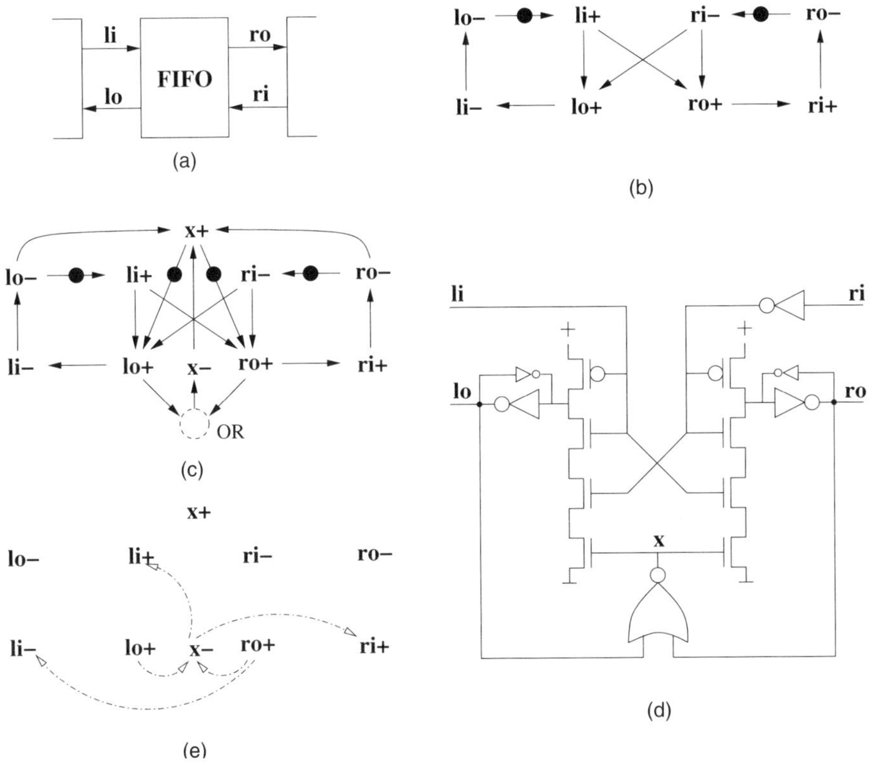

Fig. 7.14. (**a**) FIFO controller, (**b**) Specification, (**c**) Specification with state encoding signal, (**d**) relative timing implementation with gC elements, (**e**) Timing constraints sufficient for correctness

Four FIFOs were simulated by using different implementations. The cycle times of the cells were measured. The results, normalized to the delay of an inverter with fan-out of four in a given technology, are shown in Table 7.5.

The first relative timing FIFO (first row) is a circuit optimized by `petrify` using only automatic relative timing assumptions. It is depicted in Fig. 7.14d. A proper transistor sizing is required for correct operation of the circuit. No

Table 7.5. Cycle time comparison between FIFO implementations

Design	FIFO cycle time
RT	9.5
SI	11.5
RT reshuffled	5.7
SI reshuffled	7.6

user-defined assumptions on the environment are used. The timing analysis explained in Sect. 7.6 has been applied to the specification, and state encoding has been automatically solved as described in Sect. 7.5.2. With this strategy, only one additional state signal, x, was required as shown in Fig. 7.14c[4]. There are some interesting aspects of this implementation:

- The state signal x is concurrent with other activities in the circuit. This is a result of the state encoding strategy of `petrify` that attempts to increase the concurrency of new state signals until they disappear from the critical paths.
- The response time of the circuit with regard to the environment is only one event (two inverters), i.e. as soon as an output event is enabled, it fires without requiring the firing of any other internal event.
- Given that x is never triggering any output signal, the gates of l_o and r_o can be designed by having input x near Vss, thus improving their performance.

Finally, the implementation of Fig. 7.14d requires some timing constraints to be correct. Application of the method proposed in Sect. 7.7 derives five timing constraints between pairs of concurrent events, that are *sufficient* for the circuit to be correct. They are graphically represented in Fig. 7.14e.

The constraints $l_o+ < x-$ and $r_o+ < x-$ are not independent. Since the implementation of x is $x = \overline{l_o + r_o}$, it is always guaranteed that one of them will hold, whereas the other must be ensured. Since l_o+ and r_o+ are enabled simultaneously, these constraints will always hold if the delay of two gates is longer than the delay of one gate. The most stringent remaining constraint is $x- < r_i+$. In the worst case, both r_i+ and $x-$ will be enabled simultaneously by r_o+. In this case, the delay of $x-$ is required be shorter than the delay of r_i+ (from the environment). Since we assume that the environment is an identical circuit, it corresponds to requiring that the delay of $x-$ to be shorter than that of r_o+, that is easy to satisfy. In case of a very fast environment, this constraint can still be satisfied by transistor sizing or delay padding for gate x.

[4] This new specification is not strictly a Petri net, since the arcs from l_o+ and r_o+ to the *OR* place indicate an *or-causality* relation: $x-$ is triggered by the first event to fire, whereas the token produced by the latest event is implicitly consumed. An equivalent Petri Net is a bit more cumbersome and is omitted for simplicity.

The second FIFO (second row) is a speed-independent circuit derived by `petrify` with *automatic concurrency reduction* [156], and without constraining the concurrency of the input and output signals of the cell, in order to preserve performance as much as possible. The result is shown in Fig. 7.15, where CSC was obtained through state variable insertion and concurrency reduction. In comparison with the RT circuit, notice the gC elements with two p-transistors in series and the ordering between `ro+` and `lo+`. Because of concurrency reduction only one state signal is required, like in the case of the automatic RT solution. However, the state signal is on the critical cycle and the implementations of *lo* and *ro* contain additional p-transistors, which make performance of the speed-independent circuit approximately 18% worse than that of the RT one. Note that without concurrency reduction three state signals would be required to solve all state encoding conflicts, and a much larger and slower circuit would result.

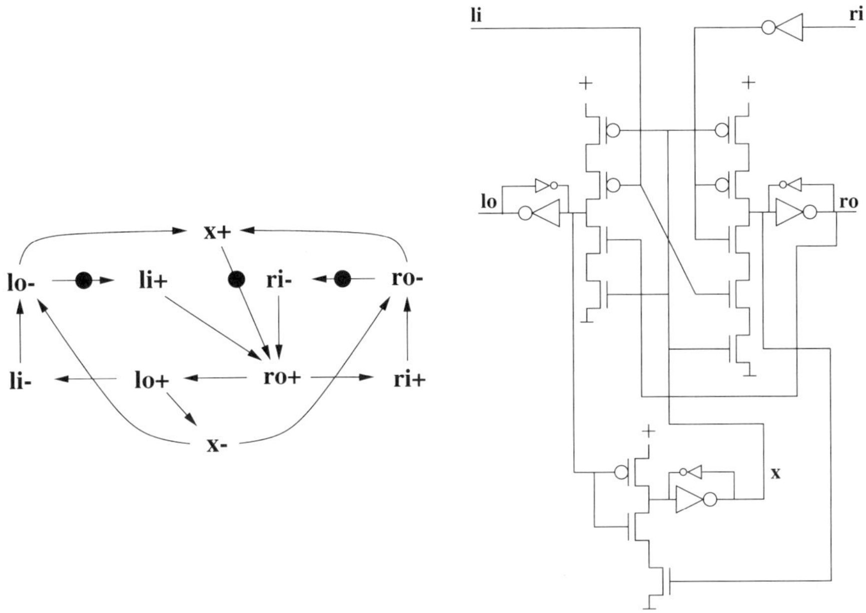

Fig. 7.15. Speed-independent specification and circuit

The third and fourth rows of Table 7.5 report results for relative timing and speed-independent circuits, further optimized for performance by applying De Morgan's laws. It can be observed again that the optimized RT circuit is approximately 25% faster than the optimized speed-independent design.

7.8.3 RAPPID Control Circuits

This section compares manually optimized RT control circuits used for RAPPID [143, 142] with those automatically derived by `petrify`. For each example, Table 7.6 reports: `m` (`manual`, obtained by applying relative timing manually), `a` (`automatic`, obtained automatically by `petrify` with relative timing) and `s` (`speed-independent`, obtained automatically by `petrify` without concurrency reduction or relative timing).

Table 7.6. Comparison for two generic representative examples (FIFO) and two control circuits from RAPPID (byte-control, tag-unit). Response time is measured in gate delays, area in transistors. `m`: manual, `a`: automatic, `s`: speed-independent

Design	Area (# tr.)			Worst case response time			Average case response time		
	m	a	s	m	a	s	m	a	s
FIFO-A	**22**	**22**	46	**3.0**	**3.0**	9.0	**2.5**	**2.5**	5.7
FIFO-B	16	**15**	46	**2.0**	**2.0**	9.0	**2.0**	**2.0**	5.7
Byte-cntr	32	**27**	71	4.0	**3.0**	5.0	3.0	**2.5**	4.1
Tag-unit	**31**	47	112	**4.0**	**4.0**	8.0	4.0	**2.7**	6.9
Summary	**101**	111	275	3.3	**2.9**	7.75	3.0	**2.4**	5.6

Results in the table show that automatic solutions are quite comparable with manually optimized RT designs. The improvement in response time by applying relative timing is about a factor of 2, substantially better than for the examples of Tables 7.3 and 7.4. This is because the designers of these circuits had a stronger interaction with the tool and provided aggressive timing assumptions on the environment that could not be derived automatically.

7.9 Summary

Lazy transition systems have been proposed as a computational model for timed circuit synthesis, where the notions of enabling and firing are distinguished for a signal switching event. We also presented implementability conditions, a synthesis algorithm and a method to derive a sufficient set of timing constraints for correctness.

The main features of the proposed design flow for synthesis with relative timing are:

- Two types of relative timing assumptions, difference (one-sided) and simultaneity (two-sided), are used.

- Timing information is defined in terms of relations among events rather than absolute delays of individual events. In this way, reasoning about the observable behavior of the system is much more efficient.
- The don't care space used for optimization is determined either by unreachability, i.e. reduction of the state space, or by laziness, i.e. expansion of the enabling region.
- Timing assumptions can be either provided by the designer or derived automatically by synthesis or analysis tools. The second feature is especially interesting for its applicability to those events that are not observable in the original specification, e.g. events of internal signals used for state encoding or logic decomposition.
- Satisfaction and verification of timing constraints (i.e. timing assumptions actually used by optimization) is left to the designer's responsibility. Some existing tools can assist in solving such task [157, 158].

This approach helps bridging two critical gaps in the synthesis of control circuits. The first gap is between the two main approaches for automated asynchronous controller synthesis, those based on fundamental mode (global timing constraints) and those based on input-output mode (fanout skew timing constraints). It also tackles the traditionally irreconcilable gap between asynchronous and synchronous circuit synthesis [159], in that it allows asynchronous circuits to exploit available timing information, rather than always making worst-case assumptions about the relative delays of gates (e.g. assuming that one gate may be slower than a sequence of three gates may be excessive in several technologies). Moreover, the exploitation of the idea of early enabling allows the synthesis process to maximize performance by increasing the effective amount of concurrency in the system.

8. Design Examples

This chapter presents a set of design examples that reflect typical cases in which the methods described in the previous chapters may be used. These examples also illustrate the practical strengths and weaknesses of the methodology, depending on the type of controllers or interfaces synthesized with it. This issue is important for two reasons. One is that real life design cases typically involve a combination of manual and automatic synthesis activities. In this respect, the reader should not overestimate the expressiveness of the language of STGs, or even PNs, as well as the power of algorithms and tools to always produce robust and efficient circuits. The class of PNs for which the corresponding STG can be implemented as a logic circuit is basically only restricted by the property of boundedness. No structural constraints are imposed on PNs. However, the method clearly has practical limitations. Those are related, firstly, to the limited size of the STGs that can be realistically implemented into circuits by the existing software tools (effectiveness aspect), and secondly, to the types of PNs and their signal interpretation, for example whether they involve regular patterns or require massive additional state coding (efficiency aspect). Some of our examples will reveal such problems.

The other reason for exposing such problematic aspects is that with their help we can show ways in which the overall Petri net-based approach can be developed further. The basic perspective is that it is not limited by the techniques based on logic synthesis from STGs through state graphs. Such complementary and novel techniques will be outlined in Chap. 10.

The major steps in the logic synthesis from STGs and the associated design flow have been described in the previous chapters. We hope that the reader has developed enough intuition on how to manipulate formal models once the STG specification has been provided. The primary goal of this chapter is to complement this knowledge by offering some guidance on how to construct synthesizable STG specifications from less formal descriptions of the design problem. Such STG specification often requires considerable effort and is by far the most time-consuming stage in synthesis, as practitioners admit [160].

Nevertheless, a situation like that exactly meets the objective that the developers of a good synthesis approach are after: to release the most valuable asset, the intellectual effort of a human designer, from complex yet formalizable and automatable model transformations to the more challenging task of constructing "good specifications". This is analogous to the synchronous domain, where despite the theoretical synthesizability of all specifications belonging to the "synthesizable subset" of a Hardware Description Language and the power of synchronous logic synthesis, a designer must still learn what constitutes a "good specification" in order to obtain satisfactory results after synthesis and physical design.

8.1 Handshake Communication

This example illustrates how an STG specification for a simple controller can be constructed from its description using timing diagrams. It then takes the reader through the main steps in the synthesis process, including making the STG implementable (bounded, consistent, output-persistent and with complete state coding), analyzing trade-offs between concurrency and state coding for logic synthesis in complex gates and generalized C-elements, decomposing the complex gate solution into simpler gates and latches, and finally, using timing assumptions for circuit optimization.

8.1.1 Handshake: Informal Specification

Fig. 8.1 shows a data processing structure consisting of two computation blocks, A and B, and a control circuit. Signals Ri and Ai are inputs, and Ao and Ro are outputs of the control. Output Ro can be seen as a latch enable signal for the data path. Output Ao is an acknowledgment signal sent to the previous control stage. The goal is, first, to synthesize a SI control circuit.

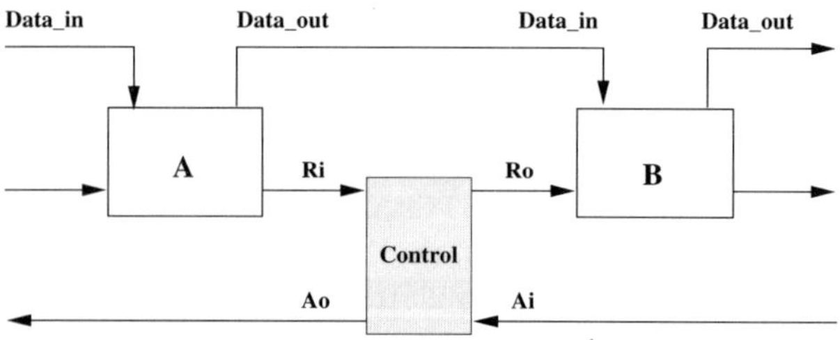

Fig. 8.1. A data processing structure

Different control disciplines are possible. Let us choose a discipline based on handshaking between adjacent stages, for instance the one described by the Timing Diagram shown in Fig. 8.2.

This discipline assumes the following:

1. The datapath includes latches which are *transparent* to input data when the control signal is low and which are *opaque*, i.e. insensitive to its data

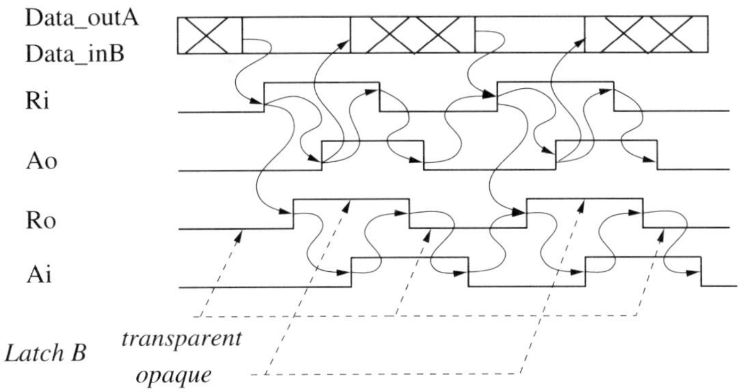

Fig. 8.2. Handshake discipline: timing diagram

input, when the control is high. e.g. the latch in B is transparent when $Ro = 0$ and opaque when $Ro = 1$.

2. The fact that Ri becomes 1 indicates that the data in the previous stage, stage A, is captured and stable (after the previous stage has become opaque).

Additional assumptions could be made if needed, depending on our knowledge of the implementation of the latches and delays in the datapath. For example, we may need to assume that there is sufficient delay between the appearance of data on the bus between stages and the rising edge on Ri, and hence on Ro, in order to guarantee the appropriate setup conditions for the latch in stage B.

8.1.2 Circuit Synthesis

STG Construction. We can construct an STG specification for the control circuit by following the causality arrows defined in the Timing Diagram. Those arrows are of the following three types. Firstly, there are arrows between the data waveform and control waveforms that reflect certain timing order assumptions between the datapath and control. Those are assumed to be guaranteed by logic or delay padding in the datapath blocks. They are not represented in the STG. Secondly, there are arrows that define the causal relationships between signal edges *within* the two handshake pairs, (Ri, Ao) and (Ro, Ai). By tracing four such arrows for each handshake in succession, it is easy to observe that the handshake satisfies the full four-phase, return-to-zero, signaling protocol in every data transfer cycle. Finally, the last group of arrows reflects the causality existing *between* the handshakes, which makes the whole protocol connected and fulfills the main purpose of transmitting data from stage A to stage B.

Fig. 8.3a illustrates the STG that specifies only the handshakes. This STG is not connected and is therefore incomplete with respect to the original

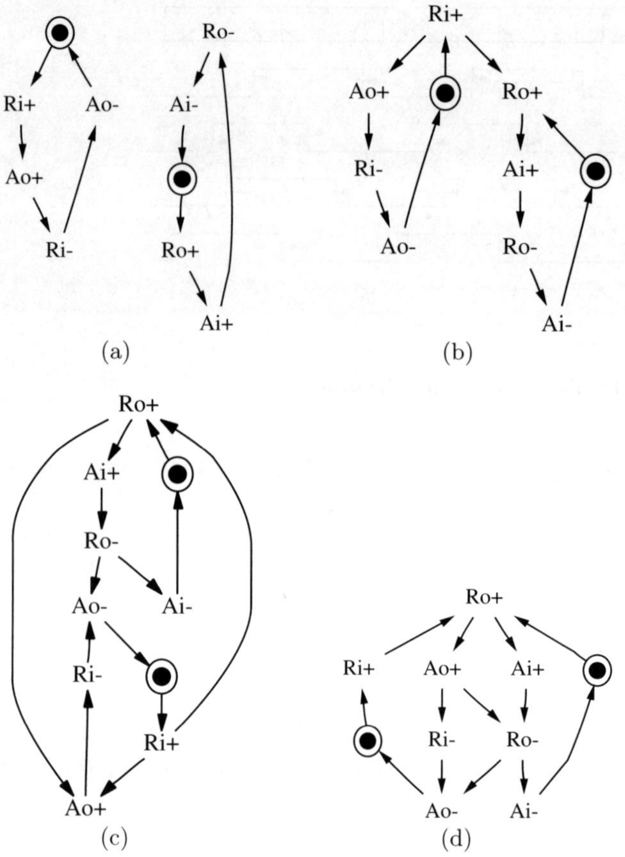

Fig. 8.3. An STG for the half-handshake: (**a**) independent handshakes at the left and right ports (**b**) unbounded version with *Ri*+ → *Ro*+, (**c**) bounded version with CSC violation, (**d**) final version with CSC

Timing Diagram in Fig. 8.2. To make it complete in that sense we should add an arc between transitions *Ri*+ and *Ro*+ as shown in Fig. 8.3b. Can this STG be used for synthesis?

The answer is "no" because its reachability analysis detects that the STG is unbounded as a PN. Its unboundedness is intuitively indicated by the fact that the left handshake cycle may produce an infinite number of tokens on the arc between *Ri*+ and *Ro*+. This would correspond to the situation where the left handshake works faster than the right handshake. In order to avoid this we should provide a reverse constraint, preventing stage A from producing another data value until the previous one has been sent to stage B. We would of course like to do it without much sacrifice in concurrency between the handshakes. Another restriction is that we cannot precondition a signal transition of the input *Ri*, since we are designing only the controller

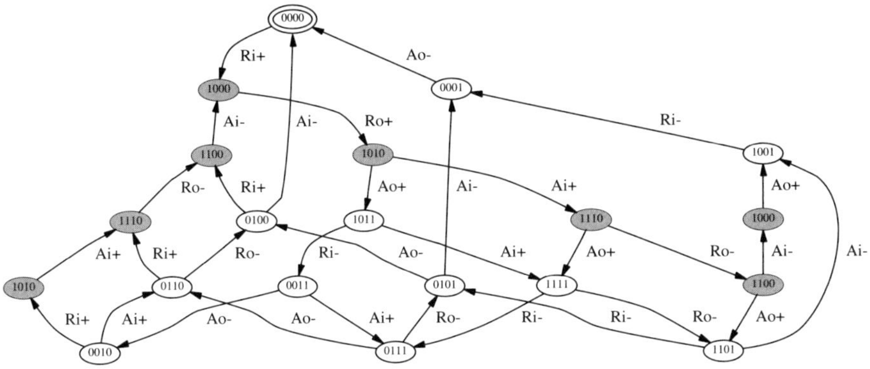

Fig. 8.4. State graph for STG in Fig. 8.3c

in Fig. 8.1, and hence we cannot change the behavior of its inputs. Thus, our additional arc will be directed from *Ro*+ to *Ao*+ as shown in Fig. 8.3c. The state graph generated by this STG is shown in Fig. 8.4. It contains 20 states. This state graph shows that the STG is bounded, consistent and signal persistent. It, however, cannot be directly implemented in a logic circuit because of CSC violations. There are four pairs of states that are in state coding conflict. They are shown shaded in the figure.

Circuit Implementation. There are two ways to move ahead with synthesis. One way could be to introduce more causality constraints to the specification, which would add more synchronization between the handshakes by *concurrency reduction*[161, 88]. For example, adding arcs $(Ro-, Ao-)$ and $(Ao+, Ro-)$, as illustrated in Fig. 8.3d, results in a state graph with 12 states, which is free from CSC violations. It is shown in Fig. 8.5.

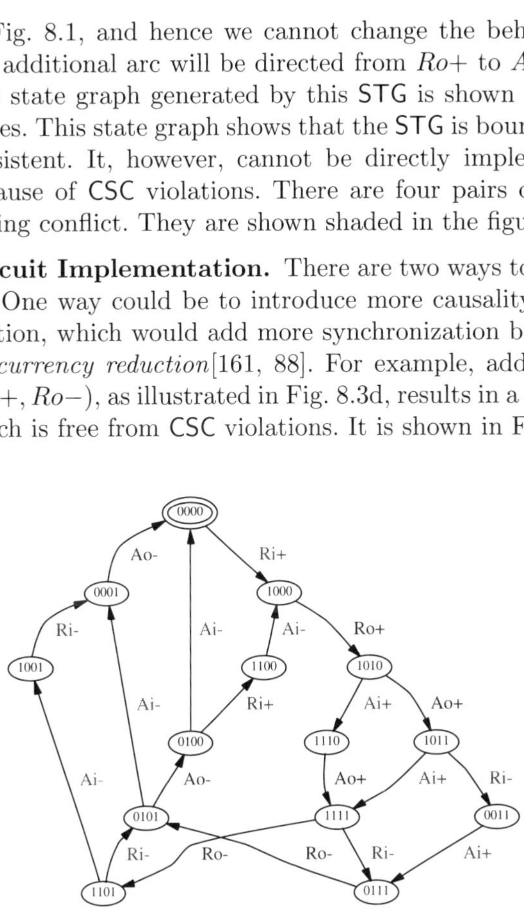

Fig. 8.5. State graph for STG in Fig. 8.3d

This graph is directly implementable with a SI circuit, described by the following complex gate equations:

$$\begin{aligned} Ro &= Ri\overline{Ao}\,\overline{Ai} + (\overline{Ai} + \overline{Ao})Ro \\ Ao &= RiAo + Ro \end{aligned} \quad (8.1)$$

The latter can be easily mapped into generalized C-elements (gC) following the technique described in Sect. 4.5. For that we simply need to derive the Set and Reset functions:

$$\begin{aligned} Set(Ro) &= Ri\overline{Ao}\,\overline{Ai}; \quad & Reset(Ro) &= AiAo \\ Set(Ao) &= Ro; & Reset(Ao) &= \overline{Ri}\,\overline{Ro} \end{aligned}$$

This circuit (called *hs1*) is shown in Fig. 8.6.

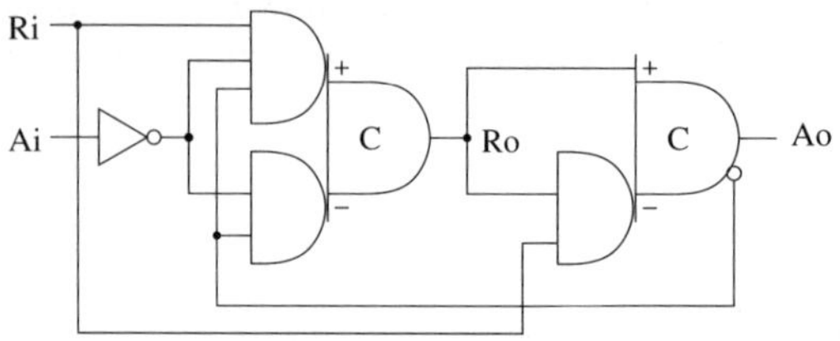

Fig. 8.6. A gC implementation of the STG in Fig. 8.3d

A second method to obtain an implementable STG from the model in Fig. 8.3c is to use the state encoding methods described in Chap. 5. The result of applying an automated procedure based on these methods is shown in Fig. 8.7a. This new STG contains two additional signals $csc0$ and $csc1$.

The complex gate implementation for this STG is defined by:

$$\begin{aligned} Ro &= csc1(Ri\,csc0 + Ro) \\ Ao &= csc0(\overline{csc1} + Ro) + RiAo \\ csc0 &= \overline{Ao}(\overline{Ro}\,csc1 + csc0) \\ csc1 &= \overline{Ai}(csc1 + \overline{csc0}) \end{aligned}$$

Similarly, a gC circuit (*hs2*) can be easily constructed.

Comparing this solution with the previous one, we can see that there has been a trade-off: a reduction of concurrency against additional logic that brings extra delay in the execution cycle (see Table 8.1).

One may try to look for an intermediate solution, for example, by adding only one concurrency-constraining arc, say $(Ro-, Ao-)$, to the STG of Fig. 8.3c. The result of this transformation leads to a state graph with 16 states, and only two pairs of states that are in CSC conflict. The CSC

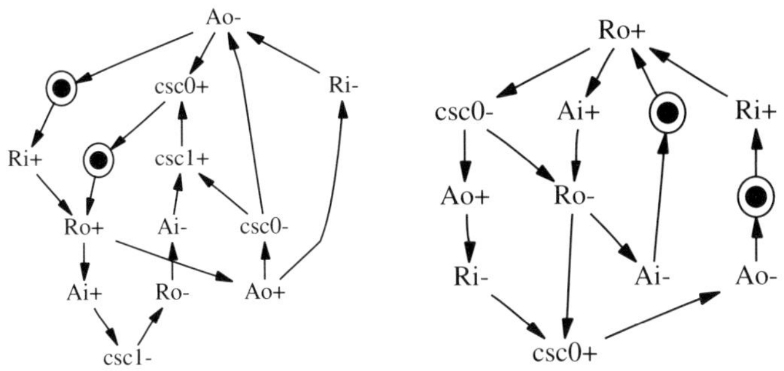

Fig. 8.7. Results of automatically solving CSC problems: (**a**) solution with two signals obtained for STG in Fig. 8.3c, (**b**) solution with one signal using an additional arc $(Ro-, Ao-)$

problem can be solved by using only one signal $csc0$, as shown in Fig. 8.7b. The resulting equations are:

$$\begin{aligned} Ro &= Ri\overline{Ai}\,csc0 + (csc0 + \overline{Ai})Ro \\ Ao &= \overline{csc0} \\ csc0 &= \overline{Ro}(csc0 + \overline{Ri}) \end{aligned}$$

It is easy to see that this solution has exactly the same complexity (but a slightly different form) as the one obtained for the STG of Fig. 8.3d and shown in Eqn. 8.1. This is simply because the other constraint added in Fig. 8.3d, namely arc $(Ao+, Ro-)$, was a constraint from an output signal Ao to another output Ro. In Fig. 8.7b, $csc0$ has exactly the same behavior (inverted) as Ao. Thus, with this order between the transitions of $csc0$ and Ao, we simply have the same effect from the insertion of $csc0$ as from the addition of arc $(Ao+, Ro-)$.

Logic Decomposition. Let us now apply decomposition techniques described in Chap. 6 to the STG in Fig. 8.3d. The result of logic decomposition using two-literal functions is as follows:

$$\begin{aligned} map0 &= Ri\overline{Ao} \\ Ro &= map0\,\overline{Ai} + (map0 + \overline{Ai})Ro \\ Ao &= RiRo + (Ri + Ro)Ao; \end{aligned}$$

This is conveniently mapped into a pair of C-elements, one AND gate and two inverters.

Timing-based Optimization. Let us consider potential timing assumptions in order to optimize the complexity and speed of our circuits, applying the techniques of Chap. 7. It is quite reasonable to assume that the delay of the environment through a handshake is greater than the delay in producing an output signal if both such actions are started at the same time. For example, the condition $Ao+ < Ai+$ corresponds to such an assumption, when both the $(Ro+, Ai+)$ handshake and output signal transition $Ao+$ have the same predecessor, event $Ro+$.

If we apply this assumption to the STG of Fig. 8.3d, the resulting logic becomes:

$$\begin{aligned} Ro &= \overline{Ai}(Ri\overline{Ao} + Ro) \\ Ao &= Ro + AoRi \end{aligned}$$

This is a simpler solution than the one given earlier by Eqn. 8.1. Further simplification, which reduces the number of transistors in the NMOS stack for the gC gate of Ro, can be achieved by adding another constraint $Ao- < Ai-$. The result is just a pair of simple (set and reset-dominant) latches and an inverter:

$$\begin{aligned} Ro &= \overline{Ai}(Ri + Ro) \\ Ao &= Ro + AoRi \end{aligned} \tag{8.2}$$

The analysis of the back-annotation results produced by `petrify` shows that both timing assumptions, $Ao+ < Ai+$ and $Ao- < Ai-$, are actually used, and thus become *timing constraints*.

The same assumptions could be added to a version of the STG in Fig. 8.3d without arc $(Ao+, Ro-)$. Recall that such an STG had some CSC violations. Now those violations are solved without adding new state signals. The solution for its logic implementation is identical to Eqn. 8.2.

Comparison between Implementations. The following table contains the summary of the area and cycle time estimates for the five gC circuit implementations of the handshake controller shown above. The area is estimated as the number of transistors. The cycle time was extracted from simulating a handshake stage with two neighboring stages and assuming that the delay in controlling the data-path is negligible, so that only the control logic delay is considered. Cycle times are normalized to the delay of an inverter with a fan-out of three.

Table 8.1. Area and cycle time comparison of handshake control circuits

Circuit	Area (transistors)	Cycle time (inverter delays)
hs1 (initial)	18	18
hs2 (max concur.)	37	23
hs3 (decomp.)	24	22
hs4 (RT)	17	17.5
hs5 (more RT)	16	16.5

It is quite clear that the best implementation among these five is the last one, with two timing constraints that help both resolve CSC and simplify logic for *Ro*.

Note also that the circuit *hs2*, obtained for the maximally concurrent specification, would have only made sense if the left and right parts of the environment of the handshake controller were significantly slower than the controller itself. This can happen, e.g. if the controller is connected to slow buses. In that case it would have been reasonable to assume that $envdel(hs1) = max(d(left+), d(right-)) + max(d(left-), d(right+)) \geq envdel(hs2) = max(d(left+) + d(left-), d(right+) + d(right-))$, where $envdel(hs1)$ and $envdel(hs2)$ stand for the delay contribution of the environment of circuits *hs1* and *hs2*, respectively. They reflect the fact that circuit *hs1* synchronizes with its left and right environments twice per cycle, whereas circuit *hs2* does it only once.

8.2 VME Bus Controller

This section illustrates how our methodology can be applied to the synthesis of a VME bus controller, which was first introduced in Chap. 2. We will look at the design of a control circuit for both read and write modes. Special emphasis will be on the practical use of relative timing assumptions.

8.2.1 VME Bus Controller Specification

Fig. 8.8 shows the I/O interface of a VME bus controller that controls the communication of a device with the bus through a data transceiver (signal *d*). On the device side there is a pair of handshake signals that follow a four-phase protocol (*lds* and *ldtack*). On the bus side there are two input signals (*dsr* and *dsw*) that follow a four-phase protocol with the output signal *dtack*. *DSr* and *DSw* indicate the beginning of a READ and WRITE cycle respectively. The timing diagram corresponding to a READ cycle was earlier depicted in Fig. 2.1b. A similar diagram for the WRITE cycle is shown in Fig. 8.8b.

For the synthesis of the VME bus controller, we need to derive a specification that includes the behavior of the READ and WRITE cycles. This can be done by using choice places, which model the nondeterminism of the environment in our Petri net formalism. In this case, the nondeterminism comes from the fact that the environment can choose to initiate a READ or a WRITE cycle after the completion of the previous cycle. An STG describing the complete behavior of the controller is shown in Fig. 8.9a. We can observe that some signal transitions, e.g. *lds*+, have multiple instances in the STG. The indices 1 and 2 have been used to distinguish events in the READ and WRITE cycles respectively.

There is also the possibility of representing the whole behavior in an STG with only one transition for each different label using the techniques of

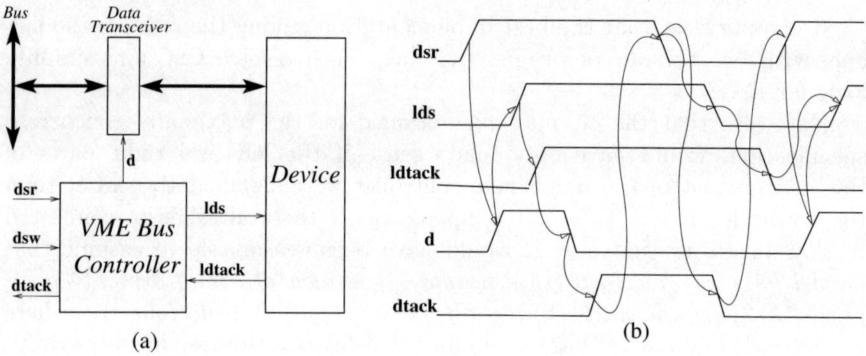

Fig. 8.8. (**a**) Interface for a VME bus controller, (**b**) timing diagram for the write cycle

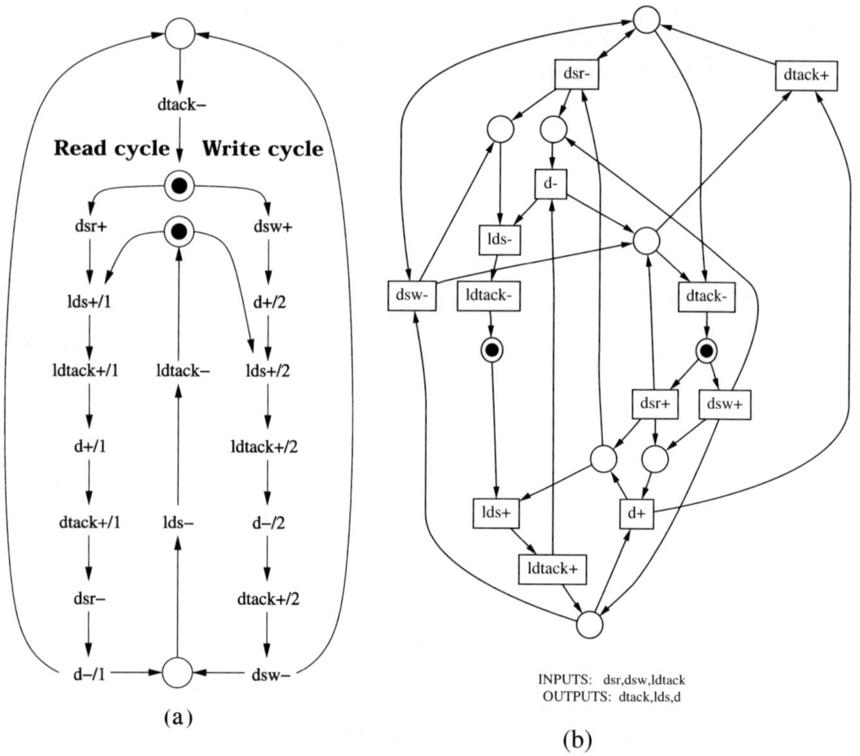

Fig. 8.9. (**a**) STG for the VME bus controller. (**b**) Equivalent STG with only one transition for each signal event

Chap. 3, which are implemented in petrify. The result is shown in Fig. 8.9b. However, this representation is much less readable than the previous one. For this reason we will use the STG of Fig. 8.9a to synthesize the circuits presented in this section. With that representation, the specification of timing assumptions will become much more intuitive, since we will be able to independently define them for the READ and WRITE cycles.

8.2.2 VME Bus Controller Synthesis

Speed-independent Solution. Let us first derive a speed-independent implementation of the VME bus controller. The state graph of its STG model is shown in Fig. 8.10. It contains 6 states that are in CSC conflict.

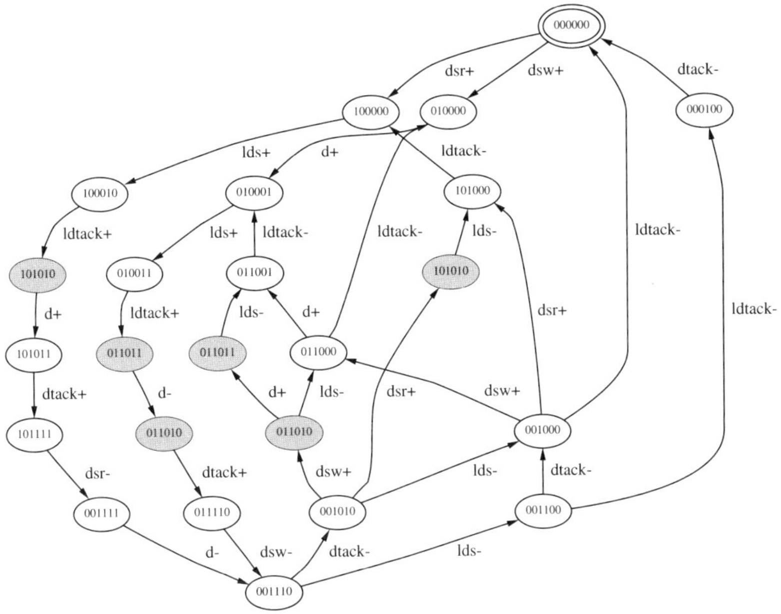

Fig. 8.10. State graph of the VME bus controller

The gC-based solution is depicted in Fig. 8.11. One internal signal, *csc*0, has been inserted to encode the states.

Assuming a Slow Environment. Let us now consider that the response time of the environment is long enough to enable the circuit to complete its internal activity. This is similar to the *fundamental mode* assumption used for synthesis of asynchronous Finite State Machines [5, 49]. Such assumption can be exploited by asking petrify to generate automatic timing assumptions after having specified the input events as "slow". This can be achieved by including the appropriate statement in the specification file and using an

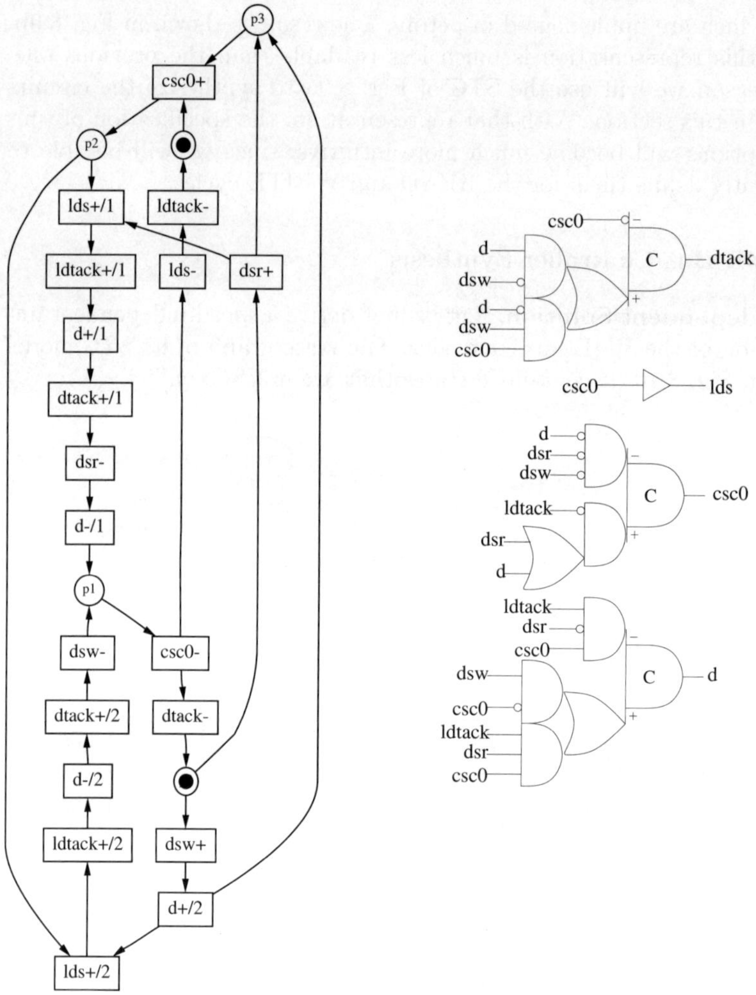

Fig. 8.11. VME bus controller: STG with a *csc*0 signal and its speed-independent implementation

option for automatic generation of timing assumptions when running petrify. The result is shown in Fig. 8.12.

Note that, in this solution, no state signals have been inserted to solve the state encoding problem. This is due to the fact that the timing assumptions make some states with conflicts unreachable. In this particular case, all conflicts disappear.

In order not to make the explanation too tedious, we will skip now the timing analysis of the solution, but we will focus on the *automatic timing assumptions* generated by petrify and shown in Fig. 8.13.

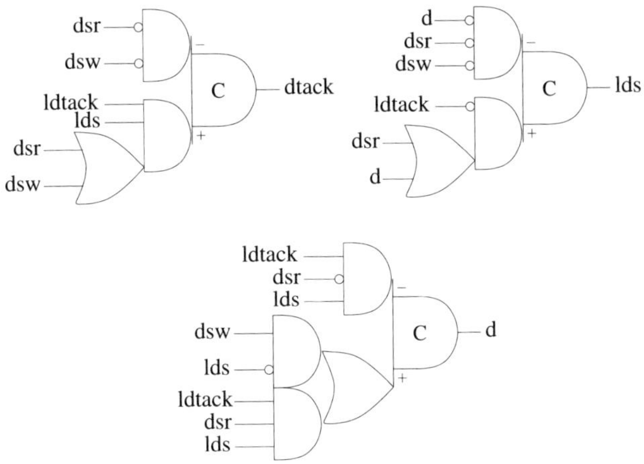

Fig. 8.12. VME bus controller: circuit under the assumption of a slow environment

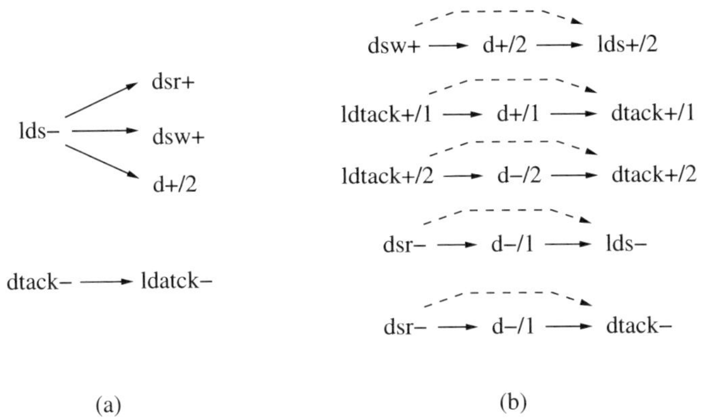

Fig. 8.13. Automatic timing assumptions: (**a**) concurrency reduction due to slow environment; (**b**) early enabling

petrify approaches the generation of automatic timing assumptions between pairs of events enabled simultaneously in the following way. Firstly, if one of the events is input, then it is assumed to be slower than the output one. Secondly, if both events are output, logic is estimated for each signal and the one having more literals in its function is assumed to be slower. It may however happen that after choosing one direction in the temporal assumption, the complexity of the logic is reversed, and delay padding may be required, as shown later. With the information provided by petrify after such an initial pre-run, the designer should then *specify* some order between the firing of non-input events. Next time, when petrify generates assumptions automatically, it will respect those given by the designer. The overall procedure of asking petrify to "pre-generate" some timing assumptions and then revising them using designer intuition is therefore step-by-step and semi-automatic.

Following this strategy, in addition to ordering input events according to the slow environment specification, petrify automatically introduces an assumption, $dtack- < lds-$, which corresponds to the ordering of two output events ($dtack-$ and $lds-$) that are enabled simultaneously. When deciding the firing order, in this case petrify chooses arbitrarily, because the number of literals for their implementation in Fig. 8.13 is the same, six.

This is one of the cases where the intervention of the designer can be useful. What would be the solution if we did not assume any firing order or another firing order?

The first possibility to explore is to add the following timing assumption in the specification: $lds- <> dtack-$, indicating that no assumptions should be made on the firing order of both events.

The second possibility can be to add the following timing assumption in the specification: $lds- < dtack-$. If such a constraint is added to the specification file, this prevents petrify from generating the opposite assumption $dtack- < lds-$.

The circuit obtained after adding $lds- < dtack-$ to the specification is shown in Fig. 8.14 and confirms the possibility of improvement by reversing the firing order of $dtack-$ and $lds-$. Again we will skip the timing analysis of this new solution.

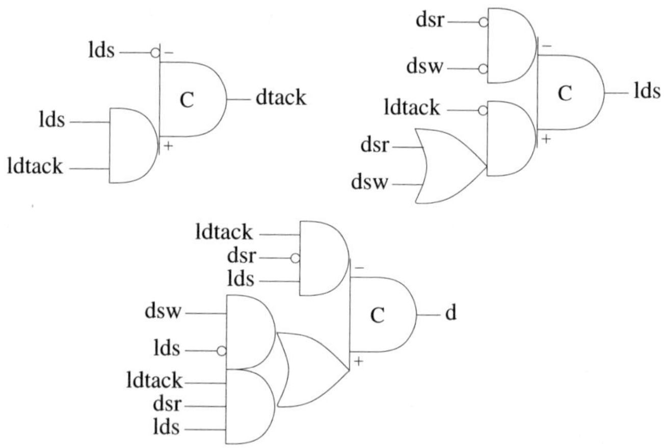

Fig. 8.14. Circuit under the assumption of slow environment and $lds- < dtack-$

Assuming a Slow Bus Control Logic. Looking at the specification of the controller in Fig. 8.9a we can observe that the return-to-zero of the protocols at both sides of the controller (bus and device) is done concurrently.

Let us now consider a case in which we know more details about the speed of the control logic on the bus side. Let us assume that this logic is so slow that no new request for a read or write cycle (*dsr*+ or *dsw*+) can arrive at the controller before the handshake with the device has completed. This can be specified by adding two new timing assumptions: *ldtack*$-$ $<$ *dsr*+ and *ldtack*$-$ $<$ *dsw*+.

The resulting circuit is shown in Fig. 8.15.

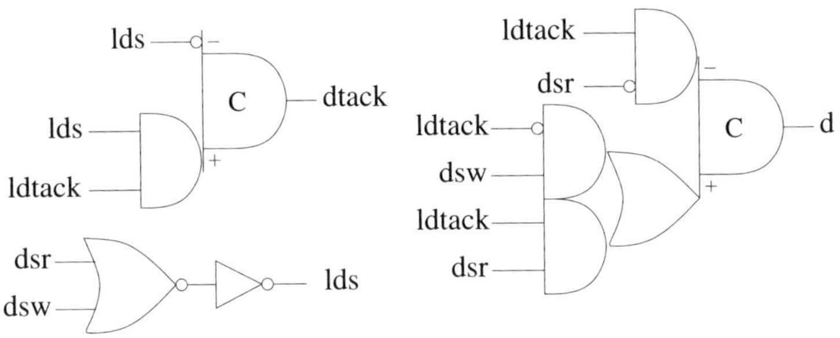

Fig. 8.15. Circuit under the assumption of a slow bus control logic

Timing Analysis. Let us now analyze the timing of the final solution depicted in Fig. 8.15. We can study the information reported by petrify, which is summarised in the timing assumptions shown in Fig. 8.16. We can distinguish three types of timing arcs in that diagram:

- **Concurrency reduction arcs** that denote the additional causality relations enforced by the logic (*lds*$-$ $\rightarrow$ *dtack*$-$ and *ldtack*$-$ $\rightarrow$ $d+/2$). These are not timing assumptions that must be verified for the circuit to be correct.
- **Firing order arcs** that denote the assumed firing ordering of concurrent events for the circuit to be correct. These are timing assumptions that must be verified or enforced by gate sizing or delay padding.
- **Early enabling arcs** that indicate the new trigger events for the early enabled events. For example, event *lds*$-$ that is triggered by $d-/1$ in the read cycle is now triggered by *dsr*$-$. This information is extracted by combining the early enabling timing assumptions and the list of trigger events reported for each solution. These are assumptions that must be verified for the circuit to be correct. In the mentioned example, it must be verified that event $d-/1$ will occur before event *lds*$-$.

As we discussed above, we derived *ldtack*$-$ $\rightarrow$ *dsr*+ and *ldtack*$-$ $\rightarrow$ *dsw*+ as assumptions about the speed of the bus control logic. Therefore, they can be considered as satisfied, or must be checked by looking at a more global picture.

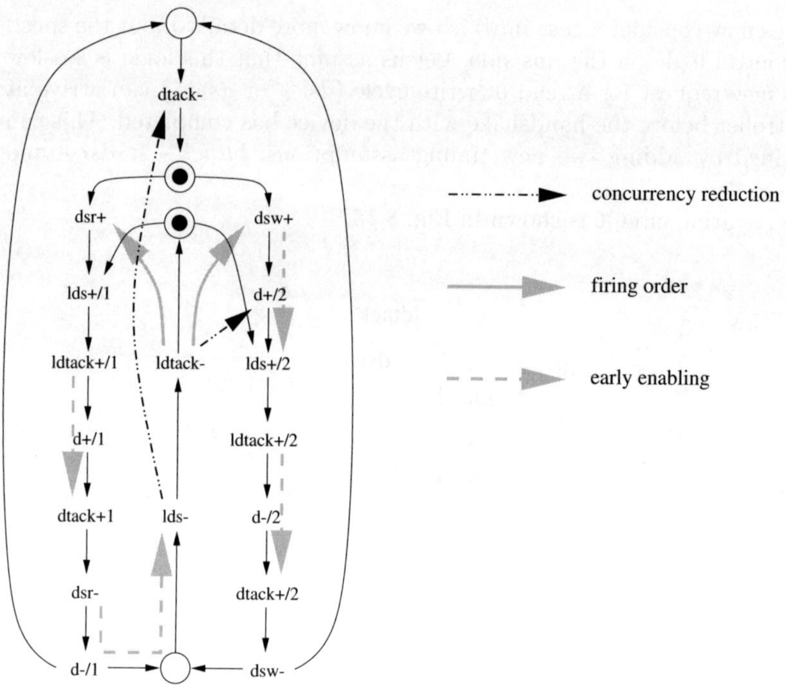

Fig. 8.16. Timing assumptions for the final solution of the VME bus

On the other hand, the assumptions on early enabling of events must be carefully analyzed, since their validity depends on the actual delays of the derived logic. All the early enabling assumptions rely on the fact that the delay of gate d is shorter than the delay of lds and $dtack$. Unfortunately, we see that gate d is more complex than the other gates, so these assumptions are generally unrealistic. Therefore, some changes must be performed on the circuit to satisfy those assumptions.

A possible solution is depicted in Fig. 8.17. The delay d_1 should ensure that $lds+/2$ and $lds-$ will fire later than $d+/2$ and $d-/1$ respectively, even though they are triggered simultaneously. The delay d_2 should ensure that $dtack+/1$ and $dtack+/2$ will fire later than $d+/1$ and $d-/2$ respectively. Note that the chosen solution only delays the triggering effect of signal $ldtack+$ on $dtack+$. Another solution would have been to add a delay at the output of gate $dtack$, but this would also delay $dtack-$, which is not necessary for the correct functioning of the circuit. Thus, we can see that the information reported by petrify allows us to finely tune the circuit so that *timing assumptions are met for individual events of a signal, rather than for all the events of the signal.*

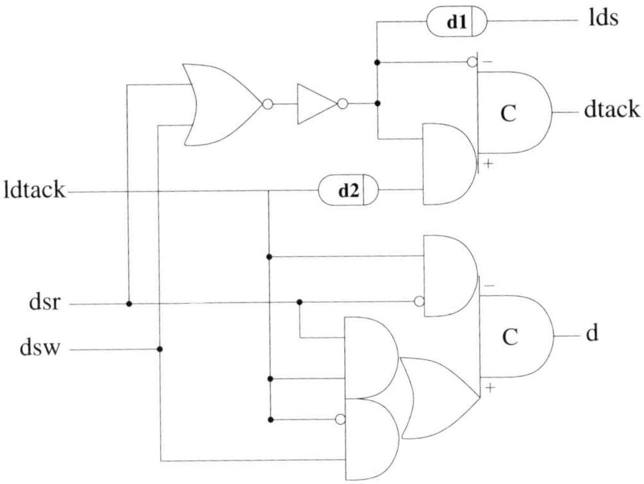

Fig. 8.17. Delay padding to satisfy timing assumptions

8.2.3 Lessons to be Learned from the Example

1. The use of timing information plays an important role in reaching efficient solutions. Timing assumptions typically originate from the knowledge of the environment of the synthesized circuit. For example, it may sometimes be possible to allow for the input signals to arrive later than the outputs (slow environment). Moreover, one part of the environment (the bus-driven interface) can be slower than another part (the local interface).
2. Timing assumptions can be either provided by the designer, together with the STG specification file, or automatically generated by the petrify tool. The latter, which are always reported back by the tool, should be verified against the actual logic implementation if they fall into the categories of *firing order arcs* and *early enabling arcs*.
3. Designers should always consider carefully whether automatically generated assumptions are reasonable or not, and if they are useful or not. In other words, they should be considered as suggestions that must be evaluated, and possibly rejected.
4. Some timing constraints must be guaranteed by appropriate gate sizing and/or delay padding in the circuit.

8.3 Controller for Self-timed A/D Converter

In this section we will consider a design example where our main focus will be on the development of STG specifications for asynchronous control circuits of a controller that requires some structural decomposition. This decomposition should be performed by the designer before he proceeds with the logic synthesis of individual blocks of the controller.

8.3.1 Top Level Description

This example comes from the design of an asynchronous A/D converter described in [162]. This converter (for a 4-bit digital value) uses a well-known successive approximation method, where the digital output is determined bit by bit, starting from the MSB, according to the result of the comparison between the input voltage and the voltage produced by an D/A converter connected to the digital output. The idea of the conversion algorithm is presented in Table 8.2.

Table 8.2. Control of Converter ("X" stands for any value, "Comp" denotes the value produced by the comparator)

Ready	State	MSB (CB3)	CB2	CB1	LSB (CB0)
1	0	X	X	X	X
0	1	1	0	0	0
0	2	Comp	1	0	0
0	3	X	Comp	1	0
0	4	X	X	Comp	1
1	0	X	X	X	Comp

The overall structure of the A/D converter is shown in Fig. 8.18. It consists of a comparator, a 4-bit register and control logic. The comparator uses analog circuitry and a bistable device which has a completion detector capable of resolving metastable states (for more on the effect of metastability in A/D converters see [162, 163]). From the point of view of the control logic, it has a special input signal, Ar (called $CLAMP$ in [162]), which is normally at 1 to keep the dual rail output, Din, at the "no valid data" or "spacer" value 00. When Ar goes to 0 the comparator is released to produce a valid dual-rail value on Din (the result of the comparison between the adjusted and input voltage), which is also acknowledged by signal Ad going to 0.

The 4-bit register is effectively a combination of a demux and a register with serial dual-rail input and parallel single-rail output. It is built from latches that are capable of storing the value even when the enable input is withdrawn. The register is controlled by five command signals arriving from the control logic in a one-hot (non-overlapping) manner. The first command signal, R, is used to reset the register to the initial state 1000 (with MSB=1), immediately before starting the conversion. For all other command signals,

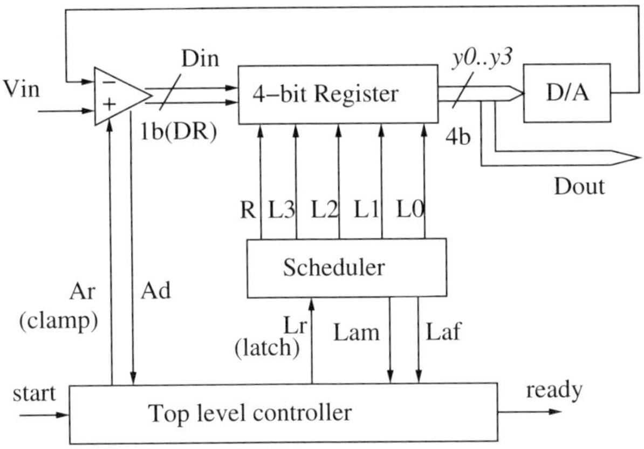

Fig. 8.18. Block diagram of A/D converter

Li, the register is supposed to latch the value from input *Din* in the appropriate bit $y(i)$ and set the lower bit, $y(i-1)$ to 1.

8.3.2 Controller Synthesis

Top Level Control. The control logic is structured to consist of a five-way scheduler and a top level controller. The latter is described by the STG shown in Fig. 8.19. Let us first ignore the dashed arrows, which stand for timing assumptions and will be considered later.

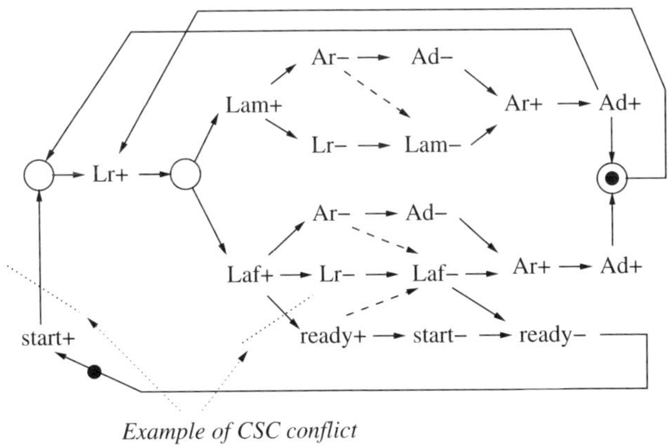

Fig. 8.19. STG specification for top level controller

The environment issues the *start*+ signal when the analog value is ready for conversion. The controller then performs the following procedure.

1. The controller issues a request to the scheduler, called *Lr*, and receives an acknowledgment, called *Lam* ("*m*" for middle), when the appropriate latching of the next bit into the register has occurred.
2. The controller executes a handshake with the comparator to force the next bit out.
3. The previous steps are repeated until the last bit has been produced, and then the scheduler issues another ack signal (mutually exclusive with respect to *Lam*), *Laf* ("*f*" from final).
4. Finally the controller generates the *ready*+ signal to the environment, indicating that the 4-bit code is available on the outputs of the register.

This STG is a free-choice PN, which is bounded. The STG is consistent and output-persistent. The only non-persistency in it is between input signals, *Lam* and *Laf*, which corresponds to a non-deterministic choice made by the environment, in particular by the scheduler. There are altogether 20 states that are in CSC conflict, with respect to all three output signals. The automatic resolution of these conflicts leads to the following logic implementation:

$$\begin{aligned} ready &= \overline{csc0} \\ Lr &= csc1(start\ csc0\ Ad + Lr) \\ Ar &= \overline{Lam}\ \overline{Laf}\ csc1 \\ csc0 &= \overline{Laf}(csc0 + \overline{start}) \\ csc1 &= (\overline{Laf}\ \overline{csc0} + \overline{Lam}\ csc0)(csc1 + \overline{Ad}) \end{aligned}$$

This solution is not quite satisfactory because of the complexity of the gate for *csc*1. It also has a disadvantage that CSC signals are inserted in sequence with output signals, which leads to the worst case path of three events between two adjacent input events.

In order to improve the size and performance of the implementation we can apply a number of timing assumptions. One such common assumption would be to suggest that output events are generated before concurrent input events activated through another handshake, provided that the outputs have the same predecessor. The dashed arcs in Fig. 8.19 illustrate such assumptions that become constraints to produce the following implementation:

$$\begin{aligned} ready &= ready\ start + Laf \\ Lr &= \overline{ready}\ start\ \overline{Laf}\ \overline{Lam} Ar\, Ad \\ Ar &= \overline{Lam}\ \overline{Laf}(\overline{Ad} + Ar) \end{aligned}$$

In this implementation all CSC conflicts have been resolved by means of the timing assumptions and concurrency reduction implied by them. As a result, the worst case path between two adjacent input events is only one event. This solution may still be not completely satisfactory because of the complexity of the AND gate for Lr. Another timing assumption, shown by the dotted arc ($start-$, $Ad+$) helps to eliminate input $\overline{ready}$. This condition is quite realistic to assume if the environment quickly resets $start$ to zero after receiving the setting of $ready$ to one. Another optimization can be achieved by logic decomposition. A NOR gate implementing the product $\overline{Lam}\ \overline{Laf}$ may be used to simplify both logic for Lr and Ar. The final circuit is shown in Fig. 8.20.

Automated Solution for the Scheduler. The STG model of the scheduler is shown in Fig. 8.21.

It consists of five identical fragments. Note the similarity of this description with the STG model of a modulo-N counter, which operates as a

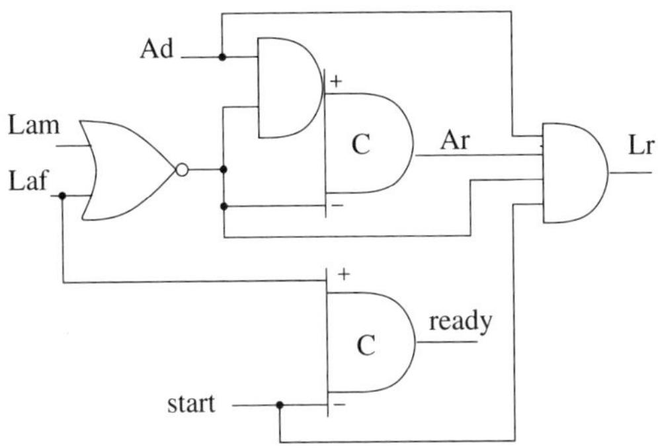

Fig. 8.20. Implementation of the top level control

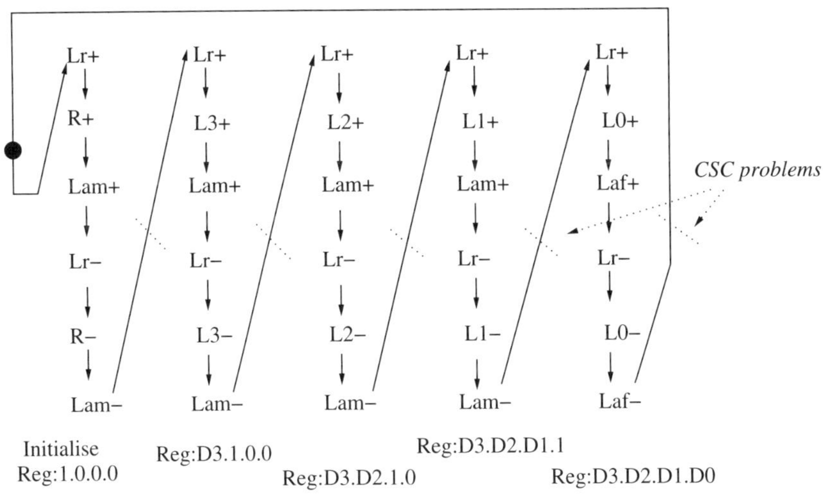

Fig. 8.21. STG specification of the scheduler

frequency divider between Lr and Laf (it produces an acknowledgment on the Lam output $N-1$ times followed by one acknowledgment on Laf). This STG has a large number of CSC conflicts. It requires internal memory to record the position of the token in this counter.

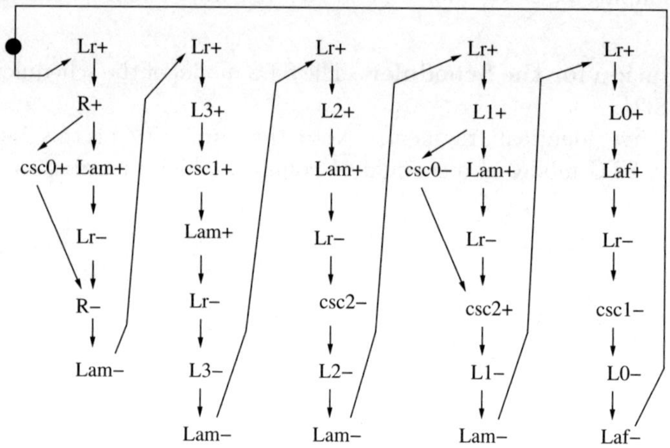

Fig. 8.22. The result of automatic CSC resolution for the scheduler STG

Giving this STG directly to the synthesis tool (`petrify`) appears to be not the best solution for this sort of logic, because the tool solves CSC conflicts by using bipartition (see Chap. 5), which indirectly implies the use of logarithmic binary encoding. Furthermore, the *csc* signals added by the tool to such an STG may be ordered with one another and with output signals. In this way, the final circuit may consist of relatively deep logic, consisting of complex gates, which would lead to large cycle time. The actual result of applying `petrify` to the STG of Fig. 8.21 is shown in Fig. 8.22.

This STG leads to the circuit described by the following equations (using complex gates):

$$
\begin{aligned}
Lam &= L3csc1 + R + L2 + L1 \\
Laf &= L0 \\
L0 &= csc1(Lr\,\overline{csc0}\,csc2 + Laf) \\
L1 &= \overline{csc2}(Lr + L1) \\
L2 &= csc2(Lr\overline{L3}\,csc0\,csc1 + L2) \\
L3 &= Lr(\overline{R}\,csc0\,\overline{csc1} + L3) \\
R &= \overline{csc0}(Lr\,\overline{csc1} + R) + LrR \\
csc0 &= \overline{L1}\,csc0 + R \\
csc1 &= csc1(\overline{Laf} + Lr) + L3 \\
csc2 &= csc2(\overline{L2} + Lr) + \overline{Lr}\,\overline{csc0}
\end{aligned}
$$

This logic is clearly irregular. Some *csc* signal transitions are indeed inserted in series with output signal transitions and, as a result, the circuit has, in the worst case, three events between two adjacent input events.

Manual Solution for the Scheduler. Unfortunately, the best way to handle such regular STG models, given the limitations of current synthesis tools, would be to insert internal memory signals by hand, in a one-hot manner. For comparison with the above automatic solution, a manual one is shown in Fig. 8.23.

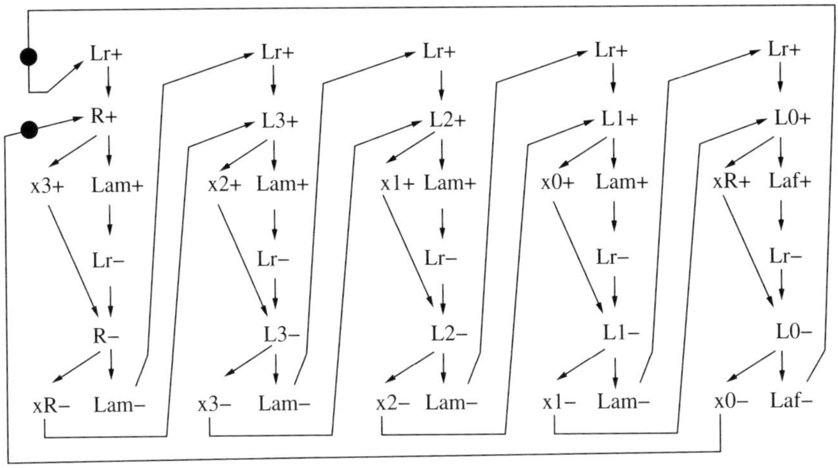

Fig. 8.23. The result of manual CSC resolution for scheduler STG

This circuit is described by the following equations:

$$
\begin{aligned}
Lam &= L3 + R + L2 + L1 \\
Laf &= L0 \\
L0 &= Lr\; x0\; \overline{x1}\; \overline{xR} + L0(\overline{xR} + Lr) \\
L1 &= Lr\; x1\; \overline{x2}\; \overline{x0} + L1(\overline{x0} + Lr) \\
L2 &= Lr\; x2\; \overline{x3}\; \overline{x1} + L2(\overline{x1} + Lr) \\
L3 &= Lr\; x3\; \overline{xR}\; \overline{x2} + L3(\overline{x2} + Lr) \\
R &= Lr\; xR\; \overline{x0}\; \overline{x3} + R(\overline{x3} + Lr) \\
x0 &= x0\; \overline{xR} + L0 + L1 \\
x1 &= x1\; \overline{x0} + L2 + L1 \\
x2 &= x2\; \overline{x1} + L3 + L2 \\
x3 &= x3\; \overline{x2} + L3 + R \\
xR &= xR\; \overline{x3} + R + L0
\end{aligned}
$$

This logic is regular. Furthermore, its state coding signals, $xR, x0, \ldots, x3$, have their transitions inserted in parallel with the handshake of the environment. It produces only two output events between two adjacent input events.

Timing Optimization. We can now reasonably assume that internal gates fire before the next event on input Lr arrives to the STG of Fig. 8.23. This leads to the following timing assumptions:

$$x3+ < Lr-, xR- < Lr+, x2+ < Lr-, x3- < Lr+, \ldots$$

As a result, the implementation is significantly simplified:

$$\begin{aligned}
Lam &= L3 + R + L2 + L1 \\
Laf &= L0 \\
L0 &= Lr\, x0\, \overline{x1} \\
L1 &= Lr\, x1\, \overline{x2} \\
L2 &= Lr\, x2\, \overline{x3} \\
L3 &= Lr\, x3\, \overline{xR} \\
R &= Lr\, xR\, \overline{x0} \\
x0 &= x0\, \overline{xR} + L0 + L1 \\
x1 &= x1\, \overline{x0} + L2 + L1 \\
x2 &= x2\, \overline{x1} + L3 + L2 \\
x3 &= x3\, \overline{x2} + L3 + R \\
xR &= xR\, \overline{x3} + R + L0
\end{aligned}$$

8.3.3 Decomposed Solution for the Scheduler

An alternative way would be to follow the idea of regularity in the scheduler and further decompose it into cells, each responsible for one command signal.

The STG model of a cell of the scheduler is shown in Fig. 8.24. This model is based on the idea of an "activity token" moving cyclically between the cells.

The logic implementation for this model, including two CSC signals introduced automatically by `petrify`, is described by the following equations:

$$\begin{aligned}
Tina &= \overline{csc1} \\
Toutr &= csc0(\overline{Tinr}\;\overline{Touta}\; csc1 + Toutr) \\
L &= Tinr(csc0 + Lr) + \overline{csc1} \\
La &= L(csc0 + La) \\
csc0 &= csc0(\overline{Toutr} + \overline{Touta}) + L\overline{La} \\
csc1 &= \overline{Lr}(\overline{csc0} + \overline{Tinr}) + \overline{Tinr}\; csc1
\end{aligned} \tag{8.3}$$

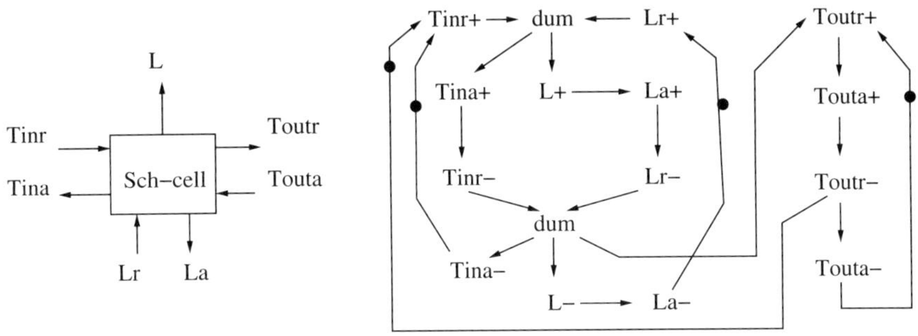

Fig. 8.24. Building the scheduler from cells: a single cell model

Modeling Complex Behavior of the Environment. Note that the STG of Fig. 8.24 models a cell in a "stand-alone" way, i.e. it does not reflect the fact that the environment of the cell may allow signal transitions on input Lr which do not activate the given cell but involve other cells. This kind of situation often happens when a large controller is decomposed into components, each component is designed separately and two or more components share some input signals. In this case it is quite easy to miss out some elements in the behavior of the environment which may at a first glance appear to be irrelevant to a specific component. In fact, to avoid inaccuracy in capturing the behavior of the environment of a scheduler cell one should specify the cell as shown in Fig. 8.25.

Note that the possibility of input Lr to change without actual involvement of the cell is shown by the sequence $(Lr = 1, Lr-, Lr = 0, Lr+)$, in which Lr may go down to 0 but only if an input token has not arrived in this cell. In other words, this can happen if the token is still traveling through other cells. Once the token is at the input of our cell, which is indicated by the firing of $Tinr+$ (the place labeled $NoToken$ is no longer marked), we assume that the environment behaves well with respect to the state of its input Lr, in order to guarantee that transition dum is not disabled. The reader may notice that in our model there is some non-persistent behavior (asymmetric conflict) between input transitions $Tinr+$ and $Lr-$, but this does not pose a problem to logic implementation because the STG remains persistent with respect to non-input transitions (as in our discussion in Sect. 4.2 and Definition 4.2.6).

An alternative way of modeling the environment where signal Lr is global to many cells, would be to precisely model a sequence of four pairs of transitions on Lr: $Lr+/1, Lr-/1, ..., Lr+/4, Lr-/4$, by substituting them for the marked arc between events $La-$ and $Lr+$ in Fig. 8.24. This would both capture the fact of multiple (irrelevant to this cell) changes on Lr and do it in a deterministic way (four such changes). However, such an approach with a fixed counter is not as elegant and modular as the one in Fig. 8.25, where we defined the behavior of a cell more or less independent of the rest of the

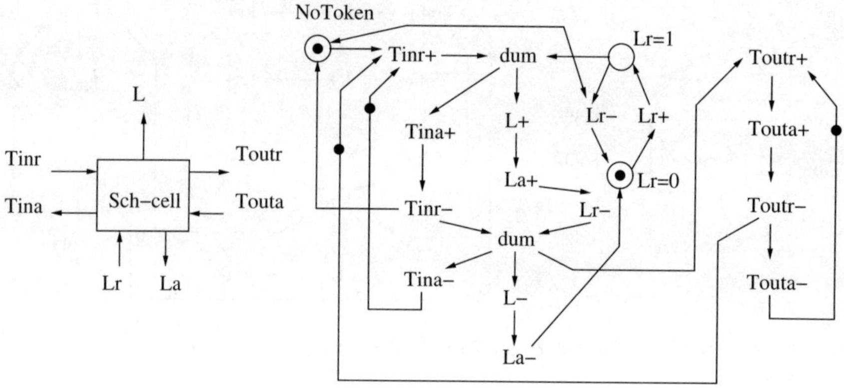

Fig. 8.25. A scheduler cell with a more accurate model of the environment

circuit, i.e. in a structure with an arbitrary number of cells. This is a good example of a situation where one pays the price of trading determinism for generality.

Verification of Decomposition. Interestingly, the logic implementation of STG in Fig. 8.25 is identical to the one described by equations 8.3, which shows that our earlier simplification of the input behavior was not critical. This is certainly not true in general and designers must be careful when performing structural decomposition of circuits and singling out components which share some global signals.

It should also be noted that in our example output signals Lam and Laf of the entire scheduler are implemented using outputs La of the five cells in the following way: $Lam = La(1) + \cdots + La(4)$ and $Laf = La(0)$. Here cell indices are associated naturally with the register bits, and hence we assume that output R of the scheduler is produced by cell number 4, i.e. $R = L(4)$.

Another issue related to the structural decomposition of the *global* specification of a control circuit into the *local* specifications of its components is that of functional verification. This issue has not been addressed at length in this book because its focus is on non-hierarchical aspects of logic synthesis (except, of course, for low-level logic decomposition). On the other hand, real life design practice cannot ignore this important issue because most designs are performed hierarchically.

The fact that our local cell model is functionally correct, i.e. that it corresponds to the overall behavior of the scheduler defined in Fig. 8.21, may be verified by composing such STGs together. We can thus explore the behavior of the composition with the internal signals, corresponding to the token handshakes between cells, considered as silent events and eventually projected away. This can be done in various ways, e.g. by using the appropriate commands in `petrify`, or by using a more appropriate tool such as

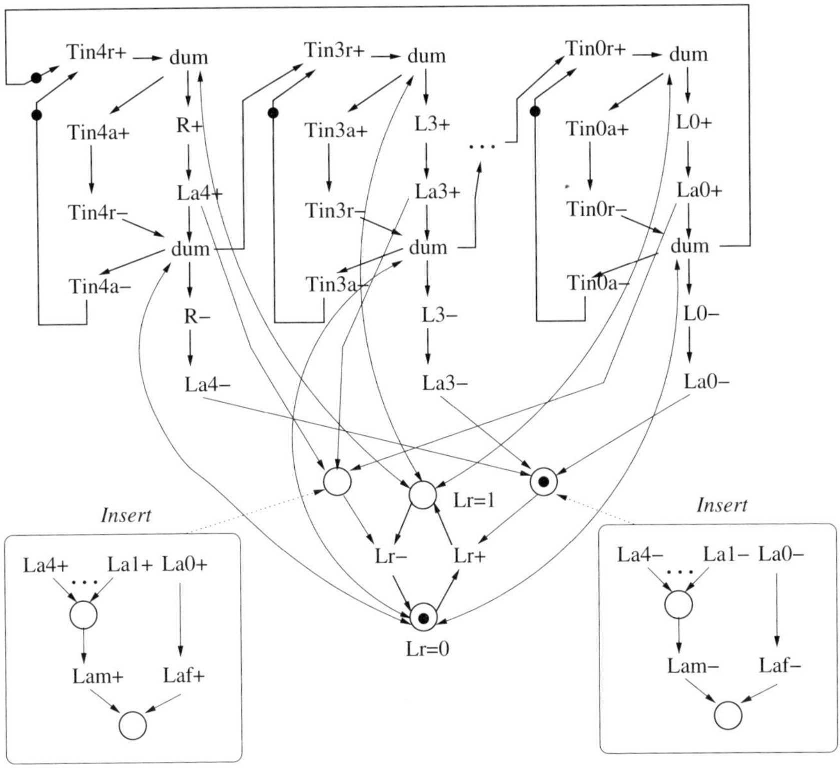

Fig. 8.26. Verifying the composition of cells

those described in [100, 164]. The composition mentioned above is shown in Fig. 8.26.

Note in this diagram how the actions on the three-wire (Lr,Lam/Laf) handshake are all merged into a single fragment, almost in a subroutine style, as compared to their unfolded instances in Fig. 8.21.

8.3.4 Synthesis of the Data Register

Finally, for completeness, we can also apply STGs to synthesize some data-path logic, namely one bit of the register (see Fig. 8.27 and 8.28). Note the use of the assignment-type notation, e.g. $y1\hat{\ }1$ (meaning $y1 = 1$), for the transitions in the STG. This actually stands for a fragment of Petri net, inserted by `petrify` automatically, which sets the value of signal $y1$ to 1 if it was at 0 initially (transition $y1+$) or keeps it the same if it was already at 1 (a dummy event). This is illustrated by the insertion box in Fig. 8.27.

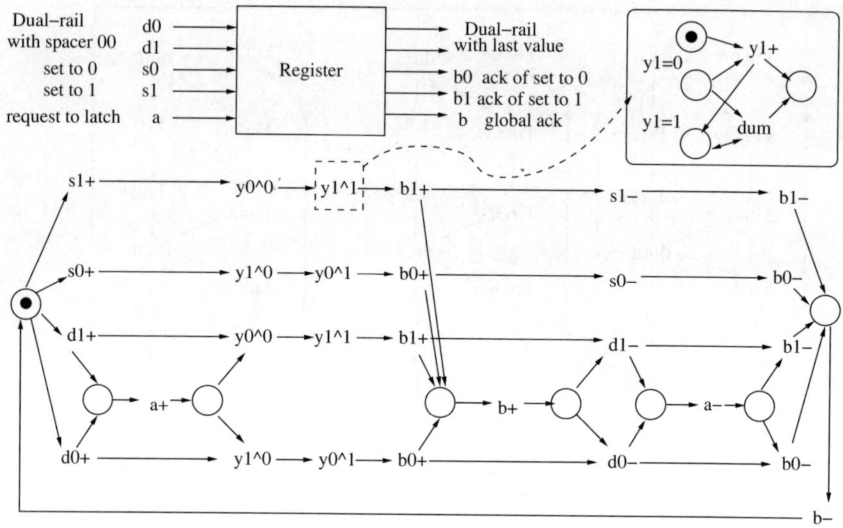

Fig. 8.27. STG model of one bit of the register

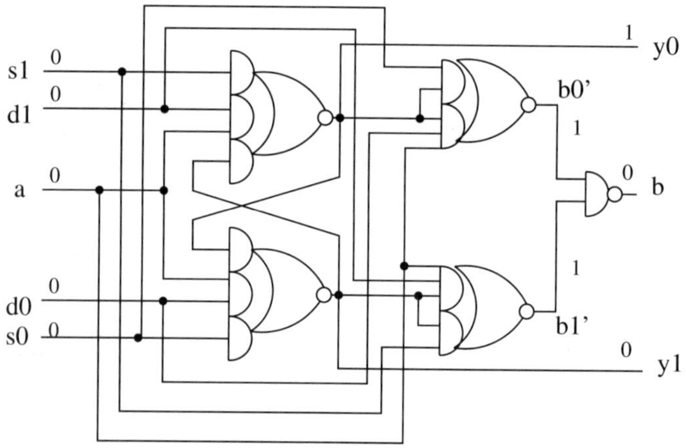

Fig. 8.28. Circuit implementation of one bit of the register

8.3.5 Quality of the Results

This hierarchical controller can be used as a first-cut design, but it involves some overhead, in that for example the signaling between the register and the comparator is done via the control logic. Additional signaling is also introduced in the controller between the environment and the scheduler.

The design published in [162] used a number of circuit optimizations. For example, it was eventually determined that a faster control logic would be obtained by allowing direct pipelining between the register and the comparator, as well as imposing the responsibility to interact with the environment on the scheduler. All such improvements were however the result of better understanding of the controller, when it was first considered as a decomposition.

8.3.6 Lessons to be Learned from the Example

1. This example shows how to *decompose* a controller into parts and construct a manageable STG model for each part. Some optimization at a later stage may be possible.
2. These STGs involve *concurrency, choice, multiple labels, dummy events, and use of short-hand assignment-type transitions*. Various ways of reshuffling handshakes could be applied in order to improve performance [156]. Relative timing assumptions, e.g. assuming that the environment changing input Lr is slower than the internal gates in the scheduler, could also be used.
3. The STGs involve *regular patterns*, which lead to many CSC conflicts. Those are better dealt with either by inserting internal memory signals by hand, or by further decomposition of the controller into cells. Currently, tools are unable to deal with regular fragments and they use a log-like state encoding policy, which may not be good for performance.
4. When performing structural decomposition of a larger circuit into parts, an *input signal is shared between these parts*. Care must be exercised when modeling the behavior of the environment for such signals, which can be more complex due to the effect of other parts. Sometimes non-determinism may appear in pursuit of generality.
5. There is no arbitration in this example. Any sort of choice in those STGs is purely *non-deterministic choice made in the environment*, often due to abstraction of its, otherwise deterministic, behavior.

8.4 "Lazy" Token Ring Adapter

The synthesis methods described in this book are targeted at logic circuits, which means that the STG specification of such circuits must satisfy the property of output persistency (Definition 4.2.6). This property guarantees that no glitches appear on the outputs of the logic gates in the synthesized circuit. While restricting the class of designs to output-persistent models is convenient for the use of logic synthesis methods, it appears that a large class of practically important interface and control circuits falls outside these conditions. For example, arbiters or interrupt controllers are precisely this kind of devices. Their specification involves conflicts between output signals. How can we handle such circuits without involving mixed-signal, analog techniques?

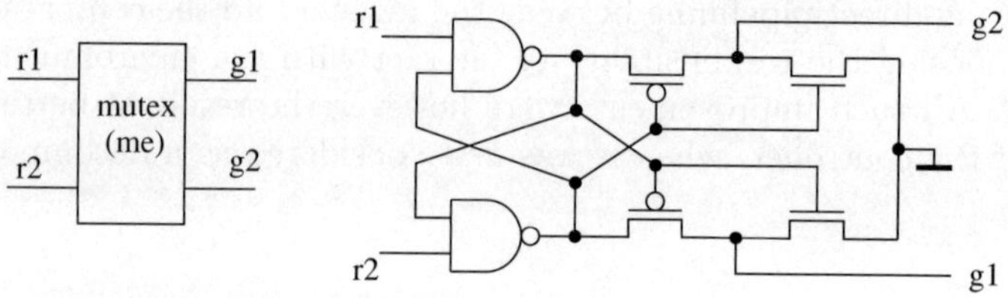

Fig. 8.29. Mutex element in CMOS

The method that we use is based on two important aspects, as discussed in [107]:

- We assume the existence of a reliable component, called a *Mutex* element, for two-way arbitration. Its well-known CMOS implementation (cf. Seitz' original NMOS circuit [91]) is shown in Fig. 8.29.
  Such an element involves analog sub-circuits, specifically aimed to deal with low thresholds and small potential differences. It confines the metastable condition within itself and resolves it to the outside world by simply "waiting long enough".
- We transform an STG specification with output non-persistency to an equivalent one, sometimes by making certain timing assumptions, in which signals which exhibit conflicts are "moved out" to the environment and declared as inputs to the Boolean part, which is subject to logic synthesis.

The design example described in this section will require us to build a circuit with arbitration.

8.4.1 Lazy Token Ring Description

Fig. 8.30 shows a distributed arbitration system with a ring architecture. Each client may want to access the shared resource (e.g. a communication channel). To gain such access the client interacts with the system via an adapter. The protocol between the client and its adapter is a simple four-phase two-wire (request-grant) handshake.

The basic idea of the distribution of the so-called "privilege token" (called P-token to avoid confusion with a Petri net place token) is depicted in the Petri net of Fig. 8.30, which describes a "basic section" in the arbitration process. Note that at any time only one adapter holds the P-token. Thus only one client can obtain the grant at a time.

If an adapter receives a request from its client, this request must be forwarded to the right, in order to fetch the P-token. This request will propagate clockwise through the ring until it reaches the adapter which currently holds the P-token. The P-token will then travel anti-clockwise until it reaches the adapter whose client has generated the request. This adapter then issues the grant to the client. When the service of the client is finished, the P-token

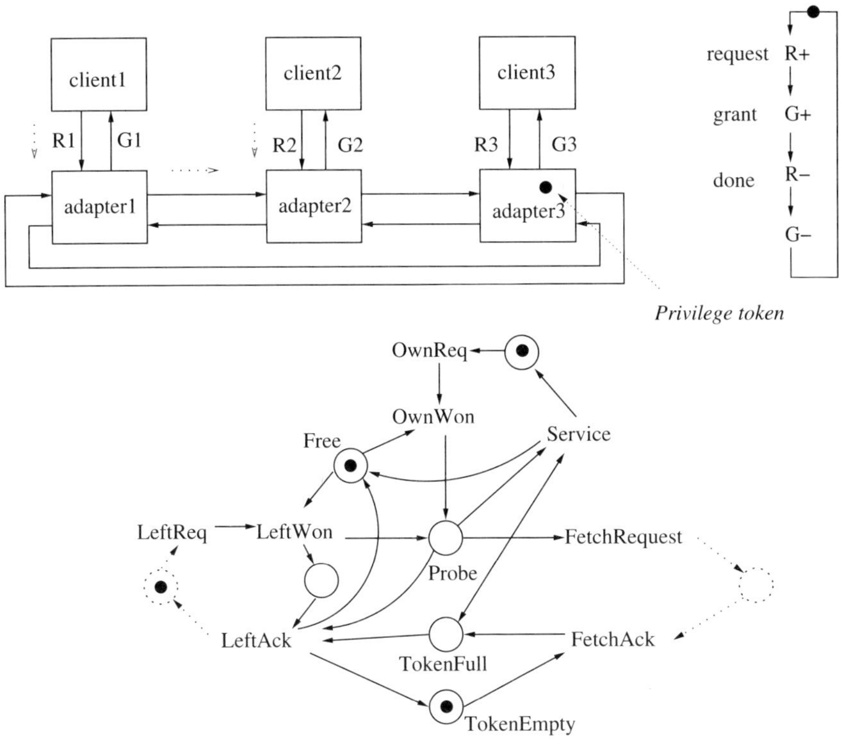

Fig. 8.30. High-level description of the "lazy ring" protocol

remains with the adapter until some other adapter at its left generates a request.

We call this arbitration mechanism "Lazy Token Ring" due to the fact that the P-token remains static until it is requested (this protocol was described in [165]). By contrast, an alternative mechanism, where the token is constantly rotating is called "Busy Token Ring". Various designs for such arbitration systems have been considered in literature. The technical report in [166] describes a number of designs using Petri nets, STGs and synthesis tools.

8.4.2 Adapter Synthesis

The STG specification for the adapter is shown in Fig. 8.31. The key element in this specification is arbitration between the request from the client (R) and the request from the left neighbor in the ring (Lr). The Petri net place labeled "me" (mutual exclusion), with a single token in it, designates the presence of dynamic (output) choice in the circuit.

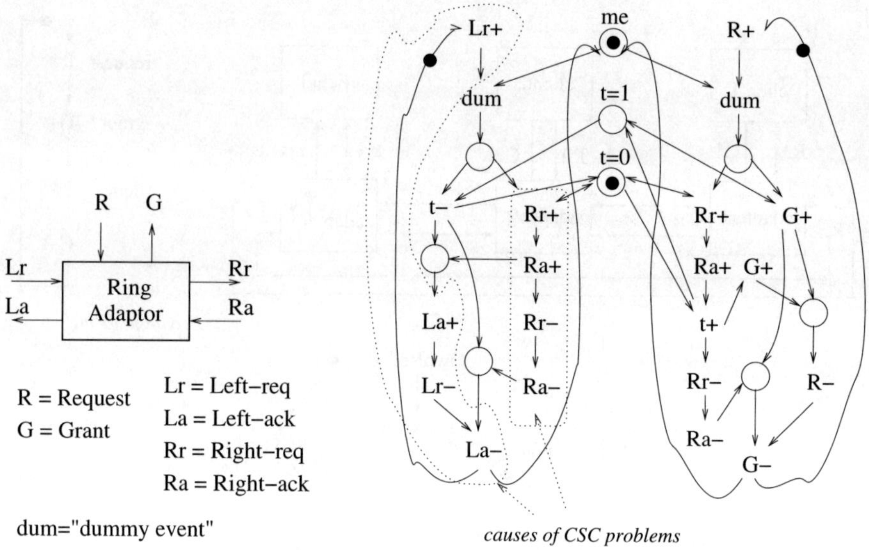

Fig. 8.31. Adapter specification

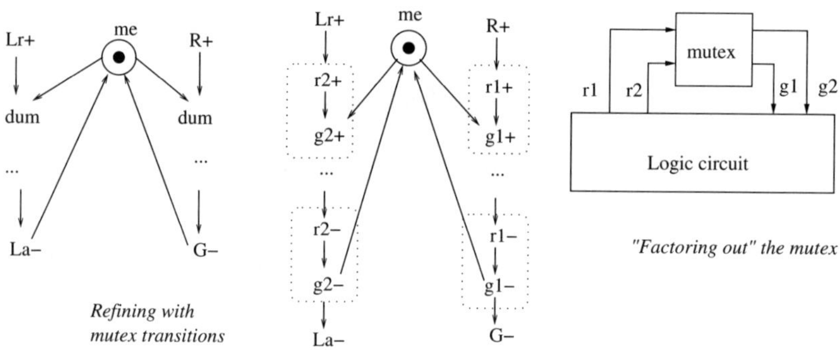

Fig. 8.32. Inserting mutex events

The specification consists of four main operation modes. These are determined, firstly, by whether the local arbitration is won by the neighbor's or the client's request. Within each of those halves, there is also choice on whether the P-token is already in the adapter or must be fetched from the right neighbor.

Note that this STG has CSC problems, because events belonging to one handshake are decoupled from other handshakes.

As we discussed before, one cannot synthesize a logic circuit with dynamic choice or conflicts between output transitions, due to the requirement of *output persistency*. Thus we should first "factor out" the conflict resolution component from the synthesized logic circuit. This is done by inserting mutual exclusion signal transitions into the STG, and declaring the grant signals of the mutex element as *inputs* to the synthesized logic circuit, as shown in Fig. 8.32. Note that the insertion of mutex events must keep the overall specification equivalent to the original STG with respect to the order between the signal transitions. This can sometimes be a problem, and requires making various assumptions, such as relative timing ones (more about this is in [107]).

The result of refinement and the circuit synthesized from the STG, which has an additional internal signal s for CSC resolution, is shown in Fig. 8.33.

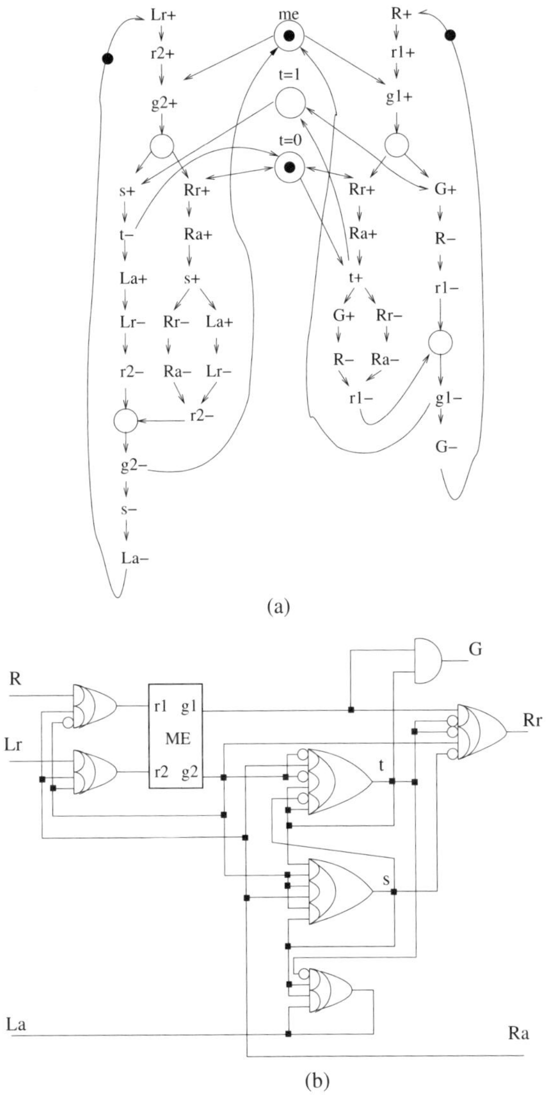

Fig. 8.33. Refined STG and circuit implementation

8.4.3 A Model for Performance Analysis

When designing circuits with arbitration, it is particularly difficult to estimate delays, due to the presence of non-determinism in their behavior.

Fig. 8.34 shows the advantages of using Petri nets in the analysis of circuit performance. The figure depicts the unfolding of the composition of the Petri net models of the adapters into a partial order graph [148], from which one can determine the maximum delay path between each pair of events of interest (in this example these are the dotted paths). For that, one should first obtain the delays of elementary actions, such as:

- $delay(Lr+ \to Rr+)$, for the case when $t = 0$,
- $delay(Lr+ \to La+)$, for the case when $t = 1$,
- $delay(Ra+ \to La+)$,
- $delay(Ra+ \to Rr-)$,

and so on, from the circuit implementation of the adapter. For example, $delay(Lr+ \to Rr+)$ is equal to the delay of two complex gates $r2$ and Rr plus

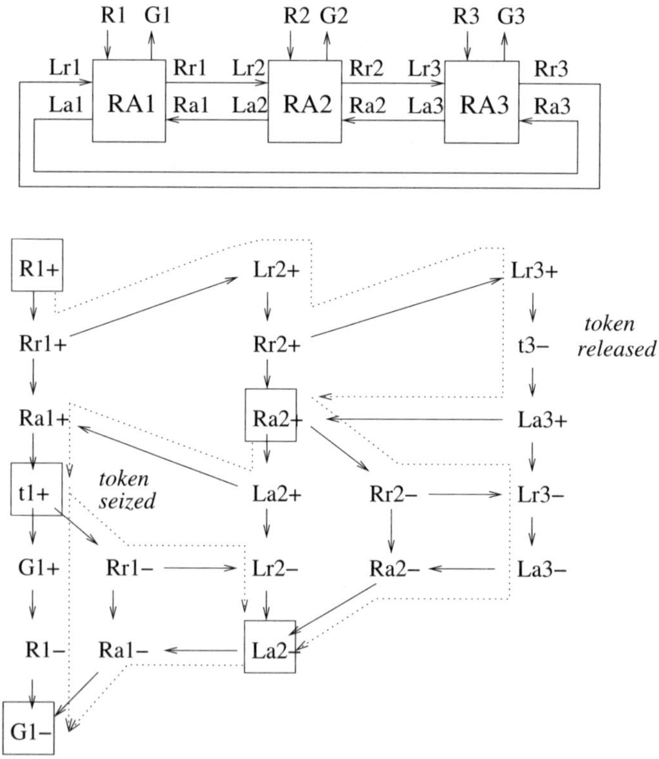

Fig. 8.34. Using partial order (unfolding) for performance analysis

the $r2$-to-$g2$ delay of the mutex, possibly involving metastability resolution (in the worst case scenario, $r1+$ and $r2+$ may arrive very closely).

The techniques for timing analysis and verification using partial order extraction are described in [158, 150]. For regular structures, such as pipelines, one can use techniques producing closed form bounds [167]. In our example, due to its regularity, one may also attempt to find a closed form bound on the maximum delay between the arrival of request $R1$ and the release of the grant $G1$ in a ring with n adapters between the requester and the adapter which currently holds the P-token. We, however, leave this task as a challenging exercise to the reader.

8.4.4 Lessons to be Learned from the Example

1. STG models of controllers with arbitration use dynamic *arbitrating choice*, as opposed to free-choice for non-deterministic selection made by the environment, and *mutex places*.
2. Designing controllers with arbitration requires handling *non-persistent* output signals. Mutex elements must be used for that, which requires inserting mutex events in the STG, and declaring the grants as inputs, i.e. *factoring out arbitration*. Sometimes, for multiple arbitration points, one may decide either to use many mutex elements or to multiplex requests on one mutex.
3. In order to evaluate the performance of asynchronous control circuits one may use the partial order semantics of the specification. The latter can be extracted from the STG, for example by means of unfolding [148].

8.5 Other Examples

Other examples of constructing Petri nets and STGs, with their subsequent synthesis using `petrify` can be found in the bibliography:

- Examples of Petri net and STG models of arbiters [168] and interrupt handlers [169]. They show various ways of inserting mutex elements.
- Deriving a Petri net model from a State Graph specification of a counterflow pipeline stage controller [170]. This example shows many tricks in constructing a synthesizable STG, including refinement of the State Graph to a form which can produce a Petri net by using the theory of regions, refinement of the Petri net into an STG with handshake signals, insertion of mutex events, and CSC signal insertion.
- Construction of Petri net models for a loadable modulo-N counter [169]. This example shows decomposition of a controller, a counter, into small parts which can be managed by the synthesis tool. It also demonstrates the effect of Petri net unfolding on understanding the behavior of the counter

and analyzing its performance, e.g. verifying that the response time is constant and independent on its size.

Finally, it is clear that a large class of controllers cannot be efficiently handled at the low (STG) level and with `petrify` [171]. These controllers require a different strategy in deriving the implementation, including the solution of the Complete State Coding problem, namely the use of direct translation techniques from Petri nets to structures like those based on David cells [172, 173, 35, 174].

9. Other Work

There are still several open issues in the design of asynchronous circuits that have not been completely covered by CAD tools. Some of them, and related research, are described in the next few sections. The interested reader is referred to a recent special issue of "Proceeedings of the IEEE" [175] for further references.

9.1 Hardware Description Languages

The design flow based on synthesis, placement and routing (SP&R) was instrumental in enabling the phenomenal productivity increase that allowed design ability to keep up with Moore's law for the last 15 years. While manual layout is still the preferred option for the datapath of high-performance micro-processors, virtually every other digital design today is performed using one of the several incarnations of the SP&R flow. This is true for the controllers and the peripherals of high-performance microprocessors, as well as for the lowly and cheap ASICs that play music when a musical greetings card is opened.

The SP&R flow is based on a few key enablers:

- the use of standardized Hardware Description Languages, VHDL and Verilog, and in particular of their "synthesizable subsets" for specification, for all design stages before layout.
- the separation between functionality and timing, that allows one to use powerful Finite State Machine and Boolean Algebra formalisms to optimize the logic implementation of multi-million gate circuits without regard to subsequent implementation choices.
- the availability of placement, routing, extraction and back-annotation tools that allow one to automatically generate the layout, and to verify the validity of this separation after layout and before fabrication.

Anybody who promotes the use of asynchronous circuits for anything but critical portions of expensive, widely used circuits (in practice, this means top-of-the-line microprocessors) must come to terms with the huge advantages of this automated, broadly standardized design flow. Any suggestion that the advantages of asynchronous circuits discussed in Chap. 1 can be achieved by requiring extensive re-training of designers to use radically new methodologies, tools and languages, or, even worse, asking them to perform a substantial amount of low-level work by hand, will encounter significant resistance from both designers and managers. Of course, things may radically change when (if) asynchronous becomes the *only* option to solve such problems ...

In the last few years, thus, we have seen a number of efforts to bring asynchronous design *flows* on a par with their synchronous counterparts, as the latest stage of a long effort to push this promising and theoretically clean design style into mainstream use. This book discussed some of the automation issues in the *logic synthesis step* for asynchronous controllers. However, *flow-related* aspects must also be taken into account.

One of the first complete flows, including both control and datapath design aspects, was developed by Martin's group at CalTech [105, 40, 149]. It was based on Communicating Sequential Processes (CSP) both as specification and as intermediate language. I.e. CSP was used to model asynchronous circuits:

- at the Register Transfer Level, where statements specify assignments, loops, as well as channel reads, writes and probes,
- at the event level, which is close to the STG level discussed in this book, where statements are sequences of assignments of Boolean values to signals and waits for signal transitions,
- at the elementary gate excitation level, which is close to Boolean equations, where statements are parallel guarded assignments of Boolean values to output signals.

This flow has several steps that are closely related to those described in this book, especially for the translation between the two last levels (e.g. insertion of state variables or concurrency reduction in case of state conflicts).

A few asynchronous microprocessors have been designed using this technique [176, 28]. Others also suggested restricting the implementation of a CSP program by using a set of small pipeline templates, which can lead to high throughput and low latency asynchronous pipelined circuits [177].

The Tangram group at Philips Research [178, 179] has long promoted the use of a flow based on a CSP-based language called Tangram, relying on a well-defined (proprietary) tool flow. Philips did a few designs using this technique [180, 181, 182, 183, 184, 19]. The Balsa group at Manchester University has developed a suite of non-proprietary tools that closely follow the Tangram methodology [185, 186]. In both cases, synthesis is performed by *syntax-directed translation* of the CSP specification. A similar step was also used by Martin's group for the translation between the first two levels.

This approach guarantees close control by the designer on the cost and performance of the resulting circuit, similar to what a good C programmer can achieve by using a relatively high-level language to optionally perform fairly low-level optimizations by hand. Thus it has been termed "VLSI programming". The approach of [187] is also similar in spirit, and so is that of [188]. The latter also introduces aspects of high-level synthesis, such as scheduling and allocation, into the pure syntax-directed approach. All syntax-directed approaches then rely on peephole optimizations to remove some of the redundancies that they inevitably introduce [106, 189, 190].

Other researchers claim that a realistic flow must deviate as little as possible from synchronous, standard HDL-based ones. A semantically faithful asynchronous implementation of the synthesizable subset of VHDL or Verilog would be an ideal solution. This is of course not possible, since at the very least, asynchronous implementations must use some form of externally visible *handshaking* for synchronization.

Renaudin's group [191] introduced standard Hardware Description Languages into a CSP-based flow, by translating a specification written in a subset of hardware-oriented CSP into VHDL processes.

Some approaches to generate and optimize asynchronous controllers with standard synchronous synthesis tools, such as the Theseus Logic approach, use a Finite State Machine-like specification style for the combinational logic [192]. However, they require registers to be manually instantiated, rather than inferred from HDL statements as in the synchronous case.

Another HDL-based approach [193] uses a high-level subset of Verilog for specification, and provides an asynchronous ASIC-style design flow. It relies on a specially developed tool only to perform the asynchronous equivalent of RTL synthesis (an initial high-level synthesis step is optional), and to manage the overall flow, by generating the appropriate scripts and calling the appropriate synchronous Electronic Design Automation tools.

All the approaches listed in this section give up some optimality, both in the choice of specification language, and in the implementation of the controller, to gain in terms of widespread applicability and ease of use. They are thus suited for fast-turnaround design cycles of low-performance circuits with low or medium production volumes, but, for example, *stringent electromagnetic emission or low-power requirements*. While this is by no means a majority of the designs around, it is a significant (growing) fraction of ASICs.

Moreover, they can produce an input suitable for the logic synthesis tools described in this book, at least for the most performance or cost-critical portions, and thus re-gain some of the lost efficiency [194].

9.2 Structural and Unfolding-based Synthesis

The synthesis techniques described in this book heavily rely on the use of the binary-encoded State Graph model, which provides:

- information for analysis of logic circuit implementability (boundedness, consistency, output-persistency and CSC), and
- a specification for the logic functions of the gates in the target circuit.

The second aspect is directly related to the two main synthesis steps discussed so far, state encoding and logic decomposition, as well as to synthesis with relative timing. Usually such an approach is normally seen as logic synthesis from a finite state machine description, for both synchronous and asynchronous design. It typically involves the optimization of the size of the synthesized circuit, which appears to be a primary quality criterion. The synthesis and analysis algorithms discussed so far however require a complete representation of the state space, which unfortunately grows exponentially with respect to the number of signals.

There are ways to avoid the complete state-space traversal of an STG specification when performing all or some steps of logic synthesis from STGs. These ways largely follow the approaches existing in the area of PN analysis and model checking which do not perform explicit state-space traversal. Those methods fall into two main categories:

- *Structural* methods. This approach checks the implementability of an STG in a logic circuit and derives Boolean covers for its implementation by means of the analysis of the structure of the underlying PN, e.g. partitioning it into State Machine components [117, 195, 196, 65]. These methods are applied to STGs whose underlying PNs belong to the class of bounded and live free-choice nets[70].
- *Partial-order* methods. This approach performs verification of the implementability properties and synthesis of the excitation and/or next-state functions for signal implementation by analysis of a finite prefix of the STG unfolding [197, 198, 199, 200, 201]. Such a prefix always exists for a bounded PN and captures (implicitly) all its reachable markings [148]. The size of the prefix (for 1-safe PNs) is in most practical cases linear with respect to the size of the original STG specification.

The above approaches are inter-related in the sense that both use the structural information, obtained either directly from the PN or its unfolding, in order to find approximations of the Boolean covers for the functions of output signals. The so-called temporal relations, precedence, concurrency and conflict, play an important role in these methods. The complexity involved in their computation and manipulation is polynomial in the size of the model (PN or unfolding prefix). The approximate nature of the methods presents a trade-off between the quality of the obtained solutions and the complexity of the optimization procedure. Indeed, in order to guarantee the best solution, these techniques use Boolean cover refinement methods, which require, albeit partial, state-space exploration.

9.3 Direct Mapping of STGs into Asynchronous Circuits

The use of structural or partial order methods, described in the previous section, helps to mitigate the complexity problem concerned with the logic synthesis approach. However, these methods are still limited in doing so due to the following issues: they are approximate in nature (exact solutions require traversing the whole state-space), the class of PNs for which they are applicable is fairly restricted (free-choice nets), and finally they exhibit polynomial complexity, rather than linear, which is still often excessive.

As an alternative to logic synthesis, one may consider the *direct translation or mapping* of behavioural specifications to circuit implementations. In this approach, which is sometimes associated with *one-hot state encoding* used in logic synthesis, the circuit acts as an object whose structure homomorphically represents the structure of the PN or STG specification. The complexity of such a syntax-based translation is linear with respect to the size of the specification. Known direct mapping techniques fall into two main categories:

- *"Place-to-latch" mapping*, which is based on the association of each place in the net with a memory element, e.g. an Sr-latch, and of each transition with appropriate logic at the inputs of the latches. The circuit in its action effectively simulates the token game of its Petri net specification [35, 64, 174]. This technique has been used in designing an asynchronous pipeline token ring interface [202]. The overall approach originates from the ideas of David about one-hot mapping of FSMs [172], advanced later by Hollaar for "parallel" FSMs [173]. The recent work of [203] incorporates the idea of direct mapping and one-hot encoding into the framework of logic synthesis.
- *Event-based two-phase circuit translation*, which is based on the correspondence between the firing of a PN transition and the switching of a logical level of a circuit signal. The firing of a transition has purely an event-based meaning, and it does not imply the direction of a transition between logic levels [35, 204]. This type of signaling is sometimes called *two-phase or non-return-to-zero*. It was used by Sutherland in his micropipeline control circuits [29]. This translation strategy originated from the work of Patil, Dennis and Furtek [205, 51, 206, 207], and was developed later by Misunas [208].

Referring to work in a similar vein, which can be collected under a generic title of "modular design", we should certainly mention the Macro-modules method of Clark et al. [209, 210, 211], the register-transfer modules of Bell and Grason[212], and the methods of Bruno and Altman [213], Keller [214], Nordmann and McCormick [215] to name but a few.

9.4 Datapath Design and Interfaces

As discussed in Chap. 1, a typical digital design is conventionally divided into controllers and datapaths. The boundaries between these two are sometimes blurred, however they typically exhibit different features: controllers are irregular and often relatively small, while datapaths are regular and large. For this reason, design tools are also different. In the synchronous case, controllers are generally specified as FSMs, then optimized using logic synthesis, and finally placed and routed using automated tools. Datapaths are often produced by interconnecting the results of module generators (e.g. adders, multiplexers, register files). These generate either directly the final layout, or gate-level netlists, subject to a limited amount of logic synthesis, placement and routing.

This book describes in detail how to implement asynchronous controllers, but provides little information on datapath design. This stems from the common belief that implementing an asynchronous datapath is much simpler than implementing an asynchronous controller. To justify this statement, let us review existing techniques for asynchronous datapath design.

- **Asynchronous-design.** The datapath is implemented using asynchronous logic design methods, constrained to preserve the usual correctness properties: hazard-freedom, delay-insensitivity, etc. In fact the problem is not much different from controller synthesis, and the very same methods could be used to design both. Note, however, that this approach shares the usual limitations of asynchronous synthesis. Thus it can be applied only to very small datapath modules. Larger blocks can be obtained by composition of smaller ones, by exploiting datapath regularity. This approach was also used in manual design methods, e.g. [105, 40].
- **Direct-mapping.** This method relies on cell-level or layout-level block generators, whose output directly satisfies the required correctness properties [191, 178, 179, 185]. The list of blocks (adders, encoders, shifters, etc.) is obviously limited, especially in the case when layout is generated. Optimization across block boundaries is virtually impossible, but the regularity of datapaths again helps to obtain relatively efficient implementations. Moreover, layout generation, or specialized placement and routing, greatly reduces the problems due to poor control of wire delays.
- **Bundled-data.** This approach [91, 29] is an elegant way of combining together an asynchronous control unit and a synchronous datapath. The control unit produces request signals that trigger the start of each computation in the datapath, and receives back acknowledge signals carrying information about computation completion. In the implementation, the acknowledgment signal is simply generated by a delay that must be *greater than the longest delay within the datapath* (matched delay). Its purpose is to provide enough time for the datapath to complete its computation before acknowledge is activated. In some sense, the request signals play the

role of local clocks for synchronous datapaths. This implementation style is very popular [18, 31, 19], because no sophisticated design techniques are required to assemble datapath sub-circuits. Moreover, conventional synchronous design tools can be used for synthesis, simulation, timing analysis, placement and routing, and so on.
Of course, this style also has the disadvantage that it cannot use true completion detection techniques in order to exploit average-case performance, since every datapath must always work at its worst-case speed. Hybrid techniques [4] use a set of matched delays, from fast to slow, attached to the datapath unit, and an appropriate delay is selected to indicate completion for each operation.

In all these approaches, the coordination between controller and datapath is done through a handshake protocol. For direct-mapping and bundled-data methods this protocol relies on explicit request-acknowledge signals, generally with a four-phase transition scheme $req+ \rightarrow ack+ \rightarrow req- \rightarrow ack-$ .... For the asynchronous-design approach, on the other hand, there need not be special signals, but handshaking proceeds through the proper sequencing of input changes and output responses.

Finally, we can conclude that datapath implementation is not really a bottleneck, as long as the designer is ready to make the correct trade-offs. For example, asynchronous-design is the most difficult and most efficient technique, bundled-data is the easiest, but is not timing-independent, and direct-mapping lies somewhere in the middle, but requires a significant library investment effort.

9.5 Test Pattern Generation and Design for Testability

Any type of integrated circuit must be tested after manufacturing, to make sure that it operates properly. A large body of literature and extensive industrial practice has been developed to test synchronous digital circuits.

Unfortunately, testing asynchronous circuits is a difficult problem, due to the following main reasons ([216]):

- All known asynchronous design methodologies ensure correct *operation* (i.e. avoid hazards on circuit outputs) by using some level of *redundancy*, i.e., by sacrificing testability.
- Asynchronous control circuits tend to have more feedback and more registers than their synchronous counterparts. This means that full-scan testing may be unacceptably expensive.

This difficulty is true despite the often heard claim that asynchronous circuits are "self-checking", i.e. that they stop in the presence of faults. This desirable property is unfortunately satisfied only for a small subset of faults, those that are functionally equivalent to delays to which the circuit is insensitive. This means that speed-independent circuits are self-checking to stuck-at-output faults [217, 218, 219] and delay-insensitive circuits are insensitive to stuck-at faults on all *observable* signals. Unfortunately no circuit can really operate without any timing assumptions. Even delay-insensitive circuits require some assumptions on the delays *within the building blocks*, much larger than standard Boolean gates, which are used for their implementation. Most faults that violate a required timing assumption will indeed cause a malfunction.

For this reason, a relatively large amount of work has focused on generating test sequences for a given set of faults in an asynchronous circuit (see [220, 221] for a physical defect-oriented evaluation of various fault models in an asynchronous context):

- either by using greedy heuristic techniques [222] to justify and propagate stuck-at faults,
- or by using manual transformations to ensure that a simple functional testing approach could test all stuck-at faults [223],
- or by using full-scan [224, 225, 226, 227] or partial-scan [228, 229] and standard combinational ATPG techniques to test the chosen *stuck-at* or *path delay* faults *robustly*, i.e. independent of the gate delays.

9.6 Verification

A reliable design flow must include verification and validation techniques. Simulation of an asynchronous circuit might require examining more traces than in the synchronous case, as one cannot rely on the fixed points reached at the end of every clock cycle. The functional verification issue, however, is often simplified by the fact that the behavior of the asynchronous implementations derived as described in this book is provably independent of component delays. Therefore any trace can be chosen for a representative simulation. The use of standard HDLs, as discussed in Sect 9.1 permits using standard simulators, originally developed mostly for synchronous circuits.

Formal verification techniques for asynchronous systems are also conceptually similar to those for synchronous systems, and are generally based on model checking. Model checking in the synchronous case is based on state space traversal of synchronous automata models, while in the asynchronous case it uses transition system models. The difference is essentially in the arc labeling, because multiple signal changes are allowed in the synchronous case, while only single signal changes are generally considered in the asynchronous case.

- Property verification for asynchronous circuits started with Muller's group, who checked semi-modularity conditions by traversing states of the circuit. The state space was represented explicitly using binary vectors. This approach was later improved in [64]. Model checking of different properties, based on symbolic traversal of the state space or on Petri Net unfoldings, was studied in McMillan's PhD thesis [230].

- Implementation verification was formulated as conformance checking in the framework of trace theory by Dill [231]. This approach was later modified by Ebergen in his verifier [232] and by other authors. Roig developed a model checking technique for verifying conformance of an implementation with respect to an STG specification model [164].
- Timing verification is required for synthesis of asynchronous circuits relying on timing assumptions (e.g. the relative timing circuits discussed in Chap. 7), as well as for verifying interfaces between synchronous and asynchronous systems. Approaches have been proposed based on timed automata [233, 234], geometric regions [152], partial orders [235], and composing untimed transition system with small timed event structures generated on demand when a failure is found [158]. Timing properties are more difficult to verify than untimed ones, and more research in this area is required.
- Methods to calculate the time separation of events have also been proposed to verify properties on asynchronous interfaces or find bounds on the performance of concurrent systems. Due to the complexity of the problem, the algorithms proposed so far work on choice-free specifications [151, 236], or specifications with non-deterministic choices [237]. Recently, symbolic timing analysis has also been proposed [238]. The calculation of time separation bounds for different instances of the same event in a concurrent system can also be used to determine the performance of cyclic behaviors [239, 2, 240].
- Verification of large systems exploits decomposition and theorem proving. The latter has been applied to asynchronous systems e.g. in [241].

9.7 Asynchronous Silicon

Only a few asynchronous chips, mostly microprocessors, have been fabricated, and more are doubtlessly needed to prove the potential benefits of this technology. Here we only provide some references to help the interested reader. More references can be found at the Asynchronous Logic Home page `http://http://www.cs.man.ac.uk/async/`

- Williams' asynchronous divider used early completion detection to speed up an iterative division algorithm [242].
- Martin's group at CalTech implemented several microprocessors [176, 243, 244, 28], using their CSP-based technique mentioned in Sect. 9.1.
- The Tangram team at Philips Research Labs implemented a Digital Cassette error corrector and a micro-controller that was used in a commercial pager [180, 181, 183, 184, 19].

- Furber's team at Manchester University designed at least three asynchronous implementations of the ARM instruction set architecture, including a bus to interface them with standard synchronous ARM peripherals [245, 246, 247, 248, 18, 249].
- Nanya's TITAC and TITAC-2 microprocessors were claimed to be the largest asynchronous microprocessors at the time [250, 251].
- Sharp and Terada's group developed an asynchronous dataflow media processor [24, 23].
- Intel's Strategic CAD Lab recently designed an asynchronous instruction length decoder, called RAPPID, for the X86 instruction set [26]. They used relative timing techniques in order to strive for maximum performance.
- IBM's interlock pipeline [252] is also an example of a very high-performance circuit structure using clock-less techniques.
- Renaudin's group at Institut National Polytechnique de Grenoble designed several micro-controllers, mostly for smartcard applications [253, 254].
- Finally Sun recently announced that an asynchronous FIFO, called GasP, was used in their latest UltraSPARC^{TM} IIIi micro-processor [13, 14].

The Sharp dataflow processor, the Philips pager micro-controller and the UltraSPARC FIFO are the only commercial application of formalized asynchronous design known to-date (of course, small manually designed asynchronous gate netlists have been used in many other instances). They were chosen for low power, low electro-magnetic emission and modularity (insensitivity to clock skew) reasons respectively. Intel and IBM are working on asynchronous circuits apparently for performance reasons.

10. Conclusions

At the time of writing, the role of asynchronous circuits in the designer's community is still unclear. Even though several advantages are claimed for asynchronous circuits, the methodologies to design and analyze them have tangible differences with respect to those used for synchronous circuits.

Moving from synchronous sequential models, e.g. Finite State Machines, to asynchronous concurrent models, e.g. Transition Systems, requires a change of mentality in the way designers must face the specification and synthesis of this type of circuits.

A similar change is required when one needs to specify some functionality that must be implemented on a distributed system. Traditional sequential programming paradigms, such as those embodied in languages like FORTRAN or C, do not lead to efficient results, because their parallelization is very hard. Thus people use concurrent specification mechanisms, such as dataflow diagrams, concurrent state machines, Communicating Sequential Processes, Kahn process networks and so on [255].

Asynchronous circuits are not far from the distributed computing paradigm. In fact, some of the formal models to specify, synthesize and analyze them, e.g. Petri nets and CSP, originated in that research area. Designers must change the way in which they reason about systems. Rather than thinking of *"what the system does in a cycle"*, they must think of *"which action must occur after another action"*. In other words, they must forget about cycles and reason in terms of events and their causality relations. So far, the main efforts to approach asynchronous circuits have been promoted by academia.

This book is an attempt to tear down the wall that separates asynchronous from synchronous design. It shows how a simple model, very close to the well-known timing diagrams, can be used to specify asynchronous systems. Moreover, it shows that modularity can be exploited, and that automated synthesis can be effectively used to derive correct and efficient circuits.

References

1. P.A. Beerel, K.Y. Yun, W.C. Chou: "Optimizing average-case delay in technology mapping of burst-mode circuits", in *Proc. International Symposium on Advanced Research in Asynchronous Circuits and Systems* (IEEE Computer Society Press, 1996)
2. A. Xie, P.A. Beerel: "Symbolic techniques for performance analysis of timed systems based on average time separation of events", in *Proc. International Symposium on Advanced Research in Asynchronous Circuits and Systems* (IEEE Computer Society Press, 1997), pp. 64–75
3. A.J. Martin: Formal Methods in System Design **1**(1), 119–137 (1992)
4. S.M. Nowick, K.Y. Yun, P.A. Beerel: "Speculative completion for the design of high-performance asynchronous dynamic adders", in *Proc. International Symposium on Advanced Research in Asynchronous Circuits and Systems* (IEEE Computer Society Press, 1997), pp. 210–223
5. S.H. Unger: *Asynchronous Sequential Switching Circuits* (Wiley-Interscience, John Wiley & Sons, Inc., New York, 1969)
6. R.K. Brayton, G.D. Hachtel, C.T. McMullen, A. Sangiovanni-Vincentelli: *Logic Minimization Algorithms for VLSI Synthesis* (Kluwer Academic Publishers, 1984)
7. R.K. Brayton, R. Rudell, A. Sangiovanni-Vincentelli, A.R. Wang: IEEE Transactions on Computer-Aided Design **CAD-6**(6), 1062–1081 (1987)
8. E. Sentovich, K. Singh, C. Moon, H. Savoj, R. Brayton, A. Sangiovanni-Vincentelli: "Sequential circuit design using synthesis and optimization", in *Proceedings of the International Conference on Computer Design* (1992)
9. I.S.I. Association: "International technology roadmap for semiconductors", see http://www.itrs.net/ntrs/publntrs.nsf
10. G.D. Micheli: *Synthesis and optimization of digital circuits* (McGraw-Hill, 1994)
11. M. Design: (2000), see http://www.montereydesign.com
12. M.D. Automation: (2000), "Gain-based synthesis: speeding RTL to silicon", see http://www.magma-da.com/articles/GBS.pdf
13. S. Microsystems: "Sun Microsystems previews UltraSPARC IIIi processor at microprocessor forum", see http://www.sun.com/smi/Press/sunflash/2001-10/sunflash.20011016.5.html
14. J. Ebergen: "Squaring the FIFO in GasP", in *Proc. International Symposium on Advanced Research in Asynchronous Circuits and Systems* (IEEE Computer Society Press, 2001), pp. 194–205
15. E. Eichelberger, T.W. Williams: "A logical design structure for LSI testing", in *Proceedings of the Design Automation Conference* (1977), pp. 462–468
16. V.S.I. Alliance: (2000), see http://www.vsi.org
17. M. Keating, P. Bricaud: *Reuse methodology manual for system-on-a-chip designs* (Kluwer Academic Publishers, 1999)

18. S.B. Furber, J.D. Garside, P. Riocreux, S. Temple, P. Day, J. Liu, N.C. Paver: Proceedings of the IEEE **87**(2), 243–256 (1999)
19. J. Kessels, P. Marston: Proceedings of the IEEE **87**(2), 257–267 (1999)
20. L.S. Nielsen, J. Sparsø: Proceedings of the IEEE **87**(2), 268–281 (1999)
21. N.C. Paver, P. Day, C. Farnsworth, D.L. Jackson, W.A. Lien, J. Liu: "A low-power, low-noise configurable self-timed DSP", in *Proc. International Symposium on Advanced Research in Asynchronous Circuits and Systems* (1998), pp. 32–42
22. K.v. Berkel, H. van Gageldonk, J. Kessels, C. Niessen, A. Peeters, M. Roncken, R. van de Wiel: "Asynchronous does not *imply* low power, but ... ", in *Low Power CMOS Design*, ed. by A. Chandrakasan, R. Brodersen (IEEE Press, 1998), pp. 227–232
23. H. Terada, S. Miyata, M. Iwata: Proceedings of the IEEE **87**(2), 282–296 (1999)
24. Sharp: "Sharp data driven multimedia processor", see http://www.sharpddmp.com
25. D.J. Kinniment: IEEE Transactions on VLSI Systems **4**(1), 137–140 (1996)
26. S. Rotem, K. Stevens, R. Ginosar, P. Beerel, C. Myers, K. Yun, R. Kol, C. Dike, M. Roncken, B. Agapiev: "RAPPID: An asynchronous instruction length decoder", in *Proc. International Symposium on Advanced Research in Asynchronous Circuits and Systems* (1999), pp. 60–70
27. I. Sutherland, S. Fairbanks: "GasP: A minimal FIFO control", in *Proc. International Symposium on Advanced Research in Asynchronous Circuits and Systems* (IEEE Computer Society Press, 2001), pp. 46–53
28. A.J. Martin, A. Lines, R. Manohar, M. Nystroem, P. Penzes, R. Southworth, U. Cummings: "The design of an asynchronous MIPS R3000 microprocessor", in *Advanced Research in VLSI* (1997), pp. 164–181
29. I.E. Sutherland: Communications of the ACM **32**(6), 720–738 (1989)
30. S.B. Furber, P. Day: IEEE Transactions on VLSI Systems **4**(2), 247–253 (1996)
31. K.v. Berkel, R. Burgess, J. Kessels, A. Peeters, M. Roncken, F. Schalij, R. van de Wiel: "A single-rail re-implementation of a DCC error detector using a generic standard-cell library", in *Asynchronous Design Methodologies* (IEEE Computer Society Press, 1995), pp. 72–79
32. T.E. Williams, M.A. Horowitz: IEEE Journal of Solid-State Circuits **26**(11), 1651–1661 (1991)
33. D. Harris: *Skew-tolerant circuit design* (Morgan Kaufmann Publishers, 2000)
34. T. Verhoeff: Distributed Computing **3**(1), 1–8 (1988)
35. V.I. Varshavsky (Ed.): *Self-Timed Control of Concurrent Processes: The Design of Aperiodic Logical Circuits in Computers and Discrete Systems* (Kluwer Academic Publishers, Dordrecht, The Netherlands, 1990)
36. S.S. Appleton, S.V. Morton, M.J. Liebelt: "Two-phase asynchronous pipeline control", in *Proc. International Symposium on Advanced Research in Asynchronous Circuits and Systems* (IEEE Computer Society Press, 1997), pp. 12–21
37. S.B. Furber, J. Liu: "Dynamic logic in four-phase micropipelines", in *Proc. International Symposium on Advanced Research in Asynchronous Circuits and Systems* (IEEE Computer Society Press, 1996)
38. G.S. Taylor, G.M. Blair: IEEE Journal of Solid-State Circuits **33**(10), 1590–1593 (1998)
39. K.Y. Yun, P.A. Beerel, J. Arceo: IEE Proceedings, Circuits, Devices and Systems **143**(5), 282–288 (1996)

40. A.J. Martin: "Programming in VLSI: From communicating processes to delay-insensitive circuits", in *Developments in Concurrency and Communication*, ed. by C.A.R. Hoare (Addison-Wesley, 1990), UT Year of Programming Series, pp. 1–64
41. A. Davis, S.M. Nowick: "An introduction to asynchronous circuit design", in *The Encyclopedia of Computer Science and Technology*, ed. by A. Kent, J.G. Williams, Vol. 38 (Marcel Dekker, New York, 1998)
42. S.M. Nowick, M.B. Josephs, C.H.K. van Berkel: Proceedings of the IEEE **87**(2), 219–222 (1999)
43. D.E. Muller, W.S. Bartky: "A theory of asynchronous circuits", in *Proceedings of an International Symposium on the Theory of Switching* (Harvard University Press, 1959), pp. 204–243
44. R.E. Miller: *Sequential Circuits and Machines*, Vol. 2 of Switching Theory (John Wiley & Sons, 1965)
45. A.J. Martin: "The limitations to delay-insensitivity in asynchronous circuits", in *Advanced Research in VLSI*, ed. by W.J. Dally (MIT Press, 1990), pp. 263–278
46. K.v. Berkel: Integration, the VLSI journal **13**(2), 103–128 (1992)
47. K.v. Berkel, F. Huberts, A. Peeters: "Stretching quasi delay insensitivity by means of extended isochronic forks", in *Asynchronous Design Methodologies* (IEEE Computer Society Press, 1995), pp. 99–106
48. L. Lavagno, A. Sangiovanni-Vincentelli: *Algorithms for Synthesis and Testing of Asynchronous Circuits* (Kluwer Academic Publishers, 1993)
49. S.M. Nowick, D.L. Dill: "Automatic synthesis of locally-clocked asynchronous state machines", in *Proc. International Conf. Computer-Aided Design (ICCAD)* (IEEE Computer Society Press, 1991), pp. 318–321
50. K.v. Berkel: (1992), "Handshake circuits: An intermediary between communicating processes and VLSI", Ph.D. thesis, Eindhoven University of Technology
51. J.B. Dennis, S.S. Patil: "Speed-independent asynchronous circuits", in *Proc. Hawaii International Conf. System Sciences* (1971), pp. 55–58
52. C.E. Molnar, T.P. Fang, F.U. Rosenberger: "Synthesis of delay-insensitive modules", in *1985 Chapel Hill Conference on Very Large Scale Integration*, ed. by H. Fuchs (Computer Science Press, 1985), pp. 67–86
53. J.T. Udding: Distributed Computing **1**(4), 197–204 (1986)
54. M.R. Greenstreet, P. Cahoon: "How fast will the flip flop?", in *Proc. International Symposium on Advanced Research in Asynchronous Circuits and Systems* (1994), pp. 77–86
55. R. Otten, R. Brayton: "Planning for performance", in *Proceedings of the Design Automation Conference* (1998)
56. H. Saito, A. Kondratyev, J. Cortadella, L. Lavagno, A. Yakovlev: "What is the cost of delay insensitivity?", in *Proc. International Conf. Computer-Aided Design (ICCAD)* (1999), pp. 316–323
57. D.A. Huffman: "The synthesis of sequential switching circuits", in *Sequential Machines: Selected Papers*, ed. by E.F. Moore (Addison-Wesley, 1964)
58. A. Davis, B. Coates, K. Stevens: "Automatic synthesis of fast compact asynchronous control circuits", in *Asynchronous Design Methodologies*, ed. by S. Furber, M. Edwards (Elsevier Science Publishers, 1993), Vol. A-28 of IFIP Transactions, pp. 193–207
59. C.A. Petri: (1962), "Kommunikation mit automaten", Ph.D. thesis, Bonn, Institut für Instrumentelle Mathematik, (technical report Schriften des IIM Nr. 3)
60. T. Murata: Proceedings of the IEEE pp. 541–580 (1989)

61. J.L. Peterson: *Petri Net Theory and the modeling of systems* (Prentice-Hall, 1981)
62. L.Y. Rosenblum, A.V. Yakovlev: "Signal graphs: from self-timed to timed ones", in *Proceedings of International Workshop on Timed Petri Nets* (IEEE Computer Society Press, Torino, Italy, 1985), pp. 199–207
63. T.A. Chu, C.K.C. Leung, T.S. Wanuga: "A design methodology for concurrent VLSI systems", in *Proc. International Conf. Computer Design (ICCD)* (IEEE Computer Society Press, 1985), pp. 407–410
64. M. Kishinevsky, A. Kondratyev, A. Taubin, V. Varshavsky: *Concurrent Hardware: The Theory and Practice of Self-Timed Design*, Series in Parallel Computing (John Wiley & Sons, 1994)
65. E. Pastor, J. Cortadella, A. Kondratyev, O. Roig: IEEE Transactions on Computer-Aided Design **17**(11), 1108–1129 (1998)
66. C.A.R. Hoare: *Communicating Sequential Processes* (Prentice-Hall, 1985)
67. J. Esparza, M. Nielsen: Petri Nets Newsletter **94**, 5–23 (1994)
68. B. E., D. R.: Theoretical Computer Science **55**, 87–136 (1988)
69. J. R., K. M.: Theoretical Computer Science **112**, 5–52 (1993)
70. J. Desel, J. Esparza: *Free-choice Petri Nets*, Vol. 40 of Cambridge Tracts in Theoretical Computer Science (Cambridge University Press, 1995)
71. J. Desel: "Basic Linear Algebraic Techniques for Place/Transition Nets", in *Lectures on Petri Nets I: Basic Models*, ed. by W. Reisig, G. Rozenberg (Springer-Verlag, 1998), pp. 257–308
72. M. Silva, E. Teruel, J. Colom: "Linear Algebraic and Linear Programming Techniques for the Analysis of Place/Transition Net Systems", in *Lectures on Petri Nets I: Basic Models*, ed. by W. Reisig, G. Rozenberg (Springer-Verlag, 1998), pp. 309–373
73. C.Y. Lee: Bell System Technical Journal **38**, 985–999 (1959)
74. R. Bryant: ACM Computing Surveys **24**(3), 293–318 (1992)
75. E. Pastor, O. Roig, J. Cortadella, R. Badia: "Petri net analysis using boolean manipulation", in *15th International Conference on Application and Theory of Petri Nets*Zaragoza, Spain (1994)
76. E. Pastor, J. Cortadella, M.A. Peña: "Structural Methods to Improve the Symbolic Analysis of Petri Nets", in *Application and Theory of Petri Nets 1999* (Springer-Verlag, 2000), Vol. 1639 of Lecture Notes in Computer Science, pp. 26–45
77. G. Hachtel, F. Somenzi: *Logic synthesis and verification algorithms* (Kluwer Academic Publishers, 1996)
78. M. Nielsen, G. Rozenberg, P. Thiagarajan: Theoretical Computer Science **96**, 3–33 (1992)
79. R. Keller: Lecture Notes in Computer Science **24**, 103–112 (1975)
80. J. Cortadella, M. Kishinevsky, L. Lavagno, A. Yakovlev: IEEE Transactions on Computers **47**(8), 859–882 (1998)
81. E. Badouel, P. Darondeau: Lecture Notes in Computer Science: Lectures on Petri Nets I: Basic Models **1491**, 529–586 (1998)
82. L. Bernardinello, G.D. Michelis, K. Petruni, S. Vigna: "On synchronic structure of transition systems", in *Proceedings of the International Workshop on Structures in Concurrency Theory (STRICT)* (1995), pp. 69–84
83. E. Badouel, L. Bernardinello, P. Darondeau: Lecture Notes in Computer Science **915**, 364–383 (1995)
84. S.R. Petrick: (1956), "A direct determination of the irredundant forms of a boolean function from the set of prime implicants", Technical Report AFCRC-TR-56-110, Air Force Cambridge Res. Center, Cambridge, MA

85. B. Lin, F. Somenzi: "Minimization of symbolic relations", in *Proc. International Conf. Computer-Aided Design (ICCAD)*Santa Clara, CA (1990), pp. 88–91
86. J. Desel, W. Reisig: Acta Informatica **33**(4), 297–315 (1996)
87. C. Ykman-Couvreur, P. Vanbekbergen, B. Lin: "Concurrency reduction transformations on state graphs for asynchronous circuit synthesis", in *Proc. International Workshop on Logic Synthesis* (1993)
88. J. Cortadella, M. Kishinevsky, A. Kondratyev, L. Lavagno, A. Yakovlev: "Automatic handshake expansion and reshuffling using concurrency reduction", in *Proc. of the Workshop Hardware Design and Petri Nets (within the International Conference on Application and Theory of Petri Nets)* (1998), pp. 86–110
89. J. Cortadella, M. Kishinevsky, A. Kondratyev, L. Lavagno, A. Yakovlev: (1995), "A region-based theory for state assignment in asynchronous circuits", Tech. Rep. 95-2-006, University of Aizu, Japan
90. T.J. Chaney, C.E. Molnar: IEEE Transactions on Computers **C-22**(4), 421–422 (1973)
91. C.L. Seitz: "System timing", in *Introduction to VLSI Systems*, ed. by C.A. Mead, L.A. Conway (Addison-Wesley, 1980)
92. T.A. Chu: (1987), "Synthesis of self-timed VLSI circuits from graph-theoretic specifications", Ph.D. thesis, MIT Laboratory for Computer Science
93. S.H. Unger: IEEE Transactions on Computers **44**(6), 754–768 (1995)
94. A. Kondratyev, J. Cortadella, M. Kishinevsky, E. Pastor, O. Roig, A. Yakovlev: "Checking Signal Transition Graph implementability by symbolic BDD traversal", in *Proc. European Design and Test Conference*Paris, France (1995), pp. 325–332
95. R.K. Brayton, F. Somenzi: "An exact minimizer for boolean relations", in *Proceedings of the International Conference on Computer-Aided Design* (1989), pp. 316–319
96. P. Siegel, G.D. Micheli: "Decomposition methods for library binding of speed-independent asynchronous designs", in *Proc. International Conf. Computer-Aided Design (ICCAD)* (1994), pp. 558–565
97. M. Shams, J.C. Ebergen, M.I. Elmasry: IEEE Transactions on VLSI Systems **6**(4), 563–567 (1998)
98. N. Starodoubtsev, S. Bystrov, M. Goncharov, I. Klotchkov, A. Smirnov: "Toward synthesis of monotonic asynchronous circuits from signal transition graphs", in *Proc. of the International Conference on Application of Concurrency to System Design 2001 (ICACSD'01)* (2001)
99. P. Beerel, T.Y. Meng: Integration, the VLSI journal **13**(3), 301–322 (1992)
100. D.L. Dill: *Trace Theory for Automatic Hierarchical Verification of Speed-Independent Circuits*, ACM Distinguished Dissertations (MIT Press, 1989)
101. J.C. Ebergen: Distributed Computing **5**(3), 107–119 (1991)
102. J.L.A.v.d. Snepscheut: *Trace Theory and VLSI Design*, Vol. 200 of Lecture Notes in Computer Science (Springer-Verlag, 1985)
103. M. Kishinevsky, J. Staunstrup: "Characterizing speed-independence of high-level designs", in *Proc. International Symposium on Advanced Research in Asynchronous Circuits and Systems* (1994), pp. 44–53
104. M. Kishinevsky, J. Staunstrup: "Mechanized verification of speed-independence", in *Proc. of the 2nd Workshop on Theorem Provers in Circuit Design* Bad Herrenalb, Germany (1994), pp. 229–248
105. A.J. Martin: Distributed Computing **1**(4), 226–234 (1986)
106. K.v. Berkel: *Handshake Circuits: an Asynchronous Architecture for VLSI Programming*, Vol. 5 of International Series on Parallel Computation (Cambridge University Press, 1993)

107. J. Cortadella, A. Yakovlev, L. Lavagano, P. Vanbekbergen: "Designing asynchronous circuits from behavioral specifications with internal conflicts", in *Proc. International Symposium on Advanced Research in Asynchronous Circuits and Systems* (1994), pp. 106–115
108. C.N. Liu: Journal of the ACM **10**, 209–216 (1963)
109. A.D. Friedman, R.L. Graham, J.D. Ullman: IEEE Transactions on Computers **C-18**, 541–547 (1969)
110. J. Cortadella, M. Kishinevsky, A. Kondratyev, L. Lavagno, A. Yakovlev: IEEE Transactions on Computer-Aided Design **16**(8), 793–812 (1997)
111. P. Vanbekbergen, B. Lin, G. Goossens, H. de Man: "A generalized state assignment theory for transformations on signal transition graphs", in *Proc. International Conf. Computer-Aided Design (ICCAD)* (IEEE Computer Society Press, 1992), pp. 112–117
112. J. Cortadella, M. Kishinevsky, A. Kondratyev, L. Lavagno, A. Yakovlev: IEICE Transactions on Information and Systems **E80-D**(3), 315–325 (1997) URL `http://www.lsi.upc.es/ jordic/petrify/refs/`
113. J. Gu, R. Puri: IEEE Transactions on Computer-Aided Design **14**(8), 961–973 (1995)
114. M.A. Kishinevsky, A.Y. Kondratyev, A.R. Taubin: "Formal method for self-timed design", in *Proc. European Conference on Design Automation (EDAC)* (1991)
115. L. Lavagno, K. Keutzer, A. Sangiovanni-Vincentelli: "Synthesis of verifiably hazard-free asynchronous control circuits", in *Advanced Research in VLSI* (MIT Press, 1991), pp. 87–102
116. K.J. Lin, J.W. Kuo, C.S. Lin: "Direct synthesis of hazard-free asynchronous circuits from STGs based on lock relation and MG-decomposition approach", in *Proc. European Design and Test Conference* (IEEE Computer Society Press, 1994), pp. 178–183
117. E. Pastor, J. Cortadella: "Polynomial algorithms for the synthesis of hazard-free circuits from signal transition graphs", in *Proc. International Conf. Computer-Aided Design (ICCAD)* (IEEE Computer Society Press, 1993), pp. 250–254
118. P. Vanbekbergen, F. Catthoor, G. Goossens, H.D. Man: "Optimized synthesis of asynchronous control circuits form graph-theoretic specifications", in *Proc. International Conf. Computer-Aided Design (ICCAD)* (IEEE Computer Society Press, 1990), pp. 184–187
119. C. Ykman-Couvreur, B. Lin: "Optimised state assignment for asynchronous circuit synthesis", in *Asynchronous Design Methodologies* (IEEE Computer Society Press, 1995), pp. 118–127
120. K.Y. Yun: (1994), "Synthesis of asynchronous controllers for heterogeneous systems", Ph.D. thesis, Stanford University
121. J. Cortadella, M. Kishinevsky, L. Lavagno, A. Yakovlev: (1995), "Synthesizing Petri nets from state-based models", Tech. rep., Universitat Politècnica de Catalunya
122. R. Brayton, G. Hatchel, A. Sangiovanni-Vincentelli: Proceedings of IEEE **78**(2), 264–300 (1990)
123. K. Keutzer: "DAGON: Technology Mapping and Local Optimization", in *Proceedings of the* 24th *Design Automation Conference* (1987), pp. 341–347
124. F. Mailhot, G.D. Micheli: IEEE Transactions on Computer-Aided Design **12**(5), 599–620 (1993)
125. S.M. Burns: "General condition for the decomposition of state holding elements", in *Proc. International Symposium on Advanced Research in Asynchronous Circuits and Systems* (IEEE Computer Society Press, 1996)

126. P.A. Beerel, W. chun Chou, K.Y. Yun: "A heuristic covering technique for optimizing average-case delay in the technology mapping of asynchronous burst-mode circuits", in *Proc. European Design Automation Conference (EURO-DAC)* (1996)
127. W.C. Chou, P.A. Beerel, K.Y. Yun: IEEE Transactions on Computer-Aided Design **18**(10), 1418–1434 (1999)
128. K.W. James, K.Y. Yun: "Average-case optimized transistor-level technology mapping of extended burst-mode circuits", in *Proc. International Symposium on Advanced Research in Asynchronous Circuits and Systems* (1998), pp. 70–79
129. M. Ligthart, K. Fant, R. Smith, A. Taubin: "Asynchronous design using commercial HDL synthesis tools", in *Proc. International Symposium on Advanced Research in Asynchronous Circuits and Systems* (2000)
130. Y.Watanabe, R. Brayton: IEEE Transactions on Computer-Aided Design **12**(10), 1458–1472 (1993)
131. P. Beerel, T.Y. Meng: "Automatic gate-level synthesis of speed-independent circuits", in *Proc. International Conf. Computer-Aided Design (ICCAD)* (IEEE Computer Society Press, 1992), pp. 581–587 URL `http://jungfrau.usc.edu/pub/iccad92.ps`
132. P.A. Beerel, C.J. Myers, T.H.Y. Meng: IEEE Transactions on Computer-Aided Design (1998)
133. A. Kondratyev, M. Kishinevsky, B. Lin, P. Vanbekbergen, A. Yakovlev: "Basic gate implementation of speed-independent circuits", in *Proc. ACM/IEEE Design Automation Conference* (1994), pp. 56–62
134. A. Kondratyev, M. Kishinevsky, A. Yakovlev: IEEE Transactions on Computer-Aided Design **17**(9), 749–771 (1998)
135. P.A. Beerel: (1994), "CAD tools for the synthesis, verification, and testability of robust asynchronous circuits", Ph.D. thesis, Stanford University
136. Y.Watanabe, L.M.Guerra, R. Brayton: IEEE Transactions on Computer-Aided Design **15**(7), 732–744 (1996)
137. E.M. Sentovich, K.J. Singh, L. Lavagno, C. Moon, R. Murgai, A. Saldanha, H. Savoj, P.R. Stephan, R.K. Brayton, A. Sangiovanni-Vincentelli: (1992), "SIS: A system for sequential circuit synthesis", Tech. rep., U.C. Berkeley
138. C.J. Myers, T.H.Y. Meng: IEEE Transactions on VLSI Systems **1**(2), 106–119 (1993)
139. D. Sager, M. Hinton, M. Upton, T. Chappell, T. Fletcher, S. Samaan, R. Murray: "A 0.18μm CMOS IA32 microprocessor with a 4GHz integer execution unit", in *ISSCC'2001* (IEEE Press, 2001), pp. 324–325
140. J. Cortadella, M. Kishinevsky, A. Kondratyev, L. Lavagno, A. Taubin, A. Yakovlev: "Lazy transition systems: application to timing optimization of asynchronous circuits", in *Proc. International Conf. Computer-Aided Design (ICCAD)* (1998), pp. 324–331
141. R. Alur: "Timed automata", in *NATO-ASI 1998 Summer School on Verification of Digital and Hybrid Systems* (1998)
142. K.S. Stevens, S. Rotem, R. Ginosar, P. Beerel, C.J. Myers, K.Y. Yun, R. Kol, C. Dike, M. Roncken: IEEE Journal of Solid-State Circuits **36**(2), 217–228 (2001)
143. K. Stevens, R. Ginosar, S. Rotem: "Relative timing", in *Proc. International Symposium on Advanced Research in Asynchronous Circuits and Systems* (1999), pp. 208–218
144. W. Coates, J. Lexau, I. Jones, I. Sutherland, S. Fairbanks: "Fleetzero: an asynchronous switching experiment", in *Proc. International Symposium on*

Advanced Research in Asynchronous Circuits and Systems (IEEE Computer Society Press, 2001)

145. J. Cortadella, M. Kishinevsky, S.M. Burns, K. Stevens: "Synthesis of asynchronous control circuits with automatically generated timing assumptions", in *Proc. International Conf. Computer-Aided Design (ICCAD)* (1999), pp. 324–331
146. T. Nanya, A. Takamura, M. Kuwako, M. Imai, M. Ozawa, M. Ozcan, R. Morizawa, H. Nakamura: "Scalable-delay-insensitive design: A high-performance approach to dependable asynchronous systems", in *Proc. International Symp. on Future of Intellectual Integrated Electronics*Sendai, Japan (1999), pp. 531–540
147. P. Vanbekbergen, G. Goossens, B. Lin: "Modeling and synthesis of timed asynchronous circuits", in *Proc. European Design Automation Conference (EURO-DAC)* (IEEE Computer Society Press, 1994), pp. 460–465
148. K. McMillan: "Using unfoldings to avoid the state explosion problem in the verification of asynchronous circuits", in *Proc. International Workshop on Computer Aided Verification*, ed. by G. v. Bochman, D.K. Probst (Springer-Verlag, 1992), Vol. 663 of Lecture Notes in Computer Science, pp. 164–177
149. S.M. Burns: (1991), "Performance analysis and optimization of asynchronous circuits", Ph.D. thesis, California Institute of Technology
150. H. Hulgaard, S.M. Burns: "Bounded delay timing analysis of a class of CSP programs with choice", in *Proc. International Symposium on Advanced Research in Asynchronous Circuits and Systems* (1994), pp. 2–11 URL `http://www.it.dtu.dk/ henrik/papers/async94.ps.gz`
151. K. McMillan, D. Dill: "Algorithms for interface timing verification", in *Proceedings of the International Conference on Computer Design* (1992)
152. C.J. Myers: (1995), "Computer-aided synthesis and verification of gate-level timed circuits", Ph.D. thesis, Dept. of Elec. Eng., Stanford University
153. B. Coates, A. Davis, K. Stevens: Integration, the VLSI journal **15**(3), 341–366 (1993)
154. S.M. Nowick: (1993), "Automatic synthesis of burst-mode asynchronous controllers", Ph.D. thesis, Stanford University, Department of Computer Science
155. B. Lin, C. Ykman-Couvreur, P. Vanbekbergen: "A general state graph transformation framework for asynchronous synthesis", in *Proc. European Design Automation Conference (EURO-DAC)* (IEEE Computer Society Press, 1994), pp. 448–453
156. A. Kondratyev, J. Cortadella, M. Kishinevsky, L. Lavagno, A. Yakovlev: "Automatic synthesis and optimization of partially specified asynchronous systems", in *Proc. ACM/IEEE Design Automation Conference* (1999), pp. 110–115
157. S. Chakraborty, D.L. Dill, K.Y. Yun: Proceedings of the IEEE **87**(2), 332–346 (1999)
158. M.A. Peña, J. Cortadella, A. Kondratyev, E. Pastor: "Formal verification of safety properties in timed circuits", in *Proc. International Symposium on Advanced Research in Asynchronous Circuits and Systems* (IEEE Computer Society Press, 2000), pp. 2–11
159. M. Horowitz: (1998), "Clocking for high performance processors", invited talk at the International Symposium on Advanced Research in Asynchronous Circuits and Systems
160. S.Furber: "Amulet experiences with petrify", in *Proc. of the Special Interest Workshop on Exploitation of STG-based Design Technology*St.Petersburg, Russia (1998), see http://sadko.ncl.ac.uk/ nay/spb/spb_report.html

161. P. Vanbekbergen, G. Goossens, F. Catthoor, H.J.D. Man: IEEE Transactions on Computer-Aided Design **11**(11), 1426–1438 (1992)
162. D.J. Kinniment, B. Gao, A.V. Yakovlev, F. Xia: "Toward asynchronous A-D conversion", in *Proc. International Symposium on Advanced Research in Asynchronous Circuits and Systems* (1998), pp. 206–215
163. T.A. Chu, L.A. Glasser: "Synthesis of self-timed control circuits form graphs: An example", in *Proc. International Conf. Computer Design (ICCD)* (IEEE Computer Society Press, 1986), pp. 565–571
164. O. Roig: (1997), "Formal verification and testing of asynchronous circuits", Ph.D. thesis, Universitat Politècnia de Catalunya
165. A.J. Martin: "Synthesis of asynchronous VLSI circuits", in *Formal Methods for VLSI Design*, ed. by J. Straunstrup (North-Holland, 1990), pp. 237–283
166. K. Low, A. Yakovlev: (1995), "Token Ring Arbiters: an exercise in asynchronous logic design with Petri nets", Tech. Rep. 537, Department of Computing Science, University of Newcastle upon Tyne
167. R. Berks, J. Ebergen: "Response time properties of linear pipelines with varying cell delays", in *Proc. International Workshop on Timing Issues in the Specification and Synthesis of Digital Systems (TAU)* (1997)
168. A. Yakovlev: "Designing arbiters using Petri nets", in *Proceedings of the 1995 Israel Workshop on Asynchronous VLSI* (VLSI Systems Research Center, Technion, Nof Genossar, Israel, 1995), pp. 178–201
169. A. Yakovlev: "Solving acid-wg design problems with petri net based methods", in *Proc. ESPRIT ACiD-WG Workshop on Asynchronous Circuit Design, TR CSN9602* (Computer Science Notes Series, University of Groningen, Groningen, 1996)
170. A. Yakovlev: Formal Methods in System Design **12**(1), 39–71 (1998)
171. I. Blunno, A. Bystrov, J. Carmona, J. Cortadella, L. Lavagno, A. Yakovlev: "Direct synthesis of large-scale asynchronous controllers using a petri-net based approach", in *Proceedings of the Fourth ACiD-WG Workshop*Grenoble (2000), see http://tima-cmp.imag.fr/tima/cis/cis.html
172. R. David: IEEE Transactions on Computers **26**(8), 727–737 (1977)
173. L.A. Hollaar: IEEE Transactions on Computers **C-31**(12), 1133–1141 (1982)
174. V.I. Varshavsky, V.B. Marakhovsky: "Asynchronous control device design by net model behavior simulation", in *Application and Theory of Petri Nets 1996*, ed. by J. Billington, W. Reisig (Springer-Verlag, 1996), Vol. 1091 of Lecture Notes in Computer Science, pp. 497–515
175. C.H.K.v. Berkel, M.B. Josephs, S.M. Nowick: Proceedings of the IEEE **87**(2), 223–233 (1999)
176. A.J. Martin, S.M. Burns, T.K. Lee, D. Borkovic, P.J. Hazewindus: "The design of an asynchronous microprocessor", in *Advanced Research in VLSI*, ed. by C.L. Seitz (MIT Press, 1989), pp. 351–373
177. A. Lines: (1998), "Pipelined asynchronous circuits", Tech. Rep. CSTR:1998.cs-tr-95-21, Caltech Computer Science
178. K.v. Berkel, J. Kessels, M. Roncken, R. Saeijs, F. Schalij: "The VLSI-programming language Tangram and its translation into handshake circuits", in *Proc. European Conference on Design Automation (EDAC)* (1991), pp. 384–389
179. J. Kessels, A. Peeters: "The Tangram framework: Asynchronous circuits for low power", in *Proc. of Asia and South Pacific Design Automation Conference* (2001), pp. 255–260
180. K.v. Berkel, R. Burgess, J. Kessels, M. Roncken, F. Schalij: "Characterization and evaluation of a compiled asynchronous IC", in *Asynchronous Design*

Methodologies, ed. by S. Furber, M. Edwards (Elsevier Science Publishers, 1993), Vol. A-28 of IFIP Transactions, pp. 209–221

181. K.v. Berkel, R. Burgess, J. Kessels, A. Peeters, M. Roncken, F. Schalij: "A fully-asynchronous low-power error corrector for the DCC player", in *International Solid State Circuits Conference* (1994), pp. 88–89
182. K.v. Berkel, R. Burgess, J. Kessels, A. Peeters, M. Roncken, F. Schalij: IEEE Design & Test of Computers **11**(2), 22–32 (1994)
183. K.v. Berkel, R. Burgess, J. Kessels, A. Peeters, M. Roncken, F. Schalij: IEEE Journal of Solid-State Circuits **29**(12), 1429–1439 (1994)
184. H.v. Gageldonk, D. Baumann, K. van Berkel, D. Gloor, A. Peeters, G. Stegmann: "An asynchronous low-power 80c51 microcontroller", in *Proc. International Symposium on Advanced Research in Asynchronous Circuits and Systems* (1998), pp. 96–107
185. A. Bardsley, D. Edwards: "Compiling the language Balsa to delay-insensitive hardware", in *Hardware Description Languages and their Applications (CHDL)*, ed. by C.D. Kloos, E. Cerny (1997), pp. 89–91
186. A. Bardsley, D.A. Edwards: "The Balsa asynchronous circuit synthesis system", in *Forum on Design Languages* (2000)
187. E. Brunvand, R.F. Sproull: "Translating concurrent programs into delay-insensitive circuits", in *Proc. International Conf. Computer-Aided Design (ICCAD)* (IEEE Computer Society Press, 1989), pp. 262–265
188. V. Akella, G. Gopalakrishnan: "SHILPA: A high-level synthesis system for self-timed circuits", in *Proc. International Conf. Computer-Aided Design (ICCAD)* (IEEE Computer Society Press, 1992), pp. 587–591
189. G. Gopalakrishnan, P. Kudva, E. Brunvand: IEEE Transactions on VLSI Systems **7**(1), 30–37 (1999)
190. A.M.G. Peeters: (1996), "Single-rail handshake circuits", Ph.D. thesis, Eindhoven University of Technology
191. M. Renaudin, P. Vivet, F. Robin: "A design framework for asynchronous/synchronous circuits based on CHP to HDL translation", in *Proc. International Symposium on Advanced Research in Asynchronous Circuits and Systems* (1999), pp. 135–144
192. M. Ligthart, K. Fant, R. Smith, A. Taubin, A. Kondratyev: "Asynchronous design using commercial HDL synthesis tools", in *Proc. International Symposium on Advanced Research in Asynchronous Circuits and Systems* (IEEE Computer Society Press, 2000), pp. 114–125
193. I. Blunno, L. Lavagno: "Automated synthesis of micro-pipelines from behavioral Verilog HDL", in *Proc. International Symposium on Advanced Research in Asynchronous Circuits and Systems* (IEEE Computer Society Press, 2000), pp. 84–92
194. T. Kolks, S. Vercauteren, B. Lin: "Control resynthesis for control-dominated asynchronous designs", in *Proc. International Symposium on Advanced Research in Asynchronous Circuits and Systems* (1996)
195. E. Pastor, J. Cortadella: (1993), "P-time unique state coding algorithms for signal transition graphs", Tech. rep., Universitat Politècnica de Catalunya
196. E. Pastor: (1996), "Structural methods for the synthesis of asynchronous circuits from signal transition graphs", Ph.D. thesis, Univsitat Politècnia de Catalunya
197. A. Kondratyev, A. Taubin: "Verification of speed-independent circuits by stg unfoldings", in *Proc. International Symposium on Advanced Research in Asynchronous Circuits and Systems* (1994), pp. 64–75
198. A. Semenov, A. Yakovlev, E. Pastor, M.P. na, J. Cortadella, L. Lavagno: "Partial order based approach to synthesis of speed-independent circuits", in *Proc.*

International Symposium on Advanced Research in Asynchronous Circuits and Systems (IEEE Computer Society Press, 1997), pp. 254–265
199. A. Semenov, A. Yakovlev, E. Pastor, M.P. na, J. Cortadella: "Synthesis of speed-independent circuits from STG-unfolding segment", in *Proc. ACM/IEEE Design Automation Conference* (1997), pp. 16–21
200. U. Kim, D.I. Lee: "Practical synthesis of speed-independent circuits using unfoldings", in *Proc. of Asia and South Pacific Design Automation Conference* (1998), pp. 191–196
201. A. Kondratyev, J. Cortadella, M. Kishinevsky, L. Lavagno, A. Taubin, A. Yakovlev: "Identifying state coding conflicts in asynchronous system specifications using Petri net unfoldings", in *Int. Conf. on Application of Concurrency to System Design* (1998)
202. A. Yakovlev, V. Varshavsky, V. Marakhovsky, A. Semenov: "Designing an asynchronous pipeline token ring interface", in *Asynchronous Design Methodologies* (IEEE Computer Society Press, 1995), pp. 32–41
203. J.C. J. Carmona, E. Pastor: "A structural encoding technique for the synthesis of asynchronous circuits", in *Proc. of the Sec. Int. Conf. on Appl. of Concurrency to System Design (ICACSD'01)*Newcastle upon Tyne, UK (2001)
204. A.V. Yakovlev, A.M. Koelmans: "Petri nets and digital hardware design", in *Lectures on Petri Nets II: Applications. Advances in Petri Nets* (1998), Vol. 1492 of Lecture Notes in Computer Science, pp. 154–236
205. S. Patil: (1970), "Coordination of asynchronous events", Tech. rep., MIT Project MAC, Tech. Rep. MAC-TR-72, Ad 711–763
206. F. Furtek: (1971), "Modular implementation of petri nets", Master's thesis, MIT
207. S. Patil, J. Dennis: "The description and realization of digital systems", in *Proceedings of the IEEE COMPCON* (1972), pp. 223–226
208. D. Misunas: Communications of the ACM **16**(8), 474–481 (1973)
209. W.A. Clark: "Macromodular computer systems", in *AFIPS Conference Proceedings: 1967 Spring Joint Computer Conference*Vol. 30 (Academic Press, Atlantic City, NJ, 1967), pp. 335–336
210. M.J. Stucki, S.M. Ornstein, W.A. Clark: "Logical design of macromodules", in *AFIPS Conference Proceedings: 1967 Spring Joint Computer Conference*Vol. 30 (Academic Press, Atlantic City, NJ, 1967), pp. 357–364
211. W.A. Clark, C.E. Molnar: "Macromodular computer systems", in *Computers in Biomedical Research*, ed. by R.W. Stacy, B.D. Waxman, Vol. IV (Academic Press, 1974), pp. 45–85
212. C. Bell, J. Grason: Computer Design (1971)
213. J. Bruno, S.M. Altman: IEEE Transactions on Computers **C-20**, 629–638 (1971)
214. R.M. Keller: IEEE Transactions on Computers **C-23**(1), 21–33 (1974)
215. B.J. Nordmann: IEEE Transactions on Computers **26**(3), 196–207 (1977)
216. H. Hulgaard, S.M. Burns, G. Borriello: Integration, the VLSI journal **19**(3), 111–131 (1995)
217. D.B. Armstrong, A.D. Friedman, P.R. Menon: IEEE Transactions on Computers **C-18**(12), 1110–1120 (1969)
218. P. Beerel, T. Meng: "Semi-modularity and self-diagnostic asynchronous control circuits", in *Advanced Research in VLSI*, ed. by C.H. Séquin (MIT Press, 1991), pp. 103–117
219. A.J. Martin, P.J. Hazewindus: "Testing delay-insensitive circuits", in *Advanced Research in VLSI*, ed. by C.H. Séquin (MIT Press, 1991), pp. 118–132
220. M. Roncken, E. Bruls: "Test quality of asynchronous circuits: A defect-oriented evaluation", in *Proc. International Test Conference* (1996), pp. 205–214

221. M. Roncken: Proceedings of the IEEE **87**(2), 363–375 (1999)
222. K.T. Cheng, V. Agrawal, E. Kuh: IEEE Transactions on Computers **39**(12), 1456–1463 (1990)
223. M. Roncken, R. Saeijs: "Linear test times for delay-insensitive circuits: a compilation strategy", in *Asynchronous Design Methodologies*, ed. by S. Furber, M. Edwards (Elsevier Science Publishers, 1993), Vol. A-28 of IFIP Transactions, pp. 13–27
224. A. Khoche, E. Brunvand: "Testing self-timed circuits using scan paths", in *5th NASA Symposium on VLSI Design* (1993)
225. K. Keutzer, L. Lavagno, A. Sangiovanni-Vincentelli: IEEE Transactions on Computer-Aided Design **14**(12), 1569–1577 (1995)
226. L. Lavagno, M. Kishinevsky, A. Lioy: "Testing redundant asynchronous circuits by variable phase splitting", in *Proc. European Design Automation Conference (EURO-DAC)* (IEEE Computer Society Press, 1994), pp. 328–333
227. S.M. Nowick, N.K. Jha, F.C. Cheng: IEEE Transactions on Computer-Aided Design **16**(12), 1514–1521 (1997)
228. A. Khoche, E. Brunvand: "A partial scan methodology for testing self-timed circuits", in *Proc. IEEE VLSI Test Symposium* (1995), pp. 283–289
229. M. Kishinevsky, A. Kondratyev, L. Lavagno, A. Saldanha, A. Taubin: IEEE Transactions on Computer-Aided Design **17**(11), 1184–1199 (1998)
230. K. McMillan: *Symbolic model checking* (Kluwer Academic, 1993)
231. D.L. Dill: "Trace theory for automatic hierarchical verification of speed-independent circuits", in *Advanced Research in VLSI*, ed. by J. Allen, F.T. Leighton (MIT Press, 1988), pp. 51–65
232. J. Ebergen, S. Gingras: "A verifier for network decompositions of command-based specifications", in *Proc. Hawaii International Conf. System Sciences*Vol. I (IEEE Computer Society Press, 1993)
233. R. Alur, D. Dill: "Automata for Modeling Real-Time Systems", in *Automata, Languages and Programming: 17th Annual Colloquium* (1990), Vol. 443 of Lecture Notes in Computer Science, pp. 322–335, warwick University, July 16-20
234. O. Maler, A. Pnueli: "Timing analysis of asynchronous circuits using timed automata", in *Correct Hardware Design and Verification Methods. IFIP WG 10.5 Advanced Research Working Conference, CHARME '95* (Springe-Verlag, 1995), pp. 189–205
235. C.J. Myers, T.G. Rokicki, T.H.Y. Meng: IEEE Transactions on Computer-Aided Design **18**(6), 769–786 (1999)
236. H. Hulgaard, S.M. Burns, T. Amon, G. Borriello: IEEE Transactions on Computers **44**(11), 1306–1317 (1995)
237. H. Hulgaard, S.M. Burns: Formal Methods in System Design **11**(3), 265–294 (1997)
238. H. Hulgaard, T. Amon: IEEE Transactions on Computer-Aided Design **19**(10), 1093–1104 (2000)
239. C.D. Nielsen, M. Kishinevsky: "Performance analysis based on timing simulation", in *Proc. ACM/IEEE Design Automation Conference* (1994), pp. 70–76
240. A. Xie, S. Kim, P.A. Beerel: "Bounding average time separations of events in stochastic timed Petri nets with choice", in *Proc. International Symposium on Advanced Research in Asynchronous Circuits and Systems* (1999), pp. 94–107
241. J. Staunstrup: *A Formal Approach to Hardware Design* (Kluwer Academic Publishers, 1994), Chap. 7, ch. 7: Self-Timed Circuits
242. T.E. Williams, M. Horowitz, R.L. Alverson, T.S. Yang: "A self-timed chip for division", in *Advanced Research in VLSI*, ed. by P. Losleben (MIT Press, 1987), pp. 75–95

243. A.J. Martin, S.M. Burns, T.K. Lee, D. Borkovic, P.J. Hazewindus: Computer Architecture News **17**(4), 95–110 (1989)
244. J.A. Tierno, A.J. Martin, D. Borkovic, T.K. Lee: IEEE Design & Test of Computers **11**(2), 43–49 (1994)
245. S.B. Furber, P. Day, J.D. Garside, N.C. Paver, J.V. Woods: "A micropipelined ARM", in *Proceedings of VLSI 93*, ed. by T. Yanagawa, P.A. Ivey (1993), pp. 5.4.1–5.4.10
246. S.B. Furber, P. Day, J.D. Garside, N.C. Paver, S. Temple, J.V. Woods: "The design and evaluation of an asynchronous microprocessor", in *Proc. International Conf. Computer Design (ICCD)* (IEEE Computer Society Press, 1994)
247. S.B. Furber, P. Day, J.D. Garside, N.C. Paver, S. Temple: "AMULET2e", in *Embedded Microprocessor Systems*, ed. by C. Muller-Schloer, F. Geerinckx, B. Stanford-Smith, R. van Riet (1996), proceedings of EMSYS'96 - OMI Sixth Annual Conference
248. S.B. Furber, J.D. Garside, D.A. Gilbert: "AMULET3: A high-performance self-timed ARM microprocessor", in *Proc. International Conf. Computer Design (ICCD)* (1998)
249. J.D. Garside, W.J. Bainbridge, A. Bardsley, D.A. Edwards, S.B. Furber, J. Liu, D.W. Lloyd, S. Mohammadi, J.S. Pepper, O. Petlin, S. Temple, J.V. Woods: "AMULET3i — an asynchronous system-on-chip", in *Proc. International Symposium on Advanced Research in Asynchronous Circuits and Systems* (IEEE Computer Society Press, 2000), pp. 162–175
250. T. Nanya, Y. Ueno, H. Kagotani, M. Kuwako, A. Takamura: IEEE Design & Test of Computers **11**(2), 50–63 (1994)
251. T. Nanya, A. Takamura, M. Kuwako, M. Imai, T. Fujii, M. Ozawa, I. Fukasaku, Y. Ueno, F. Okamoto, H. Fujimoto, O. Fujita, M. Yamashina, M. Fukuma: "TITAC-2: A 32-bit scalable-delay-insensitive microprocessor", in *Symposium Record of HOT Chips IX* (1997), pp. 19–32
252. S. Schuster, W. Reohr, P. Cook, D. Heidel, et al.: "Asynchronous interlocked pipelined CMOS circuits operating at 3.3-4.5 GHz", in *IEEE International Solid-State Circuits Conference* (2000), pp. 292–293
253. M. Renaudin, P. Vivet, F. Robin: "ASPRO-216: A standard-cell QDI 16-bit RISC asynchronous microprocessor", in *Proc. International Symposium on Advanced Research in Asynchronous Circuits and Systems* (1998), pp. 22–31
254. A. Abrial, J. Bouvier, P. Senn, M. Renaudin, P. Vivet: "A new contactless smartcard IC using an on-chip antenna and an asynchronous microcontroller", in *Proc. European Solid-State Circuits Conference (ESSCIRC)* (2000)
255. S. Edwards, L. Lavagno, E. Lee, A. Sangiovanni-Vincentelli: Proceedings of the IEEE **85**(3), 366–390 (1997)

Index

Printing: Mercedes-Druck, Berlin
Binding: Stein+Lehmann, Berlin